Editors

Lecture Notes in Chemistry

Edited by G. Berthier, M. J. S. Dewar, H. Fischer
K. Fukui, H. Hartmann, H. H. Jaffé, J. Jortner
W. Kutzelnigg, K. Ruedenberg, E. Scrocco, W. Zeil

11

Franco A. Gianturco

The Transfer of Molecular Energies by Collision:

Recent Quantum Treatments

Springer-Verlag
Berlin Heidelberg New York 1979

Author

Franco Antonio Gianturco
Institute of Physical Chemistry
The University of Bari
Via Amendola 173, Bari/Italy
and
Quantum Chemistry Laboratory
(L.C.Q.E.M.), C.N.R., Pisa/Italy

ISBN-13:978-3-540-09701-3 e-ISBN-13:978-3-642-93122-2
DOI: 10.1007/978-3-642-93122-2

Softcover reprint of the hardcover 1st edition 1979

"The abnihilisation of the etym by the grisning of the grosning of the grinder of the grunder of the first Lord of Hurtreford explodotonates through parsuralia with an ivanmorinthorrumble fragoromboassity amidwhiches general uttermosts confussion are perceivable moletons skaping with mulicules"

J. Joyce, Finnegan's wake

ABSTRACT

Volume 11: F.A. GIANTURCO

THE TRANSFER OF MOLECULAR ENERGIES BY COLLISION: RECENT QUANTUM TREATMENTS

17 figures, 327 pages, 1979

The aim of this study is to provide a summary of the currently used theoretical and computational techniques and a review of the most recent results that have dealt mainly with the simplest type of energy transfer in a molecular collision process, namely the conversion of the translational energy of a structureless atom in its ground electronic state into the internal rotational and vibrational energy of a diatomic molecule, also in its ground electronic state. This is probably one of the most common events in chemical processes and while the experimental data from molecular beam measurements have become increasingly more meaningful and reliable, only recent years have seen any thorough, ab initio approach to the various computational aspects of this problem. A brief résumé of potential scattering results is used as a preparation of the working formalism. The ab initio potential surface calculations are then examined and the various angular momentum coupling representations, those that yield the multichannel scattering formalism of vibrorotational inelastic cross sections in atom-molecule and molecule-molecule encounters, are discussed. Some of the numerical techniques that have lent themselves to the most recent applications are also reviewed together with the various decoupling schemes that are necessary when dealing with more complex cases. Finally, correlations of the state-to-state deexcitation cross sections with bulk measurements of relaxation times are also briefly presented. (498 References).

Contents: A résumé of quantum mechanical potential scattering. - Potential energy hypersurface calculations for simple systems. - Rotational and vibrational inelasticity in molecular encounters. - Dimensionality reduction methods for rotovibrational cross sections calculations.- Numerical methods for the coupled equations: a survey.- Rotovibrational relaxation models in simple gases.

FOREWORD

These Lecture Notes are intended as an introduction to the theoretical formulation and computational aspects of the molecular energy transfer processes which take place in an increasingly sophisticated range of molecular scattering experiments. They are directed to chemistry graduate students and emphasize the quantum mechanical approach, with little or no attention to classical and semi-classical treatments or to formal presentations.

Several Sections of the first Chapters are based on lectures given at the Graduate School of Physics of the University of Genoa a few years ago and I thank the students for their sense of duty in following to the end all those notation-filled blackboards and transparencies.

The kind patience of my wife Carolyn in reading the whole manuscript and improving its form is gratefully acknowledged.

Franco A. Gianturco

Bari, September 1978

CONTENTS

INTRODUCTION

The study of chemistry at a microscopic level, both from a theoretical and an 'experimental' viewpoint, is mainly the study of the numerous processes that before, during and after a chemical reaction (or in the absence of a chemical reaction) control either the transformation of one type of molecular energy into another type or its transfer into the 'environment'. Within the domain of what is commonly known as physical chemistry these studies have a very long history indeed and the many theoretical models suggested through the years for each specific case outline the progress we have made in understanding the workings of chemistry at a physical level.

When the events under study take place in the absence of external radiative fields, and when the systems are in their gaseous states, one is obviously faced with a simpler situation, albeit one that is still general enough to give rise a great variety of phenomena and processes. It is however evident that, at this level, collisions between molecules will play a very important role in bringing about such events and that our understanding of most of the phenomena will have necessarily to go through a careful examination of the collisional process itself. It has only been in recent years however that such examination has reached 'criticality' in the chemical community and that several aspects of the energy transfer which takes place during molecular collisions have been closely scrutinized.

The scope of the present Lecture Notes is therefore to provide newcomers to the subject with a compact presentation of the theoretical and computational techniques that have been implemented to yield realistic predictions for some of the possible outcomes of the molecular encounters. Because of the wide variety of energy transformations that can take place, a drastic reduction of the area examined was obviously necessary. Thus, only the simplest type of energy transfer in molecular collisions has been considered, namely that involving the conversion of the translational energy of either a structureless atom or a small molecule (both in their closed-shell, ground electronic states) into the internal rotational and vibrational energies of a diatomic molecule (also in its ground electronic state). This is probably still one of the most common events among those that happen during a chemical reaction, but while the experimental data from molecular beam measurements have become increasingly more meaningful and reliable, only recently have we been able to perform any thorough, *ab initio* approach to the various computational aspects of this problem.

From a general viewpoint, the analysis of energy transfer in molecular collisions in the gas kinetic regime can be divided into two independent stages. This is possible because the duration of a given collision between two molecules is usually much shorter than the mean time between successive collisions. This means that it is always possible to choose a time interval that is short compared to the mean time between collisions but still long enough when compared with the duration of a single collision. This allows for carrying on the treatment within a binary encounter modelling of the whole process and the system consisting of two particular colliding molecules can then be considered as isolated from all the remaining molecules over the chosen time interval.

Returning to the first stage, then, the state of the system can be described by quantum mechanical equations that only take into account the degrees of freedom of the two molecules under consideration. In this approximation the influence of all the other molecules manifests itself only through initial conditions that specify the states of the target or of both molecules before they collide. The first dynamic stage of this analysis is then completed when the appropriate solutions of the quantum mechanical problem are obtained. Such solutions in fact yield the relevant transition probabilities or state-to-state scattering cross sections which in turn provide the microscopic characteristics of the fundamental collision processes.

In the second stage of the analysis one can compute some macroscopic quantities directly accessible to experiment, namely, the rates of the different and possible relaxation processes. The aim of the theory is then to provide a microscopic interpretation of macroscopic parameters which describe a given reaction kinetics. At this essentially statistical stage of the analysis one relies on various types of kinetic equations which govern the temporal evolution (relaxation) of the distribution of molecules among various states which arises as a result of many successive binary collisions.

Such an analysis of the sequence of physical events leading to molecular energy transfer in the gas phase can thus provide a logical chapter plan for the present Lecture Notes. After a brief presentation to chemists of the most essential "tools" that are needed to deal with quantum mechanical scattering, the text continues with a discussion of potential surfaces meaning and production within the more generally used techniques of quantum chemistry (Chapters I and 2).

The detailed description of microscopic theory of binary encounters, those that lead to vibrational and rotational excitation of molecules, is the next stage of presentation (Chapter 3). It includes the SF and BF *a priori* treatments and also reports some specific, typical cases well discussed in the current literature. The natural sequel to the third Chapter is then the discussion of the most recent, simpler schemes that have been suggested to reduce dimensionality of the coupled equations. This is a much needed step either in order to upgrade the complexity of the tractable systems, or to obtain large numbers of cross sections among several internal levels with a sizeable reduction of computational effort (Chapter 4). Finally, the macroscopic quantities that have been directly accessible to bulk experiments are briefly discussed, as well as their relationship with the dynamical observables presented before. The performance of theory is thus analyzed through the examination of recent calculations in simple systems (Chapter 6).

In spite of its relative youth as a full-grown scientific subject, at least from the point of view of reliable experimental data and accurate computational treatments, molecular scattering began to diversify very early on and to form, as a consequence, several theoretical paths each of which often constitutes a whole subject in its own right. The present review has chosen to focuss its attention only on part of these developments, in the belief that an initial knowledge of the formalism that leads the coupled differential equations of the Schrödinger representation provides the most comprehensive introduction to other, more specialized or less explored, ways of theoretically discussing low-energy molecular collisions. Thus, no part of this work has attempted to bridge the gap with the more familiar classical mechanics by discussing semiclassical methods and their relative merits. Since excellent reviews have recently been published on the subjects it was considered wise to limit the size of the present Notes by excluding this interesting and alternative approach.

In the same spirit, the presentation of more formal treatments that emphasize methodology rather than the numerical aspects of yielding cross sections to compare with experiments (e.g., the optical potential approaches) has also been avoided, in spite of its increasing interest among theoreticians. The same type of choice has also been made with respect to the statistical theories that treat microscopic data to yield observables of a more general nature. The information-theoretic approach and the stochastic treatment of inelastic collisions are in fact typical examples of further handling of molecular collision problems that rely on a previous computational knowledge of the relevant cross sections: the latter was chosen instead as the main topic of these Notes.

Another excluded topic would deal with the time-dependent treatments of inelastic collisions, that rely on some specific form of each contributing trajectory and go on using some type of close-coupling expansion for describing target states. They have also become rather promising for treating high energy collisions (> 50 eV) of atom (ion)-molecule systems but have been excluded from mainly low-energy encounters discussed in the present monograph, apart from some passing references to them in Chapter 4. Finally, electronically inelastic processes or the more empirical treatment of curve-hopping, non-adiabatic transitions have also been excluded since they recently have been treated in an excellent review that has appeared in the present series of Lecture Notes.

In spite of all these limitations, however, the theoretical aspects discussed herein still remain ample in breath and scope. The close interplay of chemical and physical concepts requires in any case the bringing together of topics that are quite specialized and therefore seldom treated together in the same study. The present Lecture Notes attempt to provide, through a compact presentation, the relevant basic methods and references without lengthy digressions.

Because of the fairly large number of equations in each Chapter, one unfortunately might find a similarly large number of misprint or omissions (bits that remained in the writer's pen) together with possible accidental "degeneracies" between the notations for the symbols used in different parts of text. Whoever will be patient enough to point them out to the author will be gratefully thanked.

1. A RESUME' OF QUANTUM MECHANICAL POTENTIAL SCATTERING

1.1 - General formulation of the problem.

When discussing the gas-phase interaction of atoms and molecules as they appear in many beam experiments on reactive and subreactive encounters, the physical situation presented by the beam of projectile particles travelling across another (secondary) beam of target particles is obviously very different from the chemically more familiar situation of a stable species, where the various constituents are strongly localized in space, i.e. each elementary component is on the average rather close to the centre of the acting forces.

In a classical description, and assuming no appreciable interaction within the particles constituting the beam, each of the atoms or molecules or electrons undergoing collision will follow a trajectory that is completely determined by the initial conditions and the classical equations of motion. The former are specified by the impinging energy $E = p^2/2m$ and by the impact parameter $b(\vartheta)$.

In the simplest case of spherical and stationary force fields (to be defined later) and structureless particles, the latter are deflected by an angle with the impinging direction $\underline{z}$ that is comprised between ϑ and $\vartheta + d\vartheta$, span the solid angle $d\Omega = 2\pi \sin\vartheta \, d\vartheta$ and originate from incoming particles with impact parameters between $\underline{b}$ and $\underline{b} + \underline{db}$.

Flux conservation requirements further impose equality between the number of particles impinging per unit time on the area $2\pi b\,db$ and the corresponding number of particles scattered within the solid angle $d\Omega$ per unit time. Each scattering process therefore allows one to define a constant which, when multiplied by $d\Omega$, gives us the relative number N of particles deflected per unit time within that solid angle. Such a constant describes the Differential Cross Section of the process under study and is classifically related to the deflection function for each $\underline{b}$ and ϑ, at a given energy:

$$\sigma(\vartheta,E) = \frac{b}{\sin\vartheta}\,\frac{db}{d\vartheta} \tag{1.1}$$

The corresponding total cross section is then obviously given by:

$$\sigma(E) = 2\pi \int_0^{\pi} d\vartheta \sin\vartheta\, \sigma(\vartheta,E) = 2\pi \int_0^{b_o} b\,db \tag{1.2}$$

where b_o is the largest value of b for which it is still: $\vartheta \neq 0$.

Within a quantummechanical framework however, instead of the initial conditions specifying positions and momenta $\underline{r}(t_i)$ and $\underline{p}(t_i)$ one is dealing with the wavefunction $\Psi_i = \Psi(\underline{r}, t_i)$ which provides the corresponding initial expectation values of the variables $<\underline{r}>_i$ and $<\underline{p}>_i$. It also satisfies the time-dependent equation:

$$\mathcal{H}\Psi(\underline{r},t) = i\hbar \frac{\partial}{\partial t} \Psi(\underline{r},t) \tag{1.3}$$

where:

$$\mathcal{H} = -\tfrac{1}{2}\nabla^2 + V(\underline{r}) \tag{1.4}$$

The solution of eq.(1.3) therefore gives the time evolution of the wave packet prepared at the initial time t_i and provides the various Ψ_f's at any time t_f after collision. Whenever the initial dimensions of the wavepacket $(\Delta r)_i$ are small with respect to the range of $V(\underline{r})$ and there is no appreciable spatial 'spread' during collision, i.e. $(\Delta r)_i \sim (\Delta r)_f$, one can still adopt a classical description of the process since the quantum effects can be considered as negligible. More specifically, one can write down a condition for the linear momentum that should be satisfied for the classical picture to be valid:

$$(\Delta P_z)_i \,/\, <P_z>_i \ll 1 \tag{1.5}$$

where the z-direction is the one chosen above and $(\Delta P_z)_i$ is the initial momentum spread along that direction.

Although it is always possible to make such a packet preparation, one should however remember that, as a consequence of it, a constraint is then applied to space localizability. If r_o is the range of a spherical potential, the corresponding classical condition on the size of the wavepacket is in fact given by:

$(\Delta z)_i \ll r_o$, and since: $(\Delta z)_i\ (\Delta P_z)_i \sim \hbar$ one finds that:

$$r_o \gg \frac{h}{(\Delta P_z)_i} \gg \frac{h}{(P_z)_i} = \frac{<\lambda>}{2\pi} \tag{1.6}$$

with $<\lambda>$ being the average associated wavelength for the impinging particles. The r_o parameter corresponds to that region of interaction where the potential equates the initial average kinetic energy of the particles within the packet.

The spreading of the packet originates from the component waves dispersion within the potential field. Since the local linear momentum is given by: $\underline{P}(r) = \{2m\ [E - V(r)]\}^{\frac{1}{2}}$, such a spreading can be disregarded whenever one can write:

$$\left| \frac{r_o}{<p>} \frac{dp}{dr} \right| \ll 1 \tag{1.7}$$

which means that:

$$\left| \frac{r_o}{2[E - V(r)]} \frac{dV}{dr} \right| \ll 1 \tag{1.8}$$

within the range of the acting potential V(r).

Since eq.(1.8) diverges for E = V, which is the classical result for very small values of $\underline{b}$, the above condition breaks down and the quantum treatment becomes necessary. The quantum eigensolution can be written down in the potential--free region with the general integral form:

$$\Psi(\underline{r},t) = \int d\underline{p}'\ a_{\underline{P}'} \exp(i/h)\ |\underline{P}'\underline{r} - E't| \tag{1.9}$$

In the limit of a monoenergetic beam the coefficients in (1.9) are given by:

$$a_{\underline{P}'} = \delta(\underline{P}' - \underline{P}) \tag{1.10}$$

hence:

$$\Psi(\underline{r},t) \rightarrow \Psi_{\underline{P}}(\underline{r},t) = \exp(i/h)\ [\underline{P}\cdot\underline{r} - Et] \tag{1.11}$$

which now describes a completely delocalized plane wave. For stationary potentials one can further write:

$$\Psi_{\underline{P}}(\underline{r},t) = \psi_{\underline{K}}(\underline{r}) \exp[-(i/h)Et] \tag{1.12}$$

which, substituted into eq.(1.3), yields the familiar time-independent form of the Schrödinger equation. Whenever the beam of "structureless" particles in an elastic scattering experiment satisfies the following conditions:

(i) energetically constant for a time $\tau >> h/\Delta E$, with ΔE being the experimental dispersion in energy; and

(ii) spatially extended and uniform for radial ranges of $r >> r_0$;

it then becomes realistically valid to treat the incoming wavefunction as a planewave as in eq.(1.12). All the relevant experimental information may then be obtained from a knowledge of $\psi_k(\underline{r})$. Hence:

$$\{\nabla^2 - V'(\underline{r}) + \kappa^2\}\, \psi_k(\underline{r}) = 0 \tag{1.13}$$

with:

$$V'(\underline{r}) = \frac{2m}{\hbar^2} V(\underline{r}) \tag{1.14}$$

and:

$$\underline{k} = \underline{P}/\hbar \tag{1.15}$$

The effect of the interaction $V(\underline{r})$ is one of distorting the incoming plane wave, and this can be described as an additional contribution to the wavefunction that vanishes at large distances from the scattering centre. Hence the final asymptotic form of $\psi(\underline{r})$ can be written as:

$$\psi_k(\underline{r}) \underset{r\to\infty}{\sim} e^{i\underline{k}\cdot\underline{r}} + \frac{e^{ikr}}{r} f(\vartheta,\varphi) \tag{1.16}$$

where ϑ and φ are the polar and azimuthal scattering angles respectively, taken from the impinging direction $\hat{k}$. The $f(\vartheta,\varphi)$ is called the Scattering Amplitude.

The number of particles N that enter the volume V per unit of time is given by:

$$\mathbf{N} = \frac{\partial}{\partial\tau} \int_V \Psi^*(\underline{r},\tau)\, \Psi(\underline{r},\tau)\, dV =$$

$$= \int_V \{ \Psi^*(\underline{r},\tau) \frac{\partial}{\partial\tau} \Psi(\underline{r},\tau) + \Psi(\underline{r},\tau) \frac{\partial}{\partial\tau} \Psi^*(\underline{r},\tau) \}\, dV \tag{1.17}$$

and, via eq.(1.3), one can also write:

$$N = \frac{i\hbar}{2m} \int_V [\Psi^* \nabla^2 \Psi - \Psi \nabla^2 \Psi^*] dV \tag{1.18}$$

The hermitian operator $\mathcal{J}(\mathbf{r}_0)$ defining the current at $\mathbf{r}_0$ is in turn given by:

$$(\underline{r}_0) = \frac{1}{2m} \left\{ \underline{p}\, \delta\, (\underline{r} - \underline{r}_0) + \delta\, (\underline{r} - \underline{r}_0)\, \underline{p} \right\} \tag{1.19}$$

For a monoenergetic beam (and disregarding any asymptotic interference between incoming wave and scattered wave) the expectation value of the $\mathcal{J}$-operator yields the following expression for the (flux density) current vector:

$$\boldsymbol{j}(\underline{r}) = \frac{\hbar}{2mi} \left\{ \Psi^* \mathrm{grad}\, \Psi - \Psi\, \mathrm{grad}\, \Psi^* \right\} \tag{1.20}$$

which appears in the alternative form of eq.(1.18), form obtained by rewriting it as a surface integral using Green's Theorem:

$$N = \int_S \underline{\boldsymbol{j}} \cdot d\underline{s} \tag{1.21}$$

When one applies the operator $\underline{\boldsymbol{j}}(\mathbf{r})$ to each term on the r.h.s. of (1.16) and normalizes $\Psi_{\underline{k}}$ per unit of volume V, one obtains:

$\underline{j}_0 = \hbar k /_{mv}$ for the plane wave; (1.22a)

$\underline{j}_s = \hbar /_{mv} \left[\underline{r} \cdot \underline{k}\ \frac{|f(\vartheta,\varphi)|^2}{r^2} \right]$ for the asymptotic region of the scattered wave; (1.22b)

The total flux through the surface element $d\underline{s}$ is given by:

$$\underline{j}_s \cdot d\underline{s} = \underline{j}_s\ (\hat{r} \cdot r^2 d\Omega) = \frac{\hbar k}{mv}\ |f|^2\ d\Omega \tag{1.23}$$

Hence, from the definition of differential cross section given above, on can write:

$$\sigma(\vartheta,\varphi) = \frac{\underline{j}_s}{\underline{j}_0} = |f(\vartheta,\varphi)|^2\ d\Omega \tag{1.24}$$

while the corresponding total cross section for a cylindric-symmetry problem is:

$$\sigma(k^2) = 2\pi \int_0^{\pi} |f(\vartheta)|^2 \sin\vartheta \, d\vartheta \qquad (1.25)$$

As opposed to the classical situation, the total cross section defined above is always finite for finite-range potentials, i.e. when one can write:

$$\int_0^{\infty} |V(r)| \, r^2 dr < \infty \qquad (1.26)$$

In fact, the condition shown in eq.(1.6) for a monoenergetic beam implies, in three-dimensions, that:

$$(\Delta x)\,(\Delta p_x) \sim \hbar \quad \text{and} \quad (\Delta y)\,(\Delta P_y) \sim \hbar \qquad (1.27)$$

The beam is therefore not unidirectional and it becomes meaningless to talk of a classical deflection by an angle $\vartheta < (\Delta p_t)/_{<p_z>}$, where:

$$(\Delta P_t) = \{(\Delta P_x)^2 + (\Delta P_y)^2\}^{\frac{1}{2}} \qquad (1.28)$$

The omission of such small-angle contributions to the classical $\sigma(E)$ therefore avoids its lack of finiteness. Moreover:

$$\vartheta_{\text{limit}} >> \frac{\Delta P_t}{<P_z>} \sim \frac{\hbar}{<P_z> [(\Delta x)^2 + (\Delta y)^2]^{\frac{1}{2}}} \sim \frac{\lambda}{2\pi b} \qquad (1.29)$$

for a localized wavepacket where one can also write:

$$[(\Delta x)^2 + (\Delta y)^2]^{\frac{1}{2}} << b \qquad (1.30)$$

1.2 - Solutions of the radial equation.

In the previous section it was shown that for stationary potentials all physical information can be gathered from a knowledge of the solutions of the familiar time-independent Schrödinger equation. In spherical polar coordinates this equation assumes the well known form:

$$\left\{\frac{\mathcal{P}_r^2}{2m} + \frac{\mathcal{L}^2}{2mr^2} + V(r)\right\} \Psi(r,\vartheta,\varphi) = E\,\Psi(r,\vartheta,\varphi) \qquad (1.31)$$

where $\mathcal{P}_r$ corresponds to the hermitian linear radial momentum operator and $\mathcal{L}^2$ to the angular momentum operator commuting with $\mathcal{H}$.

Since the eigenfunctions of $\mathcal{L}^2$ (and $\mathcal{L}_z$) constitute a basis for a representation of $\mathcal{H}$, and since spherical harmonics are such eigenfunctions within the functional space of the square-integrable functions, the angular dependence of $\psi(x)$ can be determined by writing down the eigensolutions of (1.31) in the following form:

$$\Psi(\underline{r}) = f(r)\, Y_\ell^m\, (\hat{r}) \tag{1.32}$$

or

$$\Psi_\ell^m = \chi_\ell(r)\;\; Y_\ell^m(\hat{r}) \tag{1.33}$$

if one defines: $r\chi_\ell(r) = y_\ell(r)$, and since: $\mathcal{P}_r^2 = -\hbar^2\frac{1}{r}\;\frac{d^2}{dr^2}\,r$, then the previous eq.(1.31) becomes:

$$\{ -\frac{\hbar}{2m}\frac{d^2}{dr^2} + \ell(\ell+1)\frac{\hbar^2}{2m\,r^2} + V(r) - E\}\, y_\ell\,(r) = o \tag{1.34}$$

If $y_\ell(r)$ is a normalizable function in a generalized sense, preserves the $\mathcal{P}_r$ hermicity for the whole r interval and behaves at the origin as: $y_\ell\;(0) = 0$, it also constitutes a complete basis for the representation of the eigenfunctions of $\mathcal{H}$. Its continuum eigenvalues spectrum is infinitely degenerate, while the discrete spectrum is given by a set of distinct E_ℓ eigenvalues, each with a $(2\ell + 1)$-degeneracy.

Let us now briefly examine a simple example of potential form that will turn out to be useful in the following discussions of physically realistic systems.

$V(\underline{r}) = V(r) = V_0$ = const in one (or more) specific regions of the (o,∞) interval and zero otherwise. When $E > V_0$, one defines:

$$k = \frac{\sqrt{2m(E-V_o)}}{\hbar} \qquad \text{and} \qquad \rho = k\,r \tag{1.35}$$

hence eq.(1.34) can be rewritten as:

$$\left\{\frac{d^2}{d\rho^2} + \left[1-\frac{\ell(\ell+1)}{\rho^2}\right]\right\} y_\ell(r) = o \tag{1.36a}$$

or:

$$\left\{\frac{d^2}{d\rho^2} + \frac{2}{\rho}\,\frac{d}{d\rho} + \left[1-\frac{\ell(\ell+1)}{\rho^2}\right]\right\} f_\ell\;(r) = o \tag{1.36b}$$

for: $f_\ell(r) = r^{-1} y_\ell(r)$.

The general solution is given by linear combinations of spherical functions with the following properties:

(i) regular solution (Bessel function) that goes to zero as ρ^ℓ and behaves asymptotically as:

$$\left\{\frac{\pi}{2\rho}\right\}^{\frac{1}{2}} J_{\ell+\frac{1}{2}}(\rho) = j_\ell(\rho) \underset{\rho\to\infty}{\sim} \frac{\sin(\rho - \frac{1}{2}\ell\pi)}{\rho} \tag{1.37a}$$

(ii) irregular solution (Neuman function) that exhibits a pole of $(\ell+1)$-order at the origin and behaves asymptotically as:

$$(-1)^\ell \left\{\frac{\pi}{2\rho}\right\}^{\frac{1}{2}} J_{-\ell-\frac{1}{2}}(\rho) = -n_\ell(\rho) \underset{\rho\to\infty}{\sim} \frac{\cos(\rho - \frac{1}{2}\ell\pi)}{\rho} \tag{1.37b}$$

(iii) Hankel function of the 1st type:

$$h_\ell^{(+)} = j_\ell + i n_\ell \tag{1.37c}$$

(iv) Hankel function of the 2nd type:

$$h_\ell^{(-)} = j_\ell - i n_\ell \tag{1.37d}$$

The last two solutions behave asymptotically as incoming and outgoing waves respectively. When $E < V_0$, one sets: $\varkappa = \frac{\sqrt{2m(V_0 - E)}}{\hbar}$ and solutions are found as before, with k now substituted by $i\varkappa$. The only normalizable solution (hence acceptable) is given by $h^{(+)}(i\varkappa r)$ that behaves like the bound state solutions of the discrete spectrum of (1.34).

For the more restrictive case of $V_0 = 0$ there are no acceptable solutions with negative eigenvalues, while the only regular (square integrable) solutions for positive eigenvalues are given by Bessel functions. The eigenfunctions (1.33) are therefore given by the familiar expression for the free particle:

$$\Psi_\ell^m(\underline{r}) = j_\ell(kr)\, Y_\ell^m(\underline{r}) \tag{1.38}$$

with k being a continuum index. Such solutions provide a complete orthonormal set of spherical waves describing states of well-defined angular momentum and completely undetermined energy.

It is also known that the set of plane waves, like those on the r.h.s. of eq.(1.16), constitues a complete basis for a representation of the free particle and it corresponds to states of well-defined linear momentum $\hbar k$ but of undefined angular momentum.

The completeness of the above sets allows one to obtain some useful relationships. It is in fact always possible to write:

$$e^{i\underline{k}\cdot\underline{r}} = \sum_{\ell} \sum_{m=-\ell}^{+\ell} a_{\ell m}(\underline{k})\, j_{\ell}(kr)\, Y_{\ell}^{m}(\hat{r}) \tag{1.39}$$

For the somewhat simpler problem of cylindrical symmetry one has neither φ-dependence nor the summation over m (since $\langle \mathcal{L}_z \rangle = 0$) if the z-axis is chosen along $\underline{k}$. Hence:

$$e^{i\underline{k}\cdot\underline{r}} = \exp(ikr\cos\vartheta) = \sum_{\ell} c_{\ell}\, j_{\ell}(kr)\, P_{\ell}(\cos\vartheta) \tag{1.40}$$

The recurrence relation among coefficients is given by: $c_{\ell} = (2\ell+1)\, i^{\ell} c_{o}$. By putting $kr = o$ and remembering that: $j_{\ell}(o) = \delta_{o\ell}$, one finds: $c_o = 1$ and eq.(1.40) therefore becomes:

$$e^{ikz} = \sum_{\ell} (2\ell+1)\, i^{\ell} j_{\ell}(kr)\, P_{\ell}(\cos\vartheta) \tag{1.41}$$

which now provides the planewave expansion as a sum of partial waves. The more general case of $\underline{k}$ and $\underline{r}$ forming an arbitrary angle ϑ provides the equivalent expression:

$$e^{i\underline{k}\cdot\underline{r}} = 4\pi \sum_{\ell,m} i^{\ell} j_{\ell}(kr)\, Y_{\ell}^{m}(\hat{k})\, Y_{\ell}^{m}(\hat{r}) \tag{1.42}$$

1.3 - The method of partial waves.

In the case of scattering between two beams of particles interacting via a local time-independent potential, the familiar Schrödinger equation reads as follows:

$$\left\{-\frac{\hbar^2}{2m_1}\nabla_1^2 - \frac{\hbar^2}{2m_2}\nabla_2^2 + V(\underline{r}_1,\underline{r}_2) - E\right\}\psi(\underline{r}_1,\underline{r}_2) = o \tag{1.43}$$

By separating the centre of mass (c.m.) motion and the relative motion, the equation describing the latter becomes again:

$$\left\{-\frac{\hbar^2}{2\mu}\nabla_r^2 + V(\underline{r}) - E_{rel}\right\}\psi(\underline{r}) = o \tag{1.44}$$

where μ is now the reduced mass, $\underline{r}$ the relative distance vector and $E_{rel} = E - E_{c.m.}$

For local, spin independent and spherical potentials satisfying condition (1.26) and behaving asymptotically at least as r^{-2}, a general solution of (1.44) can be written down through the expansion:

$$\Psi(\underline{r}) = r^{-1}\sum_\ell A_\ell(k^2)\,P_\ell(\cos\vartheta)\,u_\ell(r) \tag{1.45}$$

which is called a partial wave expansion and where there are now two unknown to be determined, the energy-dependent coefficients $A_\ell(k^2)$ and the radial functions $u_\ell(r)$.

By substituting (1.45) into (1.44), premultiplying by $P_\ell(\cos\vartheta)$ and integrating over $\cos\vartheta$ one gets:

$$\left\{\frac{d^2}{dr^2} - \frac{\ell(\ell+1)}{r^2} - U(r) + k^2\right\}u_\ell(r) = o \tag{1.46a}$$

$$\text{with: } U(r) = \frac{2\mu}{\hbar}V(r) \quad \text{and: } k = \sqrt{\frac{2\mu E}{\hbar^2}} \tag{1.46b}$$

One has now to solve (1.46) for all angular momentum values ℓ.

In the region of small r, and for potential less singular than r^{-2}, the general solution is given by:

$$g_\ell(r) = a_\ell r^{\ell+1} + b_\ell r^{-\ell} \tag{1.47}$$

Only the first term on the r.h.s. is normalizable and hence physically acceptable, which thus provides us with a solution at the origin for all ℓ's:

$$u_\ell(r) \underset{r\to o}{\sim} r^{\ell+1} \tag{1.48}$$

In the asymptotic region of large r, provided V(r) goes to zero faster than r^{-1}, the free wave regular solution (1.37a) gives way to the following solution:

$$u_\ell(r) \underset{r\to\infty}{\sim} \sin\,(kr - \tfrac{1}{2}\,\ell\pi + \delta_\ell) \tag{1.49}$$

where δ_ℓ is an arbitrary phase factor with respect to the $j_\ell(kr)$ and depends on the nature of the potential. An alternative form of (1.49) is given by:

$$u_\ell(r) \underset{r\to\infty}{\sim} \frac{1}{2i}\left\{ e^{i\alpha_\ell}\, e^{i\delta_\ell} - e^{-i\alpha_\ell}\, e^{-i\delta_\ell} \right\} \tag{1.50}$$

where: $\alpha_\ell = kr - \tfrac{1}{2}\,\ell\pi$

Since our physical solution also needs to satisfy the asymptotic conditions (1.16), for a φ-independent symmetry one can make use of (1.41) by subtracting it from (1.45) and equating the result with the second term on the r.h.s. of (1.16). After simple manipulations one gets:

$$A_\ell\,(k^2) = \frac{1}{k}\, i^\ell\,(2\ell + 1)\, e^{i\delta_\ell} \tag{1.51}$$

and

$$f(\vartheta) = \frac{1}{2ik} \sum_\ell (2\ell + 1)\,(e^{2i\delta_\ell} - 1)\, P_\ell\,(\cos\vartheta) \tag{1.52}$$

where the scattering amplitude correctly goes to zero for a vanishing phase shift.

The problem of solving (1.44) has then been reduced to the simpler one of solving radial differential equations within the range of action of $V(r)$ and for all the contributing ℓ's in the partial wave expansion.

For all practical potential forms the equation (1.46) has to be solved numerically only for a finite number of ℓ for each given energy since: $\lim_{\ell\to\infty} \delta_\ell(k^2) = 0$. In the case of electrons scattered by atoms at a relative energy of a few eV, the typical values of contributing ℓ are $= 0, 1, 2, 3$.

One can rewrite eq.(1.49):

$$u_\ell(r) \underset{r\to\infty}{\sim} \left\{ \sin(kr - \tfrac{1}{2}\ell\pi) + \cos(kr - \tfrac{1}{2}\ell\pi)\,\tan\delta_\ell \right\} \times \cos\delta_\ell$$

$$= \sin\alpha_\ell + \cos\alpha_\ell\, K_\ell \tag{1.53}$$

where $\cos\delta_\ell$ is included in a normalization factor. One can then define: $K_\ell = \tan\,\delta_\ell$, this being the so-called K-matrix or R-matrix.

Moreover, from (1.50):

$$u_\ell(r) \underset{r\to\infty}{\sim} e^{-i\alpha_\ell} - e^{i\alpha_\ell} - S_\ell \qquad (1.54)$$

where S is now called S-matrix, defined by the asymptotic form of $u_\ell(r)$ and related to the K-matrix and the T-matrix by the relation:

$$S_\ell = \ell^{2i\delta_\ell} = \frac{1 + ik_\ell}{1 - ik_\ell} = T_\ell + 1 \qquad (1.55)$$

Which appear as diagonal matrices in this partial wave representation.

The use of real potential contains implicitely the conservation of flux during the scattering process, a fact easily related to a property of the S-matrix. In fact, from (1.55):

$$S_\ell \cdot S_\ell^{*} = 1 \qquad (1.56)$$

i.e. S_ℓ is unitary.

Since we know, from (1.24), that $|f(\vartheta)|^2 = \sigma(\vartheta)$ then the total cross section is now given by:

$$\sigma(k^2) = \frac{4\pi}{k^2} \sum_\ell (2\ell + 1) \sin^2 \delta_\ell \qquad (1.57)$$

Each partial wave contribution to (1.57) obviously satisfies the inequality:

$$\sigma_\ell(k^2) \leqslant (2\ell + 1) \times \frac{4\pi}{k^2} \qquad (1.58)$$

and the largest contribution comes for the phase shift value:

$$\delta_\ell = (n + \tfrac{1}{2})\ \pi \qquad (1.59)$$

This can be related to a classical result. Starting in fact from eq.(1.2), keeping in mind that: $\ell = \hbar\, k\, b$ and assuming unit interval of $\Delta\ell$ (in atomic units), then: $\Delta b = 1/k$ and:

$$\sigma^{cl}(k^2) = 2\pi\, b\Delta b = 2\pi\, \ell\, /_{k^2} \qquad (1.60)$$

For very large ℓ values the largest contributions to (1.57) from (1.58) become $\sim 8\pi\ell/k^2$, hence:

$$\sigma^{Q.M.}(k^2) \simeq 4\ \sigma^{C\ell}(k^2) \tag{1.61}$$

for each chosen impact parameter (angular momentum).

For real potentials the individual phase shifts are real quantities, while the scattering amplitude is complex with an imaginary part given by:

$$\mathrm{Im}\ f(\vartheta) = \frac{1}{2k}\sum_\ell (2\ell + 1)\ P_\ell\ (\cos\vartheta)\ (\cos 2\delta_\ell - 1) \tag{1.62}$$

and, from a property of the Legendre polynomials:

$$\begin{aligned} \mathrm{Im}\ f(o) &= \frac{1}{k}\sum_\ell (2\ell + 1)\ \sin^2\delta_\ell \\ &= \frac{k}{4\pi}\ \sigma(k^2) \end{aligned} \tag{1.63}$$

This very useful result is called The Optical Theorem in analogy with the behaviour of light travelling across a dense medium. In the last case, the imaginary part of the refraction index is related to the absorption coefficient, hence with the total cross section. The eq.(1.63) therefore provides a measure of the intensity loss experienced by the incoming beam of particles along the impinging direction after collision with the secondary beam. Its use in actual calculations is very important, since the evaluation of $f(\vartheta)$, in the forward direction only, allows a direct comparison of its imaginary part with measured total cross sections.

Flux conservation also requires that when the asymptotic solutions (1.49) go into the free-wave form of the Bessel functions, once outside the potential range, they exhibit continuity, as well as a continuous first derivative. If one then defines the following quantity:

$$\beta_\ell = \left\{ \frac{r_o}{u\ (r)}\ \frac{du_\ell(r)}{dr} \right\}_{r=r_o} \tag{1.64}$$

as the logarithmic derivative at the radial value where $V(r)$ vanishes, since it is also:

$$u_\ell(r) \underset{r\to\infty}{\sim} j_\ell(kr) - \tan\delta_\ell\ n_\ell(kr) \tag{1.65}$$

As seen from the definitions (1.37a,b), the following expressions can be written down:

$$e^{2i\xi_\ell} = \frac{j_\ell - in_\ell}{j_\ell + in_\ell} \tag{1.66a}$$

and:

$$\Delta_\ell + iS_\ell = kr_0 \frac{j'_\ell + in'_\ell}{j_\ell + in_\ell} \tag{1.66b}$$

where the superscripts in (1.66b) indicate first derivatives with respect to r. After simple manipulations one obtains:

$$e^{ri(\delta_\ell - \xi_\ell)} = \frac{\beta_\ell - \Delta_\ell + iS_\ell}{\beta_\ell - \Delta_\ell - iS_\ell} \tag{1.67}$$

The real phase angle ξ_ℓ coincides with δ_ℓ when $\beta_\ell \to \infty$, which happens when the potential diverges at r_0 and no partial wave can get into the target "core". ξ_ℓ is therefore called the rigid sphere phase shift and it will become useful later on when defining resonance states.

1.4 - Some properties of δ_ℓ . The Born approximation.

From (1.52) and (1.55) the scattering amplitude can be alternatively written as:

$$f(\vartheta) = \frac{i}{2k} \sum_\ell (2\ell + 1)(1 - S_\ell) P_\ell (\cos\vartheta) \tag{1.68}$$

where $(1 - S_\ell) = T_\ell$ as obtained from the previous definition of the transition matrix or T-matrix. Obviously the structure of all the matrices defined so far will become clearer when dealing with multichannel scattering formalism, although their form is particularly simple in potential scattering theory. Moreover:

$$\sigma(k^2) = \frac{\pi}{k^2} \sum_\ell (2\ell + 1)\ |1 - S_\ell|^2$$

$$= \frac{\pi}{k^2} \sum_\ell (2\ell + 1)\ |\ T_\ell|^2 \qquad (1.69)$$

If one now defines: $f_\ell (r) = kr\, j_\ell\, (kr)$, from (1.49) one can write:

$$\int_0^\infty (f_\ell \frac{d^2u_\ell}{dr^2} - u_\ell \frac{d^2f_\ell}{dr^2})\ dr = \int_0^\infty f_\ell V(r)\ u_\ell dr \qquad (1.70)$$

integrating by part and solving with respect to the phase shift:

$$\sin \delta_\ell = - \int_0^\infty r j_\ell(kr)\ V\ (r)\ u_\ell(r)\ dr \qquad (1.71)$$

which is an exact expression for δ_ℓ .

For weak potentials of finite range r_0 and at collision energies for which: $k^2 \gg |\ V(r_0)|$ it is possible to put: $u_\ell(r) \sim k r j_\ell (r)$, hence:

$$\sin \delta_\ell^B \sim \delta_\ell^B = - k \int_0^\infty V\ (r)\ [\ j_\ell(kr)\ ]^2 r^2 dr \qquad (1.72)$$

which is also valid for $\ell\to\infty$ since, whenever $\ell(\ell+ 1)\, r^{-2} \gg |\ V(r)\, |$ for the whole range of integration, the above approximate substitution is again valid for $u_\ell(r)$.

The above approximation provides the first term of the Born series. It is a positive integral and therefore δ_ℓ^B will go to zero as positive or negative according to the sign of the acting potential.

The corresponding result for the full amplitude can be obtained by substituting (1.72) into (1.52):

$$f_B(\vartheta) = \frac{1}{2ik} \sum_\ell (2\ell + 1)\ 2i\ \delta_\ell^B\ P_\ell\ (\cos \vartheta) \qquad (1.52')$$

In the low energy regimes:

$$j_\ell\ (kr) \underset{k\to 0}{\sim} \text{const} \cdot (kr)^\ell \qquad (1.73)$$

and if the finite range potential goes to an inverse-power form for $r > r_0$, one further gets:

$$\delta_\ell^B \simeq -k\int_0^{r_0} j_\ell^2 \cdot V \cdot r^2 dr - k\int_{r_0}^{\infty} V_0 \cdot j_\ell^2 \cdot r^2 dr \qquad (1.74)$$

The first integral converges if $V(r)$ is less singular than $r^{2\ell+2}$, while the second converges if: $V_0 = \frac{c}{r^n}$ and: $2-n+2\ell < 1$, i.e. if: $n > 2\ell+3$. Therefore one can also state that:

$$\delta_\ell^B \underset{k\to o}{\sim} - A_\ell \; k^{2\ell+1} \qquad (1.75)$$

Moreover, if $n = 2\ell+3$ since one can write:

$$\int_{r_0}^{\infty} \frac{dr}{r} = \lim_{k\to o} \int_{r_0}^{1/k} \frac{dr}{r} = \ln\left(\frac{1}{kr_0}\right) \qquad (1.76)$$

The phase shift energy dependence is then given more explicitely by:

$$\delta_\ell^B \underset{k\to o}{\sim} B_\ell \; K^{2\ell+1} \ln k \qquad (1.77)$$

Finally, when $n < 2\ell+3$ one obtains:

$$k \int_{r_0}^{\infty} dr \; r^{2-n} \; j_\ell^2 = k^{n-2} \int_{x_0}^{\infty} dx \cdot x^{2-n} \cdot j_\ell^2 \qquad \text{with:} x = kr \qquad (1.78)$$

when $k\to o$ then $x\to o$ and the integral (1.78) is defined at both boundaries. The dominant term becomes:

$$\delta_\ell^B \underset{k\to o}{\sim} C_\ell \cdot k^{n-2} \qquad (1.79)$$

The above derivations provide the correct energy behaviour of the phase shift even if its value is not correctly given by the approximate term (1.72).

For scattering by s-waves only:

$$u_0(r) \underset{r\to\infty}{\sim} \sin kr + \tan \delta_0 \cos kr \qquad (1.80)$$

in the low energy limit one can write:

$$\sin kr \sim kr \,, \qquad \tan \delta_0 \sim \delta_0 \,, \qquad \cos kr \sim 1 \qquad (1.81)$$

hence, via eq.(1.75) and for $\ell = 0$, eq.(1.80) becomes:

$$\lim_{k\to o} u_o(r) \sim kr + \delta_o = kr - ak = k(r - a) \tag{1.82}$$

where a is the intercept of the $u_0(r)$ at zero energy with the radial axis. Since:

$$\begin{aligned} \sigma_o (k^2) &= \frac{4\pi}{k^2} \sin^2 \delta_o = \frac{4\pi}{k^2(1 + \cot^2\delta_o)} = \\ &= \frac{4\pi}{k^2} \cdot \frac{1}{1 + \frac{1}{k^2a^2}} = \frac{4\pi a^2}{1 + k^2a^2} \end{aligned} \tag{1.83}$$

Then:

$$\lim_{k\to o} \sigma_o (k^2) = 4\pi a^2 \tag{1.84}$$

The s-wave cross section goes to a finite value at zero energy which depends on the parameter a, called the scattering length.

Now, since $f_\ell(\vartheta) \propto \delta_\ell/k$ one also has from eq.(1.75):

$$\lim_{k\to o} f_\ell (\vartheta) = \text{const} \cdot k^{2\ell} \tag{1.85}$$

The $\frac{d\sigma^\ell}{d\Omega}(k^2)$ is proportional to $|f_\ell(\vartheta)|^2$, which implies that in the low-energy limit the various partial wave contributions to $\frac{d\sigma}{d\Omega}(k^2)$ go to zero as $k^{4\ell}$ and only the s-wave contribution appears as a constant in the summation (1.24), yielding an isotropic differential cross section. This is qualitatively seen as caused by a centrifugal barrier effect at very low energies for all the partial waves except, obviously, for the s-wave which thus becomes dominant.

Equation (1.72) can be rewritten via the use of proper Bessel functions $J_{\ell+\frac{1}{2}}(kr)$:

$$\delta_\ell^B = - \frac{\pi}{2}\int_o^\infty V(r)\, J^2_{\ell+\frac{1}{2}} (kr)\, r\, d\, r \tag{1.86}$$

For small phase shifts the scattering amplitude can also be rewritten as:

$$f(\vartheta) \sim \frac{1}{k} \sum_\ell (2\ell + 1)\, \delta_\ell\, P_\ell(\cos \vartheta) \tag{1.87}$$

if one further defines the momentum transfer during collision as:

$$\mathcal{K} = 2k \sin(\vartheta/2) \tag{1.88}$$

Then the following expansion can be used:

$$\frac{\sin \mathcal{K} r}{\mathcal{K} r} = \frac{\pi}{2kr} \sum_{\ell} (2\ell + 1) P_{\ell} (\cos \vartheta) J^2_{\ell+\frac{1}{2}}(kr) \tag{1.89}$$

which allows for the rewriting of eq.(1.88) as:

$$f(\vartheta) \sim - \int_0^\infty V(r) \frac{\sin \mathcal{K} r}{\mathcal{K} r} r^2 dr \tag{1.90}$$

The accuracy of the Born approximation and its convergence rapidity needs to be checked for each special case examined. One can however try to indicate the regions of its validity. The criteria previously discussed via the r-dependence of the potential for the zero-energy case can be restated by saying that the approximation of eq.(1.72) converges if:

$$\int_0^\infty r \mid V(r) \mid dr < 2\ell + 1 \tag{1.91}$$

Or, more generally, when the potential term is dominated at all distances by the angular momentum and the energy terms, i.e. when:

$$\mid V(r) \mid \ll \left| k^2 - \frac{\ell(\ell+1)}{r^2} \right| \tag{1.92}$$

over the relevant range of r. Thus one finds that the Born approximation is good for all k when ℓ is large and for all ℓ when k is large. More specifically, from (1.86) a criterion of validity is given by:

$$\delta_\ell^B = - \frac{\mathcal{V}_0}{k} \int_0^{r_0} [kr\, j_\ell(kr)]^2 dr \ll 1 \tag{1.93}$$

where $\mathcal{V}_0$ is the potential depth and r_0 its range for a square-well form of interaction. The argument of the integral is bounded by 1 and tends asymptotically to $\sin(kr - \frac{1}{2}\ell\pi)$, hence (1.93) is certainly satisfied if:

$$\mathcal{V}_0 \cdot r_0 \ll k \tag{1.94}$$

for all ℓ values. From eq.(1.90), if $V(r) = 0$ for $r > a$, then: max $(\sin\mathcal{K}r)$ requires that: $\mathcal{K}r \sim \pi/2$ and $\mathcal{K} \sim 1/a$. Thus the region of validity of (1.94), from the definition of $\mathcal{K}$ becomes:

$$\sin\left(\frac{\vartheta}{2}\right) \leqslant \frac{1}{2ak} \tag{1.95}$$

i.e. at high energies the Born scattering is confined to deflection angles:

$$\vartheta \leqslant \frac{1}{k\,a} \tag{1.96}$$

Moreover, from (1.88):

$$\mathcal{K}\, d\mathcal{K} = k^2 \sin\vartheta \, d\vartheta \tag{1.97}$$

Hence the total cross section becomes:

$$\sigma^B(k^2) = \frac{2}{k^2} \int \mathcal{V}^2(\mathcal{K})\,\mathcal{K}\, d\mathcal{K} \tag{1.98}$$

where:

$$\mathcal{V}(\mathcal{K}) = -\frac{1}{\mathcal{K}} \int_0^\infty V(r) \sin\mathcal{K}r \; r\, dr \tag{1.99}$$

which gives an E^{-1} dependence for the cross section as the energy goes to ∞.

1.5 - Properties of the S-matrix: bound states and resonances.

Due to the solution of the basic equation (1.46) we have been able to define the S-matrix of (1.55) which is useful in the study of the more general case of k being complex.

Let us define the Jost function starting from a solution of (1.46) which exhibits the following asymptotic behaviour:

$$f_\ell(\pm k, r) \underset{r\to\infty}{\sim} \exp\left[-i\left(\pm kr - \tfrac{1}{2}\ell\pi\right)\right] \tag{1.100}$$

for k complex only k^2 changes in (1.46) and everything else stays the same. Then equation (1.56) however no longer holds and the S-unitarity breaks down. The solution $u_\ell(r)$ can be written by using the function of (1.100):

$$u_\ell(r) = \text{const} \cdot \left\{ f_\ell(k,r) + (-)^{\ell+1} S_\ell(k)\, f_\ell(-k,r) \right\} \tag{1.101}$$

Now the function in (1.100) is given by a combination of regular and irregular solutions and therefore exhibits the wrong behaviour at the origin. We thus need to define a new function that excludes the irregular component:

$$f_\ell(\pm k) = \lim_{r\to o} \frac{(kr)^\ell f_\ell(\pm k,r)}{(2\ell+1)!!} \tag{1.102}$$

This is called the Jost function (Helv. Phys. Acta, 20, 256, 1947).

Since:

$$u_\ell(r) \underset{r\to o}{\sim} f_\ell(k) + (-)^{\ell+1} f_\ell(-k)\, S_\ell(k) = o \tag{1.103}$$

This gives us a new definition for the S-matrix:

$$S_\ell(k) = (-)^\ell \frac{f_\ell(k)}{f_\ell(-k)} \tag{1.104}$$

Moreover, by definition:

$$\left\{ \frac{d^2}{dr^2} - \frac{\ell(\ell+1)}{r^2} - U(r) + (k^2) \right\} f_\ell(-k,r) = o \tag{1.105}$$

The c.c. equation has the form:

$$\left\{ \frac{d^2}{dr^2} - \frac{\ell(\ell+1)}{r^2} - U(r) + (k^*)^2 \right\} f_\ell^*(-k,r) = o \tag{1.106}$$

and

$$f_\ell^*(-k,r) \underset{r\to\infty}{\sim} \exp\left[i(-k^* r - \tfrac{1}{2}\ell\pi)\right] \tag{1.107a}$$

By comparison with (1.54) one finds that a solution of (1.107) is also given by $f^*(k,r)$, and since:

$$f_\ell(k^*,r) \underset{r\to\infty}{\sim} \exp\left[-i(k^* r - \tfrac{1}{2}\ell\pi)\right] \tag{1.107b}$$

it follows that:

$$(-)^\ell f_\ell^*(-k,r) = f_\ell(k^*,r) \tag{1.108}$$

Hence, using the definition of (1.102), one obtains a corresponding relation between

Jost functions:

$$(-)^{\ell} f_{\ell}^{*}(-k) = f_{\ell}(k^{*}) \tag{1.109}$$

In this way one can obtain a more general unitarity relation for the S-matrix, which now also holds for complex energies:

$$S_{\ell}(k)\, S_{\ell}^{*}(k^{*}) = \frac{f_{\ell}(k)}{f_{\ell}(-k)} \cdot \frac{f_{\ell}^{*}(k^{*})}{f_{\ell}^{*}(-k^{*})} = 1 \tag{1.110}$$

Moreover, from (1.104) one can also realize that S is a symmetric matrix:

$$S_{\ell}(k)\, S_{\ell}(-k) = \frac{f_{\ell}(k)}{f_{\ell}(-k)} \cdot \frac{f_{\ell}(-k)}{f_{\ell}(k)} = 1 \tag{1.111}$$

A pole in the S-matrix is a zero of the Jost function, as seen from (1.104), hence from (1.101):

$$u_{\ell}(r) = \text{const} \cdot \left\{ f_{\ell}(-k)\, f_{\ell}(k,r) - f_{\ell}(k)\, f_{\ell}(-k,r) \right\} \tag{1.112}$$

For a given energy, a zero of $f_{\ell}(-k)$ yields:

$$u_{\ell}(r) = - f_{\ell}(k)\, f_{\ell}(-k,r) \underset{r\to\infty}{\sim} \text{const} \cdot e^{ikr} \tag{1.113}$$

In the complex k-plane the above result can be used to associate some physical property of the system with the imaginary axis. In fact, at a pole of $S_{\ell}(k)$, if $\text{Im}\, k > 0$, one can write:

$$\begin{aligned} u_{\ell}(r) \text{ at the pole} &\sim e^{i(\text{Re}k + i\,\text{Im}k)r} \\ &\sim e^{i\text{Re}kr}\, e^{-\text{Im}kr} \end{aligned} \tag{1.114}$$

The first term on the r.h.s. of this equality is a phase factor of unit amplitude, whereas the second one describes an asymptotically decaying bound state. Since one is dealing with hermitian operators, the bounded, integrable wavefunctions must correspond to real eigenvalues or real energies. Hence it must be that: $E = \frac{1}{2\mu} k^2 =$ real quantity. Thus k^2 can be either positive or negative, i.e.: $k^2 = (k^*)^2$, which implies that: $\text{Re}\, k = 0$. Therefore the poles of the S-matrix in the upper-half of the complex plane are restricted to the imaginary axis and correspond to negative energy eigenvalues, i.e. to bound states of the particle under the influence of V(r). Obviously the strength of the potential will control how far the poles will be from the real axis.

If Im $k < 0$, the corresponding wavefunction is no longer bounded and the energy eigenvalues are no longer negative. It is easy to show that there is an infinity of such poles related to the structure of the continuum spectrum (see below).

Obviously, a completely similar analysis can be done in terms of zero of the Jost function, for which the bound states will correspond with points in the lower-half of the complex plane and along the imaginary axis, as seen from (1.104).

Let us now look at another interesting set of poles of the S-matrix that lay close to the real axis but are not on the real axis. From the general properties of the unitary and symmetric nature of the S-matrix one assumes that, if there is a pole at k there must be a corresponding zero at (-k) and another pole at $(-k^*)$, with another zero at k^*. Therefore the poles are symmetrically placed about the immaginary axis (conjugate poles). The zeros are symmetric with respect to the real axis. The poles with Re k = 0 and Im $k < 0$ are quite often called virtual states poles.

Conversely, if one looks for the Jost function in the upper-half of the complex k-plane, these states will correspond to its zeros since such a function is analytic only in the lower-half plane. Whenever Im k is small, i.e. close to the real axis, then $k = i\gamma$ lies in regions where $f_\ell(k)$ is analytic and the following Taylor expansion is applicable:

$$f_\ell(k) = f_\ell(i\gamma) + a_1(k - i\gamma) + a_2(k - i\gamma)^2 + \dots \qquad (1.115)$$

Since $f_\ell(i\gamma) = 0$ by the assumption of the existence of a virtual state and only the linear term can be retained for sufficiently small values of $|k - i\gamma|$, the corresponding form of the S-matrix can therefore be written as:

$$S_\ell(k) \sim (-)^\ell \left(\frac{k - i\gamma}{-k - i\gamma}\right) \qquad (1.116)$$

with the corresponding cross section:

$$\sigma_\ell(k^2) = \frac{4\pi(2\ell + 1)}{k^2 + \gamma^2} \qquad (1.117)$$

The corresponding expression for a weakly bound state can be obtained with a similar procedure and is given by:

$$\sigma_\ell(k^2) \sim \frac{4\pi\,(2\ell+1)}{k^2 + 2E_n} \tag{1.118}$$

for $\mu = 1$ and for a binding energy of E_n.

The different energy dependence of the (1.117) could then be a way of detecting virtual state effects in low energy scattering.

Going now into the complex energy plane, let us consider a pole in the S-matrix with a small imaginary part Γ, i.e. a pole at energy $E = E_r - \frac{1}{2}i\Gamma$.

The associated probability function will necessarily exhibit a time dependent behavior:

$$|\,\psi_\ell\,(\underline{r},t)|^2 = |\,\psi_\ell(r)|^2 \;\; e^{-\Gamma\; t/\hbar} \tag{1.119}$$

While the real part provides only a phase factor, one then sees that the imaginary part behaves like a decaying non-stationary state, called a resonant state.

Taking advantage of the small imaginary parts for non-overlapping poles, it is possible to derive a simple expression for the form of the S-matrix at the resonance by performing the analitic expansion of the relevant Jost functions:

$$S \sim S_o \frac{E - E_r - \frac{1}{2}i\Gamma}{E - E_r - \frac{1}{2}i\Gamma} \tag{1.120}$$

where S_o represent the effects of 'distant' poles away from the real axis. It is therefore a slowly-varying quantity across the resonance state pole. Hence:

$$S_\ell\,(k) = e^{ri\delta_\ell} = e^{ri\delta_o} \cdot e^{ri\delta_r} \tag{1.121}$$

where:

$$e^{2i\delta_r} = \frac{E - E_r - \frac{1}{2}i\Gamma}{E - E_r + \frac{1}{2}i\Gamma} = \frac{e^{i\delta_r}}{e^{-i\delta_r}} \tag{1.122}$$

$$= \frac{\cos\,\delta_r + i\,\sin\,\delta_r}{\cos\,\delta_r - i\,\sin\,\delta_r}$$

Hence:

$$\tan\,\delta_r = \frac{\frac{1}{2}\,\Gamma}{E_r - E} \tag{1.123}$$

which allows one to write for a given partial wave phase shift:

$$\delta_\ell(k^2) = \delta_{\ell,o}(E) + \delta_{\ell,r}(E)$$
$$= \delta_{\ell,o}(E) + \tan^{-1}\frac{\frac{1}{2}\Gamma(E)}{E_r - E} \qquad (1.124)$$

which is the well known Breit-Wigner one-level formula for a one-level resonance case. This treatment can be generalized to a many levels system. Across the resonance energy E_r the total phase shift varies very rapidly because of the $\tan^{-1}$ dependence of the resonance shift and since the background term is only a slowly varying function of energy.

For realistic interactions in heavy particle collision phenomena the 'mixed' nature of the relevant potentials, i.e. their short-range repulsive nature and long range attractive form, makes it rather interesting to look at those resonances which occur in single-channel scattering (potential scattering) and are called <u>potential or shape resonances</u>. In such cases the systems exhibit a repulsive barrier in the outer region of the interaction and the position of the resonance is determined within the experimentally observed width, since the corresponding pseudo-bound state can decay by barrier penetration.

The corresponding scattering amplitude within the resonance region can be written as follows:

$$f_\ell(\vartheta) = \frac{(2\ell+1)}{k}\, e^{ri\delta_{o,\ell}} \left\{ \frac{\Gamma/2}{E_r - E - i(\Gamma/2)} + e^{-i\delta_{o,\ell}} \sin\delta_{o,\ell} \right\} P_\ell(\cos\theta) \qquad (1.125)$$

Whenever the background phase shift is either zero (as it often is in neutron scattering experiments) or negligible, the total cross section is controlled mainly by the resonant part and the corresponding angular distribution depends on the Legendre polynomial of the resonant partial wave (eq.(1.125)):

$$\sigma_\ell(k^2) = \frac{4\pi}{k^2}(2\ell + 1)\sin^2\delta_\ell =$$
$$= \frac{4\pi}{k^2}(2\ell + 1)\frac{\Gamma^2/4}{(E - E_r)^2 + \Gamma^2/4} \qquad (1.126)$$

In nuclear physics, the outer barrier of the potential resonance is caused by the

Coulomb repulsion. In atomic and molecular physics, the origin of the barrier lies in the repulsive nature of the centrifugal angular momentum term associated with each partial wave equation like in (1.46a).

The radial wavefunction for such a state will exhibit large oscillatory amplitude in the inner region, decaying behaviour across the barrier and oscillations with much smaller amplitudes further outside:

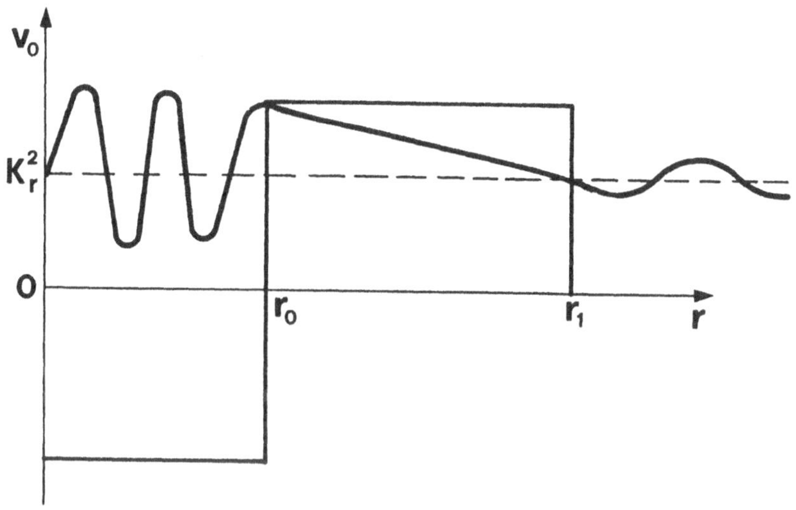

Fig.1 - A schematic representation of centrifugal barrier leakage.

In the limit of a very narrow resonance with nearly infinite lifetime, the radial function will go into the bound-state behaviour. For very large ℓ values, the typical potentials acting between atomic and molecular systems will exhibit very narrow shape resonances but only at high collision energies.

If one now goes back to the time-dependent representation as in eq.(1.9):

$$\Psi_{\text{incoming}} = \int a(\underline{k}' - \underline{k}) \exp\left\{ i(\underline{k}' \cdot \underline{r} - \omega' t) \right. \; d\underline{k}' \qquad (1.127)$$

where the energy is distributed at about the same mean value as the resonant energy. If: $\underline{q} = \underline{k}'-\underline{k}$ then the asymptotic form of (1.127) becomes:

$$\Psi_{incoming} \sim e^{i\underline{k}'\cdot\underline{r}} \int a(\underline{q})\, e^{i(\underline{q}\cdot\underline{r}-\omega' t)} d\underline{q} \quad (1.128)$$

where: (1.129)

$$\omega' = \frac{\hbar}{2\mu} k'^2 = \frac{\hbar}{2\mu}(k^2 + 2\underline{q}\cdot\underline{k} + q^2)$$

For a case where: $|\underline{q}| << |\underline{k}|$ and assuming the width of the wavepacket to be large when compared with the particle associated λ (see condition (1.6)):

$$\omega' \sim \omega + \frac{\hbar}{\mu}\underline{q}\cdot\underline{k}$$

and (1.130)

$$\Psi_{incoming} \sim e^{i(\underline{k}\cdot\underline{r}-\omega t)} \int a(\underline{q})\, e^{i\underline{q}(\underline{r}-\frac{\hbar k}{\mu}t)} d\underline{q}$$

which can be written further as:

$$\Psi_{incoming} \sim e^{i(\underline{k}\cdot\underline{r}-\omega t)}\, a(\underline{r} - \underline{v}t) \quad (1.131)$$

where the last term on the r.h.s. describes a wave packet moving along the incident direction with velocity $\underline{v} = \hbar\underline{k}/\mu$.

The corresponding scattered wave is given by:

$$\Psi_{scatt} \sim \int a(\underline{k}' - \underline{k})\, e^{-i\omega' t} \cdot f_{\underline{k}'}(\hat{r}) \frac{e^{ik'r}}{r} d\underline{k}' \quad (1.132)$$

which describes the scattering due to a particular momentum $\underline{k}'$ and averaged by $a(\underline{k}'-\underline{k})$ over the whole packet momentum distribution.

At a collision energy where there is no resonance the quantity $f_{\underline{k}'}$ is a slowly-varying function of energy and one can rewrite (1.132) as:

$$\Psi_{scatt} \sim \frac{f^{o}_{k}(\hat{r})}{r} e^{i(kr-\omega t)} \int a(\underline{q})\, e^{i\underline{q}(\hat{k}\cdot r - \underline{v}t)} d\underline{q} \quad (1.133)$$

This corresponds to a localized wave at the position $\hat{k}\cdot r = \underline{v}t$, i.e. there is no time delay with respect to the incident wavepacket of (1.131).

Let us now assume a resonance for a specific partial wave ℓ:

$$f_{res}(\vartheta) = \frac{2\ell+1}{rik}\,(e^{2i\delta_\ell} - 1)\, P_\ell\,(\cos\vartheta) \tag{1.134}$$

Then, to a 1st approximation, one can write:

$$\delta_\ell \sim \delta_{\ell,0} + \frac{\partial\delta_\ell}{\partial k}\, dk \tag{1.135}$$

where the resonance width has been assumed to be much larger than the wavepacket width. The quantity $f_{\underline{k}'}(\hat{r})$ can no longer be taken out of (1.132):

$$\Psi_{scatt} \sim \frac{f^{0}_{\hat{k}}\,(\hat{r})}{r}\, e^{i(kr - \omega t)} \int a(\underline{q})\, e^{i\underline{q}(\hat{k}\cdot r - \underline{v}t + 2\hat{k}\, d\delta_\ell/\partial k)}\, d\underline{q} \tag{1.136}$$

which corresponds now to a localized wave at:

$$\hat{k}\cdot r = \underline{v}t - \hat{k}\cdot\frac{2d\delta_\ell}{dk} \tag{1.137}$$

The additional term on the r.h.s. of (1.137) represents a <u>time delay</u> $\Delta t = \frac{2}{v}\frac{d\delta_\ell}{dk} = \frac{2\mu}{\hbar}\frac{d\delta_\ell}{dk}$. By using the definition of δ_ℓ from eq.(1.124) to perform the derivative and by taking the average of the time delay over the whole resonance, one obtains a very useful relationship between resonance width and delay time:

$$(\Delta t)_{av.} = \frac{4\hbar}{\Gamma} \tag{1.138}$$

Therefore the resonance lifetime is inversely proportional to the resonance experimental width, a result which will often be used in the following Chapters. The time delay can in turn be related to the nature of the interaction potential, since it can be shown that the energy gradient of the relevant phase shift is related to the range r_0 of the potential:

$$\frac{\partial\delta_\ell}{\partial k} \geqslant -r_0/2 \tag{1.149}$$

which is a way through which we will be able to show the effect of coupling matrix elements of the radial equations on the structure of the multichannel S-matrix.

If one assumes that it is possible to define $\delta_\ell(k^2)$ as a continuous function of energy, then it results that as $E\to\infty$ the $\delta_\ell(E) = 0$ for local, short-ranged potentials of the type discussed thus far. By further remembering that the Jost function had zeros but no poles in the lower-half of the complex k-plane, one can integrate the following expression within an area C of that lower plane and with contours following the real axis:

$$I_\ell = \int_C dk \left\{ \frac{d}{dk} \ln f_\ell(k) \right\} \tag{1.140}$$

Along the real axis one obtains:

$$I_\ell = \lim_{k\to\infty} \left\{ \ln f_\ell(-k) - \ln f_\ell(+k) \right\} + + \ln f_\ell(0_+) - \ln f_\ell(0_-) \tag{1.141}$$

and the integration along the infinite semicircle does not contribute for the common, real potentials of interest. Hence, from (1.104):

$$\begin{aligned} I_\ell &= \ln S_\ell(k\to o) - \ln S_\ell(k\to\infty) \\ &= 2i\left\{ \delta_\ell(o) - \delta_\ell(\infty) \right\} \end{aligned} \tag{1.142}$$

On the other hand, from Cauchy residue theorem one can also write:

$$\frac{1}{2\pi i} I_\ell = \sum_{p=1}^{n} r_p - \sum_{q=1}^{m} s_q \tag{1.143}$$

The two summations being over the p zeros of order r_p and the q poles of order s_q respectively. Since f (k) is analytic in the lower-half of the complex plane, the second sum on the r.h.s. of (1.143) goes to zero. Moreover, the zeros of the Jost function correspond in the lower-half plane, to the bound states n_b of the system. Hence:

$$I_\ell = 2\pi i \cdot n_b = 2i\left\{ \delta_\ell(o) - \delta(\infty) \right\} \tag{1.144}$$

which allows us to write down an important result for potential scattering via finite, local potentials:

$$\delta_{\ell}(o) - \delta_{\ell}(\infty) = \pi \cdot n_b \tag{1.145}$$

which is called the Levinson Theorem (Kgl. Danske Vid. Sels. Mat. Fys. Medd. 29 (1949) 9).

This result can be qualitatively depicted as the need to preserve the orthogonality of the lowest continuum state ($k \sim o$) within the inner-core ($r < r_o$) region of the potential to the bound state eigensolutions of the problem under study. Since the outermost, n_bth bound state will exhibit (n_b-1) nodes, then the corresponding continuum amplitude will have to force its $\tan \delta_{\ell}$ to have $(n_b-1)+1 = n_b$ distinct zeros, i.e. δ_{ℓ} will reach a corresponding multiple of π, as stated by eq.(1.145).

For physical situations where one is dealing with a non-local interaction, then the following equality holds:

$$U(r)\, u_{\ell}(r) = \int_0^{\infty} \mathbf{K}(r,r')\, u_{\ell}(r')\, dr' \tag{1.146}$$

In such cases the Kernel $\mathbf{K}$ of eq.(1.146) may in certain circumstances support bound states which are normalizable and which lie in the continuum. An alternative way of putting it is to remember that non-local terms arise mainly from Pauli principle requirements, which apply an extra orthogonality constraint to $u_{\ell}(r)$ over the whole r region.

Hence a generalization of (1.145) reads:

$$\delta_{\ell}(o) - \delta_{\ell}(\infty) = \pi \cdot \{ n_b + n_p \} \tag{1.147}$$

where n_p is now the number of states excluded by the Pauli principle.

For a given scattering problem, the potential in the one-dimensional wave equation for each one partial wave is obviously the same, and hence a strong correlation between the properties of the associated partial wave S-matrices is to be expected. In particular, if ℓ is allowed to vary continuously, the movements of the S-matrix poles can be considered as a function of ℓ as well. As ℓ is increased, in fact, the effective potential attractive strength is correspondingly decreased so that a bound state pole would be expected to move down the positive imaginary k-axis towards the origin. As the bound state moves into the continuum it splits into two poles, which move symmetrically into the fourth and third quadrant to give rise to resonance and conjugate poles respectively. It is only in the special case where there exists a pole at the origin for $\ell = 0$ that only a virtual state pole is formed. The 'trajectory' exhibited by the $S(k,\ell)$ in the complex k-plane as a function of ℓ gives us some information on the resonant states. The angle that each trajectory (with Im $k < 0$) makes with the real axis depends on the corresponding value of its ℓ when the pole goes to the origin (Re $k = 0$). This angle decreases as ℓ increases. Thus, a set of bound states and resonances can be attributed to a single pole in the S-matrix, in cases where this is treated as a function of ℓ. Further, the sharpness of the corresponding set of resonances will increase with an increase in the strength of binding of the associated bound state in the potential. The resonances in a given set will in fact increase in width with increasing ℓ, since the deeper bound states need a greater centrifugal barrier term to lift them to the same energy. This larger barrier will in turn inhibit the bound state decay and will give rise to a sharper resonance. Thus the higher the ℓ-value in a given energy region the sharper the resonance.

In the multichannel case there is no longer any point in choosing a specific k-plane, and a 2^N sheeted Riemannian surface in the energy plane should be considered (for the N channel case). The branch cuts are conventionally taken along the positive real axis. The movement of the poles as ℓ is varied is similar to the potential scattering (single channel) case, except that all poles now move off the real axis into the negative Im E plane portion at the first threshold and at each new threshold. It further moves rapidly away from the resonance region, a new pole taking over at each incoming threshold. This causes in general new threshold effects in the resonance structure of the cross sections, although they are often too narrow and superimposed to each other to be practically detected experimentally.

1.6 - Classical and semiclassical scattering: a set of definitions.

The scattering of two heavy particles (atoms and molecules) by their mutual interaction involves, except at very low energies, significant contributions from very-many partial waves. Furthermore, the variations from one phase shift to the next are often small even though the individual phase shifts are large, thus allowing the summation over partial waves to be replaced by an integration. In these circumstances it is then possible to use a classical path method to compute the cross sections. More specifically, the limiting conditions outlined in section 1.1 on linear momentum and impact parameter and contained in eq.s (1.29) and (1.96) indicate the classical approximation to be valid in most cases except at very small angles. Now the conservation of angular momentum and energy give respectively:

$$\ell = m\, r^2\, \dot{\vartheta}_1 = m\, v\, b \tag{1.148}$$

and

$$E = \tfrac{1}{2}\mu \left\{ \dot{r}^2 + r^2\, \dot{\vartheta}_1 \right\} + V(r) = \tfrac{1}{2}\omega\, v^2 \tag{1.149}$$

hence:

$$E = \tfrac{1}{2}\mu\, \dot{r}^2 + \tfrac{1}{2}\,\frac{\ell^2}{\mu r^2} + V(r) = \text{const} \tag{1.150}$$

or:

$$\dot{r} = \left\{ \frac{2}{\mu} \left[E - V(r) - \frac{\ell^2}{2\mu r^2} \right] \right\}^{\frac{1}{2}} \tag{1.151}$$

with the usual meaning of the symbols as defined in the Figure below:

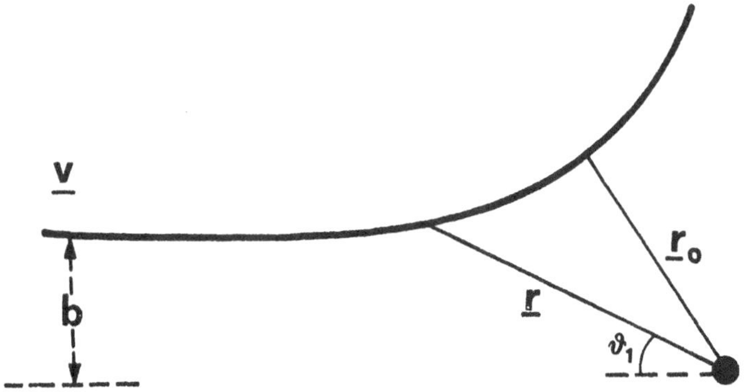

Fig.2 - A schematic representation of classical scattering from a centre of forces.

From (1.148) one gets that: $d\vartheta_1 = \ell\, dt/\mu r^2$, hence:

$$d\vartheta_1 = \ell\ dr\ /_{\mu}\ r^2 \left\{ \frac{2}{\mu} [E\ .\ v(r) - \ell^2/_{2\mu}\ r^2] \right\}^{\frac{1}{2}} \qquad (1.152\text{ a})$$

or:

$$d\vartheta_1 = b\ .\ d\ r\ .\ r^{-2} \left\{ 1 - \frac{b^2}{r^2} - \frac{V(r)}{E} \right\}^{-\frac{1}{2}} \qquad (1.152\text{ b})$$

By integrating eq.(1.151) one obtains the scattering angle ϑ. Thus:

$$\vartheta = \pi - \vartheta_1 \qquad (1.153\text{ a})$$

$$\vartheta(b,\ E) = \pi - 2b \int_{r_0}^{\infty} dr\ .\ r^{-2} \left\{ 1 - \frac{b^2}{r^2} - \frac{v(r)}{E} \right\}^{-\frac{1}{2}} \qquad (1.153\text{ b})$$

where r_0 is the outermost zero of the curly brackets term in the integrand (classical distance of closest approach).

From the last eq.s the differential cross section is thus given by (1.1), under the conditions (1.96). For simple inverse power potentials that are only attractive, the angular dependence is given analytically by:

$$\frac{d\sigma}{d\Omega}\ \alpha\ \vartheta^{-[(S+1)/3]} \qquad (1.154)$$

where S is the power of the r dependence. Eq. (1.154) presents a result which is often valid over a wide angular range.

The small angle collisions correspond to large values of b, where it is then possible to use the WKBJ approximation for the phase shifts. This method was applied to quantum mechanical scattering by Jeffreys (Proc. Lond. Math. Soc. (2), 23 (1923) 428) and then developed simultaneously by G.Wentzel, H.A.Kramers and L.Brillouin in 1926.

Provided that the variation in wavelength is small compared with λ itself, i.e.:

$$\text{grad}\ ƛ \ll 1 \qquad (1.155)$$

a condition already presented in eq.(1.5), it is possible to develop a solution of the Schrödinger equation in powers of $\hbar$. The first term in the expansion gives the classical limit. Thus, if we expand the solution of eq.(1.44) after writing it in the following form:

$$u(r) = e^{is(\underline{r})/\hbar} \tag{1.156}$$

where the classical limit is found by taking the limit $\hbar \to 0$ in the following expansion:

$$s = s_o + \hbar s_1 + \hbar^2 s_2 + \ldots.. \tag{1.157}$$

By substituting the full eq.(1.157) into (1.44) one gets:

$$\frac{i}{\hbar}\left\{ s''_o + \hbar s''_1 + \hbar s''_2 + \ldots \right\} - \frac{1}{\hbar^2}\left\{ s'_o + \hbar s'_1 + \hbar s'_2 + \ldots \right\}^2 + k^2 = o \tag{1.158}$$

where: $k^2 = \frac{2\mu}{h^2}\left[E - V(r) \right]$

Now, by equating the various powers of $\hbar$ one obtains:

$$s'^2_o = \hbar^2 k^2 \tag{1.159a}$$

$$i\, s''_o = 2s'_o \,.\, s'_1 \tag{1.159b}$$

The first of the above equations corresponds to the limit of taking only s_0 in the expansion (1.157) and it is explicitely written as:

$$\frac{1}{2\mu}\left\{ \nabla_{\underline{r}}\, s_o \right\}^2 = E - V(r) \tag{1.160}$$

The above expression becomes the classical Hamilton-Jacobi equation if $s_0(\underline{r})$ is identified with Hamilton's characteristic function. By analogy with optics, eq. (1.160) is termed the eikonal equation. Its integration determines the orbits and, if $s_0(\underline{r})$ is substituted into (1.156), the resulting wavefunction is defined as the eikonal wavefunction. The use of this wavefunction in the integral equation for the scattering amplitudes forms the basis of the eikonal approximation.

For many cases of interest in low energy collisions between heavy particles, the scattering is confined to small angles and the trajectories differ little from straight lines parallel to the incident direction.

The reference frame can be chosen with the $\hat{k}$ versor parallel to the z-axis and defining $\underline{b}$ as a vector orthogonal to $\hat{k}$. The classical **orbit** is then:

$$\underline{r} = \underline{b} + z \, . \, \hat{k} \tag{1.161}$$

Eq. (1.160) then reduces to:

$$\frac{1}{\mu} \left(\frac{\partial s_0(z)}{\partial z} \right)^2 = E - V(r) \tag{1.160 b}$$

with solution:

$$s_0(z) = \int \left\{ 2\mu \left[E - v(r) \right] \right\}^{\frac{1}{2}} dz + c \tag{1.162}$$

as z goes to $-\infty$ the wavefunction must approach a plane wave, thus $s_0(z) \sim kz$. Using this result in (1.162) one gets, for large values of $\hbar k$, with $\hbar = 1$:

$$s_0(z) = kz - \frac{\mu}{k} \int_{-\infty}^{z} v(r) dz \tag{1.163}$$

The scattering amplitude is then:

$$f.(\hat{r},\underline{k}) = - \frac{1}{2\pi} \int d\underline{r} \, e^{-i\underline{k}'.\underline{r}} \cdot \mu \, V(r) \, e^{i\left[\underline{k}.\underline{r} - \frac{\mu}{k} \int_{-\infty}^{z} V(r) \, dz\right]} \tag{1.164}$$

The momentum transfer $\underline{q}$ is: $\underline{k} - \underline{k}'$, with $q = 2k \sin \vartheta/2$. Since $|\underline{k}| = |\underline{k}'|$, $\underline{q}$ is almost perpendicular to $\underline{k}$ for potential scattering at small scattering angles, hence:

$$\underline{q} \cdot \hat{k} = k(1 - \cos\vartheta) \sim \frac{1}{2} k \, \vartheta^2 \tag{1.165}$$

Moreover, we can write:

$$\exp i(\underline{k} \, . \, \underline{r} - \underline{k}' \, . \, \underline{r}) = \exp \, i\underline{q}(\underline{b} + \hat{k}z) \tag{1.166}$$

for angles such that: $\vartheta^2 k \, r_m << 1$, with r_m being the largest r contributing to the integral (1.164), then $\exp(i\underline{q}\cdot\hat{k}z)$ may be replaced by 1. Therefore eq.(1.164) becomes:

$$f(\hat{r},\underline{k}) = -\frac{1}{2\pi}\int_0^{2\pi} d\phi \int_0^{\infty} b\,db \int_{-\infty}^{\infty} dz\; e^{i\underline{q}\cdot\underline{b}}\; \mu\, V(r)\; x$$

$$x \exp\left\{ -\frac{i\mu}{k}\int_{-\infty}^{\infty} V(r)dz'\right\} = \qquad (1.167)$$

$$= \frac{k}{2\pi i}\int_0^{2\pi} d\phi \int_0^{0} b\,db\; e^{i\underline{q}\cdot\underline{b}}\left[e^{-i\frac{\mu}{k}\int_{-\infty}^{\infty} V(r)dz} - 1\right]$$

for spherical potentials the integration over ϕ is immediately obtained:

$$\int d\phi\; e^{iqb\cos\phi} = 2\pi \quad J_0(qb) \qquad (1.168)$$

that yields, for eq.(1.167):

$$f(\vartheta,\underline{k}) = \frac{k}{i}\int_0^{\infty} J_0(qb)\left\{e^{i\chi(b)} - 1\right\} b\,db \qquad (1.169)$$

where the eikonal phase term is defined as:

$$\chi(b) = -\frac{\mu}{k}\int_{-\infty}^{\infty} V(r)\; dz \qquad (1.170)$$

For small ϑ values and large ℓ one can write: $P_\ell(\cos\vartheta) \sim J_0(2\ell\sin\vartheta/2)$, hence from the partial wave expansion of the scattering amplitude:

$$f(\vartheta,\underline{k}) = \sum_\ell (2\ell + 1)k^{-1}\; P_\ell(\cos\vartheta)\;\frac{1}{2i}(e^{2i\delta_\ell} - 1) \qquad (1.169b)$$

The large-ℓ region allows to approximate the P_ℓ's and to integrate (1.169b) :

$$f(\vartheta,\underline{k}) = \frac{1}{ik}\int_0^{\infty} b\,db \quad J_0(2\ell\sin\vartheta/_2)\;(e^{2i\delta_\ell} - 1) \qquad (1.169c)$$

Hence a very interesting relationship is established, under semiclassical conditions, between the eikonal phase and the phase shift:

$$2\,\delta_\ell = \chi(b) \qquad (1.170b)$$

where b and ℓ are related by: $\ell = kb$.

The solutions of eq.s (1.159a) and (1.159b) are given by:

$$s_0 = \pm \int \sqrt{k^2}\, dr \qquad (1.159c)$$

$$s_1 = \tfrac{1}{2} i \log k \qquad (1.159d)$$

where we have written: $\hbar = \mu = 1$. If one now uses these results with the correct solution of (1.44) as given in (1.156):

$$u(r) \sim \frac{1}{\sqrt{k(r)}} \exp \{\pm i \int \frac{\sqrt{2(E-V)}}{k} dr\} \qquad (1.171)$$

where only the first term in the expansion (1.157) has been used. We see then that it presents a wider range of applicability than the purely classical solutions, since it is valid even in regions where $E < V(r)$, which are just the regions forbidden to classical particle trajectories. One is then forced to develop connection formulae between these two regions, as extensively discussed in some of the suggested reading of this chapter(ref.7). In these treatments one can show that it is more accurate to introduce the angular momentum term in the form of $(\ell + \frac{1}{2})^2$ rather than in the more usual $\ell(\ell + 1)$ form. From eq.(1.171) it then follows that, to the first order in $\hbar$, the phase introduced by 'switching on' the potential interaction is:

$$\delta_\ell \sim \int^\infty \left\{2\mu \left[E - V(r)\right] - (\ell + \tfrac{1}{2})^2/r^2 \right\}^{\frac{1}{2}} dr$$
$$- \int^\infty \left\{2\mu E - (\ell + \tfrac{1}{2})^2/r^2\right\}^{\frac{1}{2}} dr \qquad (1.172)$$

the lower limit of integration being the zero of the integrand in the classical path integration. Whenever $V(r) << k^2$ the expansion of the square root term of the first integral on the r.h.s. of (1.172) gives:

$$\delta_\ell \sim \int_{(\ell+\frac{1}{2})/k}^\infty \frac{V(r)dr}{\left\{k^2 - (\ell+\frac{1}{2})^2/r^2\right\}^{\frac{1}{2}}} \qquad (1.173)$$

If we compare it with (1.153b) we find the following relationship between the JWKB phase and the classical deflection function:

$$\vartheta(\ell) = 2\,\frac{\partial \delta_\ell}{\partial \ell} \tag{1.174}$$

Hence, for the cross section one can choose two partial waves regions, the one with $\ell < L$ where the phase is large and the outer one with $\ell > L$ where the phase is small and can be approximated; thus the cross section becomes:

$$\sigma \sim \frac{4}{k^2}\left\{ \int_0^L 2\ell \sin^2 \delta_\ell \, d\ell + \int_L^\infty 2\ell \sin^2 \delta_\ell \, d\ell \right\} \tag{1.175}$$

If one assumes in the first integral that $\sin^2\delta_\ell$ is on the average $\sim \frac{1}{2}$, i.e. there are no regions of stationary phases (random phase approximation) and uses in the second integral the result of eq.(1.173), then eq.(1.175) becomes:

$$\sigma(E) \sim \frac{2L^2}{k^2} + \frac{4}{k^2}\int_L^\infty 2\ell\, \delta_\ell^2 \, d\ell \tag{1.176}$$

For purely attractive potentials of the r^{-s} type and by choosing the outer limit for the ℓ-value to be that L for which: $|\delta_L| = 1$ one obtains:

$$\sigma = \left\{ 2 + \frac{4}{s-2} \right\} \times \left\{ \frac{F(s)}{k} \right\}^{2/s-1} \tag{1.177}$$

where $F(s)$ is a function of s that depends on its value (Mott and Massey Proc.Roy. Soc. A114 (1934) 188).

If one considers a situation where two particle trajectories have classical angular momenta ℓ_1 and ℓ_2 corresponding to two impact parameters b_1 and b_2 with the same scattering angle ϑ_1, the corresponding classical deflection function, at a given energy, is associated with two different values of b. The total scattering amplitude in the semiclassical approximation is then given by:

$$f_{s.c.}(\vartheta) = \sum_j \left\{ \sigma_{c\ell}(\vartheta) \right\}_j^{\frac{1}{2}} e^{i\beta j} \tag{1.178}$$

where the phase β_j depend on the classical path taken for each trajectory (with a mixed potential). For the two branches discussed before one has that:

$$\sigma_{s.c.}(\vartheta) = \left| \left[\left\{ \sigma_{c\ell}(\vartheta) \right\}_1^{\frac{1}{2}} + \left\{ \sigma_{c\ell}(\vartheta) \right\}_2^{\frac{1}{2}} \right] \exp\{i(\beta_2 - \beta_1)\} \right|^2 \qquad (1.179)$$

Hence the semiclassical differential cross section exhibits interference effects as ϑ is varied.

Furthermore, eq.(1.1) can be cast in an equivalent form by using $\ell = \mu vb$:

$$\sigma(\vartheta) = \frac{1}{\mu^2 v^2 \sin \vartheta} \, \ell \, \frac{d\ell}{d\vartheta}$$

$$= \frac{1}{\mu^2 v^2 \sin \vartheta} \, \ell \times \left(\frac{d\vartheta}{d\ell} \right)^{-1} \qquad (1.180)$$

which shows that $\sigma(\vartheta)$ has a pole at the points where:

$$\left(\frac{d\vartheta}{d\ell} \right) = 0 \quad ; \quad \sin \vartheta = 0 \qquad (1.181)$$

They correspond to very important experimental effects which will be more extensively discussed in later Chapters. The first discontinuity is called Rainbow scattering while the second one correspond to a Glory scattering (forward, backward).

Finally, another interesting potential scattering phenomenon is seen for those collision energies for which the denominator in eq.(1.153b) repeatedly vanishes, hence the corresponding integral diverges at the lower limit. The corresponding deflection function therefore exhibits a pole for the particular angular momentum involved. Physically this corresponds to the mutual capture of the colliding particles, a situation that can arise for any attractive potential with long range terms, provided that the collision energy is low enough to prevent the maximum in the effective potential: $V(r) + \ell_1(\ell_1 + 1)/r^2$ to be replaced by a point of inflection. For the angular momenta that are lower than ℓ_1 the corresponding barrier is lower and the incident particle penetrates the barrier, thus getting trapped in one of a series of resonance levels that makes it go by π repeatedly. The first one is called an orbiting phenomenon and it is a classical effect for which a semiclassical approximation for the phase shift cannot be derived. The second set of resonance effects occurs at a further decrease of ℓ that will in turn cause the turning point to make a sudden jump inwards since V_{eff} is now less than the collision energy. This produces a discontinuity in the semi-classical phase shift. To remove such an unphysical behaviour one thus has to take full account of the quantum mechanical tunnelling through the potential barrier for large ℓ.

A further important influence of the classically inaccessible region arises from the presence of virtual energy levels. Considering in fact the variations of V_{eff} with r at a fixed ℓ_1 for which an inner attractive well exists, if this well is deep and wide enough there may exist a number of levels supported by it and which, for $E > 0$, are not stable due to leakage through the barrier. These are virtual levels that appear broadened because of their finite lifetime, controlled by eq.(1.138). In general the width of each of them is so small that the energy resolution available in practice will be insufficient to detect the effect. If the virtual levels occur however near the top of the wall, the resonance may be sufficiently long-lived to be observed (core-excited resonances).

SUGGESTED READING FOR CHAPTER 1

[1] N.F. Mott and H.S.W. Massey: The Theory of atomic collisions, 3rd Ed, Oxford U.P. (1965).

[2] R.G. Newton, Scattering Theory of waves and particles, Mc-Graw-Hill, N.Y. (1966).

[3] M.L. Goldberger and K.M. Watson, Collision Theory, J. Wiley (1967).

[4] E.G. Phillips, Functions of complex variables, Oliver & Boyd, London (1957).

[5] J. Heading, Phase integral methods, Methuen, N.Y. (1962).

[6] H. Goldstein, Classical Mechanics, Addison Wesley N.Y. (1959).

[7] K.W. Ford and J.A. Wheeler, Ann. Phys. 7 (1959) 259, 287.

[8] B.H. Brandsen, Atomic Collision Theory, Benjamin Inc., N.Y. (1970).

[9] M.S. Child, Molecular Collision Theory, Academic Press, London (1974).

[10] R.K. Nesbet, Adv. Quantum Chem. 9 (1975) 215.

[11] D.R. Bates, Atomic and molecular processes, Academic Press, New York (1962).

[12] R. Glauber, Lectures in Theoretical Physics (ed. W.E. Britten and L.G. Dunham) 1 (1959) 315.

[13] P.G.Burke, Potential Scattering in Atomic Physics, Plenum Press, New York and London (1977).

2. POTENTIAL ENERGY HYPERSURFACE CALCULATIONS FOR SIMPLE SYSTEMS

2.1 - Kinematics considerations.

As soon as the collision partners acquire the complexity of the real systems, any evaluation of the potential $V(\underline{r})$ becomes a topic in itself and simple analytic expressions have to give way to actual calculations that aim at rigorously taking into account all the forces at play.

First of all, one readly recognizes that the effects of the interactions between the electrons belonging to the scattering partners and those corresponding to the interactions between the nuclei are 'switched on' within different time scales due to the different velocities involved when molecular collisions at near-thermal energies are examined. Thus, if one considers that a typical velocity for a valence electron is of the order of 10^8 cm $\cdot$ sec^{-1} the corresponding velocity for two colliding atoms will reach similar values only for collision energies larger than 100 KeV. Moreover we have already seen that in the simplest case of structureless partners the motion is classically separated into one associated with a fixed laboratory system (see eq. (1.43)) and another associated with the 'inertial' system of the centre of mass (c.m.) (eq. (1.44)). Any actual measurement will provide the scattering angle Θ in the laboratory system, while θ will be the scattering angle in the c.m. system as in eq. (1.164 a). They are in turn related by the well known transformation:

$$\tan \Theta = \frac{\sin \theta}{\cos \theta + (M_1/M_2)} \qquad (2.1)$$

so that, when for instance the partners are an electron and an atom, then: $M_1/M_2 \simeq$

$\simeq 10^{-4}$ and one can assume that the scattering centre is fixed in the laboratory system (i.e. $M_1/M_2 \simeq 0$). In the cases of interest here, however, the ratio is often close to unity and the transformation (2.1) has to be taken into account when comparing measured Θ's with the computed θ's in the c.m. system of relative motion.

Any evaluation of the interactions between realistic systems starts therefore with detailed analysis of the internal structure of the partners and of the relevant coordinates to be used in either of the above reference systems.

In the laboratory, space-fixed (SF) reference system the corresponding time independent Schrödinger equation can be written as:

$$\{-\tfrac{1}{2}\sum_{i=1}^{n} \nabla^2_{\underline{r}_{oi}} - \frac{1}{2M_A}\nabla^2_{\underline{R}_A} - \frac{1}{2M_B}\nabla^2_{\underline{R}_B} + V(\underline{r}_{oi},\underline{R}_A,\underline{R}_b) - E_{tot}\}\,\Psi_{tot} \equiv 0 \qquad (2.2)$$

where M_A and M_B are the masses of the two atomic (or molecular) partners in atomic units, $\underline{r}_{oi}$ are the coordinates of the ith electron (for an n-electron total system) with respect to the SF origin and $\underline{R}_A$, $\underline{R}_B$ are the atomic nuclei (or molecular c.m.'s) vector positions with respect to the same origin. More precisely, one should have used for each atom (or molecule) the coordinate of the (electrons + nucleus) or (electrons + nuclei) system. The low mass ratios, however, allow without much loss of accuracy [1] the substitution of these coordinates with the ones defined in eq. (2.2).

Because of the large velocity differences between electrons and nuclei (molecules) at near thermal energies, it is more useful to employ the relative coordinate of their centres of mass for the heavier particles and to express the electron coordinates relative to their respective nuclei. Thus eq. (2.2) becomes (in a.u.):

$$\{\frac{1}{2M}\nabla^2_{\underline{R}_{cm}} - \tfrac{1}{2}\sum_{i=1}^{n_a}\nabla^2_{\underline{r}_{Ai}} - \tfrac{1}{2}\sum_{j=1}^{n_b}\nabla^2_{\underline{r}_{Bj}} - \frac{1}{2\mathcal{M}}\nabla^2_{\underline{R}'} + V - E_{tot}\}\,\Psi_{tot} = 0 \qquad (2.3)$$

where: $\underline{r}_{Ai} = \underline{r}_{oi} - \underline{R}_A$, $\underline{R}_{Bj} = \underline{r}_{oj} - \underline{R}_B$ and: $i = 1,2\ldots n_a$, $j = n_a + 1,\ldots n$. Moreover: $M\underline{R}_{cm} = (M_A + n_a m)\,\underline{R}'_A + (M_B + n_b m)\,\underline{R}'_B$, with m being the electron mass and $\underline{R}'_A$, $\underline{R}'_B$ satisfy the equalities:

$$(M_A + n_a m)\,\underline{R}'_A = M_A\,\underline{R}_A + m \sum_i^{n_a} \underline{r}_{oi} \tag{2.3'}$$

$$(M_B + n_b m)\,\underline{R}'_B = M_B\,\underline{R}_B + m \sum_{j=n_a+1}^{n} r_{oi} \tag{2.3''}$$

the meaning of the other vectors of eq. (2.3) involves the overall system centre-of-mass: $M = M_A + M_B$ and $M\underline{R}_{cm} = (M_A + n_a m)\cdot\underline{R}'_A + (M_B + n_b m)\cdot\underline{R}'_B$. The relative coordinate: $\underline{R}' = \underline{R}'_A - \underline{R}'_B$ and $\mathcal{M}$ is the total reduced mass. In analogy with eq. (1.44) one can then write:

$$\Psi_{tot} = e^{i\underline{K}_{cm}\cdot\underline{R}_{cm}} * \Psi(\sum_i r_{Ai}, \sum_j r_{Bj}, \underline{R}') \tag{2.4}$$

which yields the following equation for the relative motion:

$$\{-\tfrac{1}{2}\sum_i^{n_a} \nabla^2_{\underline{r}_{Ai}} - \tfrac{1}{2}\sum_j^{n_B} \nabla^2_{\underline{r}_{Bj}} - \frac{1}{2\mathcal{M}} \nabla^2_{\underline{R}'} + V - E'_{tot}\}\Psi = 0 \tag{2.5a}$$

where:

$$E'_{tot} = E_{tot} - \frac{2}{M} K^2_{cm} \tag{2.5b}$$

When the two partners are molecules it often becomes more useful to choose for all the bound electrons a set of coordinates that are centred on the centre-of-mass of all the nuclei alone, i.e. to disregard the small displacing effects of the electronic masses in eq.s (2.3') and (2.3"), thus following a "supermolecule" picture for the electronic part. Then: $\underline{r}_i = \underline{r}_{oi} - \underline{R}_{cm}$, with: $M\,\underline{R}_{cm} = M_A\,\underline{R}_A + M_B\,\underline{R}_B$ and: $\underline{R} = \underline{R}_A - \underline{R}_B$. All the electron-electron, electron-atom and atom-atom interactions are included in the potential term $V(\underline{r}, \underline{R})$ with: $\underline{r} = \sum_i^n \underline{r}_i$. The linear transformation thus applied to the kinetic energy of the nuclei and electrons provides cross terms of the type: $\nabla_{\underline{r}_{Ai}} \cdot \nabla_{\underline{r}_{Bj}}$ that arise from the finite masses of the nuclei. However, they are of the order of $m/(M_A + M_B)$ when compared with the $\nabla^2_{\underline{r}_i}$ terms and hence can be disregarded. The above eq. (2.5a) therefore becomes:

$$\{-\tfrac{1}{2}\sum_i^n \nabla^2_{\underline{r}_i} - \frac{1}{2\mathcal{M}}\nabla^2_{\underline{R}} + V(\underline{r}, \underline{R}) - E'_{tot}\}\,\Psi = 0 \tag{2.6}$$

For slow molecular velocities one can look for an approximate solution of (2.6) by disregarding $\nabla^2_{\underline{R}}$ and writing instead an equation with a separable potential $V'(\underline{r}; \underline{R})$:

$$\{-\tfrac{1}{2}\sum_i^n \nabla^2_{\underline{r}_i} + V'(\sum_i \underline{r}_i ; \underline{R})\}\,\Psi_m(\underline{R}, \sum_i^n \underline{r}_i) = E_m^{el}(\underline{R})\,\Psi_m \tag{2.7}$$

where the $\Psi_m(\underline{R}, \underline{r})$ are eigensolutions for the n-electron problem in a field of stationary nuclei at fixed $\underline{R}$. For the simplest case of two interacting atoms the various $E_m^{el}(R)$ obviously provide the corresponding potential for the nuclear internal motion that yields in turn the following time-independent equation:

$$\{-\frac{1}{2\mathcal{M}}\nabla^2_{\underline{R}} + V_N(\underline{R}) + E_m^{el}(\underline{R})\}\,\chi_\gamma^m(\underline{R}) = E'_{tot}\,\chi_\gamma^m(\underline{R}) \tag{2.8}$$

where $V_N(R)$ contains the nucleus-nucleus interactions and γ is a collective quantum number describing the rotational and vibrational states of the diatom in the mth electronic state. Since the eigensolutions of eq.s (2.7) and (2.8) provide two complete sets of orthonormal functions, we can write the total wavefunction for the relative motion as:

$$\Psi\,(\underline{R}\,,\,\underline{r}) = \sum_{m,\gamma} \Psi_m\,(\underline{R}\,;\,\underline{r})\;\chi^m_\gamma\,(\underline{R}) \tag{2.9}$$

By substituting (2.9) into (2.6), premultiplying it by Ψ^*_m and integrating over $\underline{r}$ one obtains:

$$\{ -\frac{1}{2\mathcal{M}}\,\nabla^2_{\underline{R}} + V_N\,(\underline{R}) + E^{el}_m\,(\underline{R}) - E'_{tot}\,\}\,\chi^m_\gamma\,(\underline{R}) =$$

$$= \frac{1}{2\mathcal{M}} \sum_{m',\gamma'} C_{mm'}\,(\underline{R})\,.\,\chi^{m'}_{\gamma'}\,(\underline{R})\,\} \tag{2.10}$$

where:

$$C_{mm'}\,(\underline{R})\;\chi^{m'}_{\gamma'}\,(\underline{R}) = A_{mm'}\,(\underline{R})\;\nabla_{\underline{R}}\;\chi^{m'}_{\gamma'}\,(\underline{R}) + B_{mm'}\,(\underline{R})\;\chi^{m'}_{\gamma'}\,(\underline{R}) \tag{2.11}$$

and with the following meaning of the A's and the B's:

$$A_{mm'}\,(\underline{R}) = 2\int d\underline{r}\;\Psi^*_m\,(\underline{r}\,,\,\underline{R})\;\nabla_{\underline{R}}\;\Psi_{m'}\,(\underline{r}\,,\,\underline{R}) \tag{2.12a}$$

$$B_{mm'}\,(\underline{R}) = \int d\underline{r}\;\Psi^*_m\,(\underline{r}\,,\,\underline{R})\;\nabla^2_{\underline{R}}\;\Psi_{m'}\,(\underline{r}\,,\,R) \tag{2.12b}$$

This final form of the stationary equation can be approximated at different levels. In what is called the Born-Oppenheimer (B-O) approximation all the r.h.s. terms in eq. (2.10) are ignored. One then obtains a definite potential surface

E_m ($\underline{R}$) and the coupling terms between nuclear and electronic motion are assumed to be negligible. If the diagonal term only is included from (2.11), one still has a potential surface, while the extra terms can be considered corrections allowing for the coupling of the nuclear and electronic motions. This is called the adiabatic approximation. Finally, if all terms are included in (2.10) the coupling between different electronic states becomes allowed and the definition of a potential surface is then more problematic.

The validity and justification of the expansion (2.9) was originally put forward in a classic paper by Born and Oppenheimer [2], who derived this result by essentially applying a perturbative treatment of the molecular eigenvalue problem. They thus introduced a parameter λ which was given by the fourth root of the ratio of the electronic mass and a mass of the order of magnitude of the nuclear masses. The molecular Schrödinger equation was expanded in terms of the λ parameter and the ensuing 'fixed nuclei' eigenvalue problem for the electronic part (eq. (2.7)) was developed in powers of the internuclear displacement, provided the latter were assumed to be small. They thus showed via the above two expansions [3] that when no degeneracy in the electronic energy occurs, i.e. for most ground-state electronic configurations of molecules, then the molecular wavefunctions are given, within an error of the order of λ^3, by a product of electronic and nuclear factors. Moreover, the corresponding molecular energy levels can be obtained, with an error of the order of λ^6, from the eigenvalue equation satisfied by the nuclear factors. The above treatment gives therefore a systematic ordering in powers of λ of the terms contributing to the energy and to the wavefunction, hence it throws some light on the order of magnitudes of the errors involved.

These conclusions cannot, however, be applied to molecular collisions. First of all, because the assumption made by them, that nuclear motions are confined to a small vicinity of an equilibrium configuration, is obviously not met in collisions. Moreover, the accuracy of the computed total energy is not a problem in scattering calculations, since it is generally fixed by the experimental conditions and it is not a fundamental observable.

Several recent derivations have however dealt with the adiabatic approximation for collision states and with the order of magnitude of the errors involved [4,5,6,7]. It appears from their treatment that such an approximation yields a transition T-matrix correct up to the order of λ^3, while the total wavefunction is correct up to order of λ^2. These results are subject to the condition that the electronic energy $E_m^{el}(\underline{R})$ is not degenerate.

Thus the physical reasons for applying such an approximation turn out to be the same for both the bound state and the scattering situation: the electronic kinetic energy has to be much larger than the nuclear kinetic energy [3]. It is in fact the excitation of electronic levels in atoms or molecules by transfer of nuclear kinetics energy that causes the breakdown of the approximation.

Let us now briefly examine the atom-atom interaction between closed-shell systems. The $E_m^{el}(\underline{R})$ has been computed in (2.7) without the inclusion in the separable potential $V'(\underline{r}_i ; \underline{R})$ of the nuclear interactions: $(Z_A \cdot Z_B) \cdot \underline{R}^{-1}$. Hence the B-O version of eq. (2.10) in the absence of electric and magnetic fields can be rewritten as follows:

$$\left\{ -\frac{1}{2\mathcal{M}} \nabla_{\underline{R}}^2 + \frac{Z_A \cdot Z_B}{R} + E_m^{el}(R) - E_m^{el}(\infty) - \left[E'_{tot} - E_m^{el}(\infty)\right]\right\} \chi_\gamma^m(\underline{R}) \equiv 0 \tag{2.13}$$

where: $E_m^{el}(\infty)$ is the electronic energy of the two partner atoms at infinity in the absence of interaction and hence it is given by the sum of their isolated electronic energies.

The effective central potential:

$$V(R) = \frac{Z_A \cdot Z_B}{R} + E_m^{el}(R) - E_m^{el}(\infty) \tag{2.14}$$

provides the Born-Oppenheimer potential curve for the diatomic system once the centre of mass motion is taken out. It can be readily extended to groups of atoms when representing interacting molecular systems within their nondegenerate total electronic state Γ :

$$V(\underline{R}, \underline{X}) = \sum_{i,j}^{N} \frac{Z_i Z_j}{|\underline{X}_i - \underline{X}_j|} + E_\Gamma^{el}(\underline{R} ; \underline{X}) - E_\Gamma^{el}(\infty ; \underline{X}) \tag{2.15}$$

where the vector $\underline{X}$ now stands for all the nuclear coordinates of the two molecular partners, i.e. it describes the (3N - 3) internal degrees of freedom of the supersystem, while $\underline{R}$ is the relative position vector of their centres-of-mass.

For the diatomic systems that form stable molecules, eq. (2.14) obviously provide bound states corresponding to roto-vibrational states of the system within a certain range of $\underline{R}$ values. More in general the potentials of (2.15) for the cases that we will be examining here will involve either a very small number of loosely bound 'complexes' or none at all. A detailed knowledge of $V(\underline{R}, \underline{X})$ is however of central importance for studying energy transfer and energy disposal effects during collisions, whereby the kinetic energy of the relative motion goes into or gains from some energy-storing degrees of freedom of a portion of the supersystem within the electronic state Γ, i.e. the rotations or vibrations of one or both of the molecules involved. The methods for acquiring this knowledge will be the subject of the following sections in this Chapter.

2.2 - Theoretical development of *a priori* methods.

From the discussion of the previous section, we have estabilished that the total Hamiltonian for our problem can be conveniently written as:

$$\mathcal{H} = -\frac{1}{2\mathcal{M}}\nabla^2_{\underline{R}} + \sum_i \mathcal{H}^{(i)}_{int}(\underline{X}_i) + V(\underline{R}, \underline{X}) \qquad (2.16)$$

where the potential V is the one given in eq. (2.15) and defined to vanish at infinity: $\lim_{\underline{R}\to\infty} V = 0$. The internal Hamiltonians $\mathcal{H}^{(i)}_{int}$ depend only on a group of internal variables for each subsystem and yield stationary eigensolutions at $\underline{R}\to\infty$:

$$\mathcal{H}^{(i)}_{int}(\underline{X}_i)\varphi^{(i)}_j(\underline{X}_i) = E^{(i)}_j\varphi^{(i)}_j(\underline{X}_i) \qquad (2.17)$$

Here $E^{(i)}_j$ will correspond to the sum of the isolated vib-rotational energies either for each of the two molecules, or for one molecule only or is simply the reference level $E_m^{el}(\infty)$ of eq. (2.13). For the cases to be discussed in the following Chapters, the E_j's could be given by rigid rotor eigenfunctions or by Morse oscillator eigenfunctions or by other, more sophisticated descriptions of simple diatomics. The expression (2.17) is however applicable in all cases and therefore provides a good starting point for constructing the total scattering wavefunction.

We are interested here in the calculation of a $V(\underline{R}, \underline{X})$ that one can obviously rewrite as:

$$V(\underline{R}, \underline{X}) = E'_{\Gamma}(\underline{R}, \underline{X}) - E'_{\Gamma}(\infty, \underline{X}) \tag{2.18}$$

where $E'_{\Gamma}(\underline{R}, \underline{X})$ is now the total electronic energy (including nuclear repulsion) of the supersystem at arbitrary geometries specified by $\underline{R}$ and $\underline{X}$ and $E'_{\Gamma}(\infty, \underline{X})$ is the total electronic energy (+ nucl. repulsions within each fragment) at infinite separation of the colliding partners. For a general system of n electrons at position $\underline{r}_i$, N nuclei of masses M_K, charges Z_K at positions $\underline{R}_K$ in a SF frame of reference (taken usually at the centre of mass of the supersystem, as discussed before), the time-independent Schrödinger equation is written as (in atomic units):

$$\Big\{ -\tfrac{1}{2} \sum_i^n \nabla^2_{\underline{r}_i} + \sum_{i<j} \frac{1}{|\underline{r}_i - \underline{r}_j|} - \sum_{i,k} \frac{Z_K}{|\underline{r}_i - \underline{R}_K|} + \sum_{k<l} \frac{Z_K Z_l}{|\underline{R}_K - \underline{R}_l|} \Big\} \Psi_{el}^{\Gamma} = E'_{\Gamma}(\underline{R})\, \Psi_{el}^{\Gamma}(\underline{r}, \underline{R}) \tag{2.19}$$

where $\underline{r}$ and $\underline{R}$ indicate collective indices for all the electronic and nuclear positions respectively.

While the problem of solving eq. (2.19) is well defined in principle, in practice not even for the simplest reaction systems is the potential surface $E'_{\Gamma}(\underline{R})$ available accurately enough that it can be regarded as known. The reasons for this stem from a number of computational difficulties:

(i) The multicentre problem. All interactions within simple molecular systems require solution of at least a three-centre problem, which in turn demands the evaluation of many time-consuming multicentre integrals over the electrostatic operators of (2.19).

(ii) The functional $E'_{\Gamma} = E'_{\Gamma}(\underline{R})$. In contrast to the study of molecules in their equilibrium geometries $\underline{R}_{eq}$, chemical reactions or subreactive encounters require the knowledge of the functional over a large region of $\underline{R}$ values. For a three--atom case, for instance, at least 200 points are needed to determine the surface, and even more as the number of atoms increases. This implies very extensive computational work and the further need of transforming the point-like form of $E'_{\Gamma}(\underline{R})$ into an analytic expression for use in scattering problems (see next Chapter).

(iii) Accuracy thresholds. For chemical reactions, or for weakly interacting nonchemical systems, a very high accuracy is essential; e.g. a change in the barrier height by $\simeq$ 1.5 Kcal can change computed rate constants at room temperature by a factor of $\simeq$ 10. Since the total energies are of the order of $10^5 - 10^6$ Kcal this implies that an accuracy of 0.001% or better is desiderable for problems where potential wells or repulsive barriers are important. Of course, certain aspects of the non-reactive systems can be studied with cruder surfaces but a word of caution is always needed to go beyond the semiquantitative level.

(iv) Tests with observables. Again in constrast with calculations involving stable molecules, for which available dissociation energies, equilibrium distances and measured force constants or electric permanent multipoles can be used to provide checks for approximate treatments, there is very little direct knowledge on the various regions of a potential surface required in a chemical reaction. The information available from bulk kinetic data in the gas phase or from molecular beam experiments cannot as yet yield quantitative values for all the features of a potential surface, albeit being very useful for certain specific features of a mixed potential [8].

It is the purpose of this Section to survey briefly the methods of computation of molecular wavefunctions and potential curves that provide direct applications of the scattering problems we are dealing with to specific, realistic systems that have been experimentally analyzed in recent years [8,9]. The numerous but often closely related formalisms that have been put forward in the literature to carry out the above computations will therefore not be examined in detail. It is in-

-tended, however, to indicate different levels of performance and success and to discuss the common ground of physical approximations. Only ab initio methods will be presented in some detail, but only passing reference will be made to the many empirical reductions of complexity. We will mention however, the difficulty raised by the electronic operator $(|\underline{r}_i - \underline{r}_j|)^{-1}$ that is often referred to by the theoretical chemists as the correlation problem, of which interesting recent reviews have been given by Kelly [10] and Nesbet [11,12].

All the calculations that have been made so far by treating all terms of the Hamitonian (2.19) and that are usually defined as being of the ab initio type, have involved an expansion of the total wavefunction for the supersystem. It was in fact recognized quite early [13,14] that it is convenient to write the Ψ_{tot} in the form:

$$\Psi_i = \sum_j C_{ij}\, \phi_j\, (\underline{r},\underline{R},\alpha) \tag{2.20}$$

where the ϕ_j are the N-particle composite functions of an elementary basis set $\{\chi_k\}$, depending on the electronic and nuclear coordinates and on a set of parameters $\{\alpha\}$. The ϕ_j's are antisymmetric functions of electronic space-spin coordinates and are usually (but not necessarily) eigenfunctions of the operators which commute with the electrostatic Hamiltonian of (2.19). The χ_k can in turn be one-particle (orbital) functions, two-particles (geminal or pair) functions, or even (less commonly) multi-particle functions. A complete set of trial functions would of course lead to exact results. For computational reasons, however, the expansion is usually truncated and only a finite (usually small) number of terms can be included in eq. (2.20). If the ϕ_j,s do not contain interelectronic distance coordinates correctly, the convegence of the expansion is slow. If, however, such coordinates are directly used the corresponding integrals of the following equations are much more difficult and have 3N dimensionality, where N is the number od electrons. The way of determining the functions ϕ_j and the linear coefficients C_{ij} is therefore very important.

The main method usually employed is based on the Rayleigh-Ritz variation principle; that is one searches for a minimum of the functional:

$$E_{i,true} \leq \frac{\langle\Psi_i|\mathcal{H}|\Psi_i\rangle}{\langle\Psi_i|\Psi_i\rangle} \tag{2.21}$$

The use of a function Ψ_i given by (2.20) generates the well-known secular equation:

$$|\mathbb{H}_{nm} - E_i\,\mathbb{S}_{nm}| = 0 \tag{2.22}$$

where:

$$\mathbb{H}_{nm} = \langle\phi_n|\mathcal{H}|\phi_m\rangle \tag{2.23a}$$

$$\mathbb{S}_{nm} = \langle\phi_n|\phi_m\rangle \tag{2.23b}$$

and where both matrix elements are a function of $\underline{R}$ and α. The minimum value for the functional of the above is an upper bound to the true value of the energy for the ground electronic state of the surface. Accurate lower bounds are much more difficult to achieve and have been used much less in calculations [15].

An alternative way to approach the problem is to minimize the variance $U_i^2(E)$ over a grid of points in configuration space , where:

$$U_i^2 = \frac{\int(\mathcal{H}\Psi_i - E_i\Psi_i)^2\,d\tau}{\int\Psi_i^2\,d\tau} \tag{2.24}$$

the substitution of the expansion (2.20) into (2.24) now yields another form of the secular equation:

$$|\mathbb{H}^2_{nm} - E_i \mathbb{H}_{nm} + (E_i^2 - U_i^2)\,\mathfrak{S}_{nm}| = 0 \qquad (2.25a)$$

where:

$$\mathbb{H}^2_{nm} = \int \phi_n \mathcal{H}^2 \phi_m \, d\tau \qquad (2.25b)$$

and the energy E_i of (2.25a) is a prior estimate for the true energy. This approach, coupled with an extrapolation procedure for E_i [16], appears to have the advantage over the previous variation treatment that cruder values for the required matrix elements are already satisfactory for estimating the variance. It has however the disadvantage that the energy obtained is not a bound to the true energy.

Finally, a further alternative technique has been suggested by Boys [17,18] starting from the general form of the equations of the method of moments [19] that can be written:

$$< X_r | \mathcal{H} - E | \sum_i C_i \psi_i > = 0 \qquad r = 1, 2, \ldots n \qquad (2.26)$$

here the functions X's are essentially arbitrary and separated from the set $\{\psi_i\}$ which form a specific approximate solution, $\phi = \sum_i C_i \psi_i$, of the Schröedinger equation. Boys and Handy choose:

$$X_r = \frac{\partial}{\partial \beta_R} (C^{-1} \phi') \qquad (2.27a)$$

and:

$$\psi_i = \frac{\partial}{\partial \beta_r} (C \, \phi') \qquad (2.27b)$$

where C is a correlation function $C = \prod_{i>j} f(r_{ij}, r_i, r_j)$, and ϕ' is an independent particle trial determinantal function. The β_r's are parameters upon which both C and ϕ' depend. It is possible to choose C such that the most difficult integrals involve integrations of not more than nine dimensions, and usually only six dimensions, instead of 3N, where N is the number of electrons. The electron-electron singularities are removed from the relevant integrals appearing in the matrix elements and they can be evaluated by direct numerical integration over six dimensions [20].

The significance of the method, called by the authors the "transcorrelated function" method [18], is that very complicated and accurate forms for Ψ can be used in the eq. (2.26) as long as a set of functions X can be found to cancel the complications, so to speak, in order to render the required matrix elements relatively simple to handle. The primary drawbacks of such functions are, however, the absence of an upper bound relation for the energy eigenvalue and the necessity for dealing with unsymmetric matrix problems which often have undesiderable stability specifications.

In most studies of systems with more than two electrons the trial functions of (2.20) are expressed in a determinantal form over the elementary basis set $\{\chi\}$:

$$\phi_j = ||\chi_1(1)\chi_2(2) \; \ldots\ldots \; \chi_n(n)|| \qquad (2.28)$$

where the χ functions are one-electron spinorbitals. For the surface calculation problem we can distinguish between two limiting choices for the single particles χ's: (i) atomic functions that are in some way centred on one atom of the "super--molecule" system; and (ii) molecular functions in which the χ's are distributed over some or all the centres of the polynuclear structure. The most common examples of the first type are: the Slater-type orbitals or STO's, the Gaussian-type functions called GTO's and a mixture of the two given by the so-called 'contracted GTO functions' in which a linear combination of GTO's with coefficients and exponents fixed by atomic calculations are used to provide a way of overcoming the poor behaviour of the GTO's near the nuclei [21].

For the case of elementary basis of molecular functions, suffice it here to mention the united-atom or one-centre expansion, where the origins of the functions (like the types above) are fixed on a single centre although their non-linear parameters are chosen to represent the distribution over the entire molecule. Convergence of expansions in single-centre basis sets are so slow, however, as to severely limit the utility of this practice for making accurate calculations [22]. Single-centre wavefunctions can, however, yield fairly accurate values of the multipolar static moments of the target molecule charge distributions [23] and have been used to obtain potentials for the scattering of electrons by simple molecules [24,25].

The most common method for obtaining reasonably well-behaved expansion functions as defined in eq. (2.28) is provided by the Hartree-Fock (HF) self-consistent field (SCF) procedure [26] . For closed shell systems one can write:

$$\phi_j = ||u_1\alpha(1)\ u_1\beta(2)\ \ldots\ldots\ u_n\beta(2n)|| \tag{2.29}$$

where the u's are now spatial orbitals and the $\alpha(i)$, $\beta(i)$ are the spin functions. In the above, restricted, Hartree-Fock approximation (RHF), 2n electrons are accomodated in n orbitals and a single Slater-determinant [26] is used to represent each expansion wavefunction ϕ_j. Moreover only one ϕ_j function is used in (2.20).

Each of the $u_j(k)$ is an eigenfunction of the Fock operator F(k):

$$F(k)\ u_j(k) = \varepsilon_j\ u_j(k) \tag{2.30}$$

where the ε_j's are called orbital energies and:

$$F(k) = H(k) + \sum_i \{2J_i(k) - K_i(k)\} \tag{2.31}$$

with H(k) being the kinetic energy and nuclear attraction operator in (2.19). The $J_i(k)$'s are the Coulomb operators and the $K_i(k)$'s are the exchange operators defined as follows:

$$J_i(k) = \int (r_{k\ell})^{-1} u_i^*(\ell)\, u_i(\ell)\, d\tau_\ell \tag{2.32a}$$

$$K_i(k)\, u_j(k) = \int (r_{k\ell})^{-1} u_i^*(\ell)\, u_j(\ell)\, d\tau_\ell \tag{2.32b}$$

Hence the eq. (2.30) is termed a pseudo-eigenvalue equation since the Fock operator depends on the lowest n eigenfunctions $u_j(k)$. Such an equation is therefore solved by using an iterative method that first guesses some trial orbitals $u_j(k)$ and then used these to evaluate the Fock operator itself. In the end one uses this operator to calculate a new set of orbitals in (2.30), which are in turn used to calculate an improved Fock operator and so on. According to the quality of the initial guess, this procedure usually converges fairly rapidly for closed-shell molecules.

The best possible function of the form (2.29) is called the HF wavefunction. For molecules it is difficult to solve eq. (2.30) numerically. On the other hand, the SCF procedure outlined above has a number of advantages: its orbitals are eigenfunctions of a known Hamiltonian, like the one defined in eq. (2.19), even if the total wavefunction is an approximate solution of the N-electron problem. Moreover, the diagonal expectation values of one-electron operators are estimated correctly to second order. The most widely used procedure was the one suggested by Roothaan [27], who proposed a method of obtaining approximate SCF orbitals by an expansion in analytic functions that was called a linear combination of atomic orbitals (LCAO) and that applied to systems which could be represented by all closed-shell molecular orbital wavefunctions:

$$u_i(k) = \sum_p b_{pi}\, \chi_p(k) \tag{2.33}$$

and the χ's could now be the elementary set of functions discussed before. The problem of solving the pseudo-eigenvalue equation (2.30) has thus been transformed into the one of varying the coefficients in the expansion (2.33) so as to find the best possible solutions within the limit of the level of truncation used in the expansion itself. As the size and flexibility of the basis set is increased, the SCF orbitals and energy approach the true Hartree-Fock ones. In eq. (2.30) there is a set of occupied u_i's ($i = 1,2,\ldots n$) plus a set of unoccupied orbitals ($i > n$) that form a complete set of eigenfunctions associated with the Fock operator. This basic method in several modifications was coupled to numerous early molecular integral programs that provided the basis for the bulk of molecular electronic energy calculations until the middle 1960's [28,29,30].

The advantages of the HF method are that it provides a conceptually attractive orbital picture of the molecular wavefunctions. There are, however, a number of deficiencies in this approach that have prevented its wide-scale application to the calculation of reliable basis sets that could then be used in molecular scattering problems. Closed-shell Molecular Orbital (MO) wavefunctions, for instance, cannot represent all of the eigenfunctions for some of the useful operators that commute (or nearly commute) with $\mathcal{H}$ and thus serve to characterize and differentiate the molecular states. Moreover, closed-shell MO wavefunctions cannot in general correlate well with the separated atom states that couple to form a given molecular state. The correlation energy tends to be, for strongly overlapping systems, a fairly strong function of intermolecular separation, hence the single-determinant approach contains a built-in error when used to yield potential surfaces.

Examples of this failing are found in SCF calculations of diatomic potential curves for H_2 [32], F_2 [33], LiF [34] and many other systems.

Furthermore, the HF method is designed to yield a good approximation for the ground electronic state of molecular systems. It has been used to examine excited states in those cases where there exists a lower state of the some symmetry, but one has to keep in mind that there is no justification on theoretical grounds to ignore the strong requirement of checking that the excited state in question is orthogonal to all lower states of the same symmetry [35].

Many systems of interest in inelastic or reactive molecular collisions have electronic structures with unpaired electrons which cannot be described by a single determinantal w.f. like in (2.29). The orbital pseudo-eigenvalue equations that arise for such open shell systems are, in fact, considerably more complicated than those presented by the closed shell case. The essence of the Hartree-Fock model is, in fact, the intent to replace the detailed and complicated description of the repulsions between every pair of electrons in the system by the average field that each electron exerts on every other. Moreover, in its restricted form (RHF) the Molecular Orbitals of eq. (2.29) are obtained with two mayor constraints placed on them. The first is that each u_i transforms according to one of the irreducible representations of the point group to which the molecule belongs. The second restriction is that the space functions of (2.33) come in identical pairs, one with one spin function and the other with the opposite spin function. This latter equivalence restriction [36] can be removed and a single determinant wavefunction is usually obtained. Since this method allows the introduction of a "Coulomb hole" to supplement for electrons of opposite spins (the "Fermi hole")already accounted for in the RHF model [37]) a lower value of the total electronic energy is usually obtained.

It is however possible for the so called unrestricted HF wavefunction (UHF) obtained in this way to start going beyond the HF approximation limit and thus include some of the correlation energy. The latter energy is of fundamental importance within the HF framework since it indicates the source of an inherent error in the HF wavefunctions, i.e. the fact that the pair probability of electrons with opposite spins is given in this theory by a simple product wavefunction, while the same pair probability for electrons with identical spins contains a 100% negative correlation of motion via the exchange terms, as required by the Pauli principle [38]. Since the Hamiltonian of eq. (2.19) does not include relativistic effects and hence cannot yield the true total energy of a system, one defines the correlation error as the difference between the electronic energy obtained in the HF limit and the exact electronic energy of the system, thus including in the above error both the correlation proper and the correct accounting of the relativistic correction [39].

While the UHF model retains the obvious advantage of the single-determinant wavefunction, its main disadvantage is that it leads to a wavefunction which is not an eigenfunction of the spin operator $\langle S^2 \rangle$. If the $\langle S^2 \rangle$ is thus found to diverge from its anticipated value, a projection operator can be applied in order to recover a spin-pure or a nearly spin-pure state. This can be done in several ways, from a method which cancels out only the contaminant of next higher spin multiplicity hence assuming it to be the mayor contaminant [40], to a method of complete projection that yields a wavefunction as a linear combination of Slater determinants [41] but which requires the projecting to be done before optimization on each iteration in order to obtain the correctly optimized wavefunction.

Since the correlation energies are normally at least as large as energies of chemical interest, such as dissociation energies or potential energy barriers, the potential energy surfaces calculated by the HF model can therefore be useful only if the correlation correction does not vary a great deal over the spatial regions of interest of the surface. On the other hand, one knows that the Hartree-Fock potential energy surface increases too steeply along a coordinate that leads to dissociation of the equilibrium structure, as confirmed by the too-short bond lengths and too-large force constants usually obtained via the HF description of the equilibrium geometry of a molecule [42]. An important example of an exception to this general behaviour is however given by the HF wavefunctions in which the electronic structures of the dissociation products corresponds to closed-shell systems, i.e. each spatial orbital is doubly occupied in its MO description. The single-determinantal w.f. reduces then properly, in the asymptotic region, to a product of wavefunctions which describe the correct dissociation products. In such systems the number of electron pairs remains constant in the various regions of the surface, while in the cases where the asymptotic parts of the interaction region correspond to dissociation leading to a decrease in the number of electrons pairs, a single-determinantal w.f. is unable to describe the products properly.

In conclusion, one can anticipate that mayor variations of the correlation energy will occur in the potential surface only when the number of electron pairs predicted by the MO picture changes in going from one region of the surface to another. The reactions involving closed-shell reactants and products should therefore be reasonably well described by the HF model of their B.O. potential surface [43]. This prediction has been confirmed by several calculations of heats of reaction [44] and of energies of reactions [45] and has been thoroughly examined in a recent review [46].

Another important part of the potential surface that affects scattering calculations stems from the dispersion forces operating between systems at large distances and coming out of small but significant correlative interaction of the motions of the electrons in the partners. The resulting polarization of the two systems and the consequent small changes in energy are not, however, accounted for in the single-determinantal HF model and need to be treated in a more sophisticated description going beyond the independent particle picture [47] . A great deal of effort has thus gone into the development of formalism that retain the advantages of the SCF model while overcoming the defects of the closed-shell LCAO-MO theory. If the electrons have parallel spins they are in fact kept correctly apart in the HF model by the antisymmetric nature of the wavefunction and produce what is usually known as a Fermi hole. Electrons of opposite spin, on the other hand, should also avoid each other but this is not correctly allowed by the single-determinantal wavefunction. The method of configuration interaction (CI) provides one of the earliest techniques used to overcome this problem [48] and it is based on an extensive mixing of the states arising from different spinorbital configurations. In theory one may expand the exact solution (2.20) in terms of the complete (infinite) set of determinantal wavefunctions which are in turn constructed from a complete set of one-electron spinorbitals [42].

The total wavefunction must be an eigenfunction of the operators which commute with the chosen Hamiltonian of the systems and whose eigenvalues approximately characterize the molecular states of interest. The form of a symmetry-adapted Ψ_i depends on the method selected for evaluating matrix elements and on the method chosen for generating eigenfunctions of $\mathcal{S}^2$. The determinantal approach thus uses:

$$\Psi_i = \sum_k a_{ik} D_{ik} \tag{2.34}$$

where the D_{ik} are Slater deteminants of space-spin orbitals. The mixing coefficients and the energies of the resulting CI states are found by solving the familiar set of secular equations:

$$||\mathbb{H} - \mathbb{E}(CI)\mathbb{S}\,|\mathbb{C} = 0 \qquad (2.35)$$

where $\mathbb{C}$ is the coefficients matrix and the matrix elements of the Hamiltonian are given by:

$$H_{k\ell} = \langle D_{i\ell}|\mathcal{H}|D_{ik}\rangle \qquad (2.36a)$$

with:

$$S_{\ell k} = \langle D_{i\ell}|D_{ik}\rangle \qquad (2.36b)$$

An RHF calculation over an n-electron, closed-shell, system and via N-orbitals expansion will thus produce $^n/_2$ doubly-occupied MO's and $(N-^n/_2)$ vacant MO's. A standard CI procedure will therefore form the other D's by systematically promoting electrons from the occupied MO's of the ground state D_{io} to the virtual MO's. The number of configurations which can be formed in this way is of the order of n^N, thus making the full CI approach possible only for very small systems even with today's high speed computers [46].

Another method which follows the same approach but is an improvement over the well known slow convergence of the expansion (2.34) is the multiconfiguration SCF (MC-SCF) method, which can be described as one which forces both the CI coefficients of eq. (2.34) and the MO coefficients of eq. (2.33) to be varied to minimize the individual CI energies an the total electronic energy. A doubly iterative process is therefore needed and it yields, upon convergence, the optimum orbitals and the optimum configuration mixing coefiicients for the basis sets used. In general one will need fewer configurations as compared to a conventional CI calculation, since the virtual orbitals are always forced to be within the same physical space as the occupied orbitals. Whal, Das and coworkers [49] have developed a special form of MC SCF formalism that omits the changes in atomic core as a function of R. This has been called the optimized valence configuration (OVC) method, and, while mainly implemented for equilibrium configuration studies, appears to have interesting capacities for applications to special regions of a potential surface [50].

Much study and effort has also gone into the extensions of the independent particle models which have as their aim the direct computation of correlation energies. One branch of this work originates from the chemical definition of "localized" bonds and "separated" electron pairs [51,52], while another branch of developement arises from the many-body theory techniques [53]. In the former treatments, one is after some criterion to reduce the number of configurations to be included in eq. (2.34) and to choose just how many of them are important. The slow convergence of a standard CI calculation originates from the nature of the virtual orbitals used to form the excited configurations, which do not occupy the same physical space as do the occupied MO's.

One can begin then by relying on Brillouin's theorem and disregard those configurations (singly-excited) which make no first-order contribution to the total energy [54]. Alternatively one can disregard all doubly-excited configurations that again do not mix in first-order CI calculations. Finally, one can disregard all excited configurations with energy E_k for which $|E_k - E_o|$ is very large and contribute very little from perturbation theory criteria. An interesting example is provided by the natural spin orbitals (NSO), first introduced by Löwdin [55], which have been shown to produce the most rapidly convergent expansion of the first-order density matrix, thus greatly reducing the number of needed determinants for any required accuracy in the wavefunction whenever they are constructed from natural orbitals at each stage [56]. This unfortunately becomes a not very practical advatage as soon as one moves to large systems, since the NO's first require a full CI before producing the NO's of the next interaction. The PSNO, or pseudo-natural orbital Method, tries then to perform an NO calculation on selected pairs of electrons only in the HF field of the (n-2) electron core. These orbitals can thus form a smaller basis for a CI calculation [57]. In contrast to the features of virtual and occupied orbitals, the NO's are all localized in the same region of physical space, hence the strong coupling that is produced in this way by the overlap of ground-state and excited orbitals effectively lowers the electronic energy of the whole system.

Finally, another method of including electron correlation within the framework of "chemical" bonds is the "independent electron pair approach" (IEPA). This method treats the correlation as a sum of the correlation energies of pairs of electrons in the field of other electrons [58]. It can then be shown that the total energy of the system can be written as:

$$E_{tot} \simeq E_{HF} + \sum_{\alpha} \eta_{\alpha\alpha} + \sum_{\alpha<\beta} \eta_{\alpha\beta} \qquad (2.37)$$

where $\eta_{\alpha\alpha}$ is the energy associated with the correlation between electrons in the same orbital (intrapair correlation) and $\eta_{\alpha\beta}$ represents the energy resulting from the correlation of electrons in different orbitals (interpair correlation). This method has yielded satisfactory results for potential surfaces like Li-H_2^+ with various H_2 bond distances and Li^+ in the long-range region of interaction [58] and for the similar ionic interaction of Li^+ with CO molecules at different bond distances [59] (see last Section of this Chapter).

In conclusion one can say that the correlation error in the HF method may be associated with two different physical causes:
(i) the increase in correlation energy due to the formation of new electron pairs, thus bringing more electrons into closer proximity with each other; and
(ii) the inability of the single determinantal configuration to describe correctly the dissociation of the super-molecule (when coming from a reactive encounters) into the component partner molecules or atoms.
If one considers, as a general way of correcting for both the above errors, the use of expansion (2.34) in a limited form then the criteria for selecting the added configurations are different for error (i) or error (ii). Thus, the addition of a few configurations that allows for the correct dissociation of the products makes use of the ones that are not important when describing the close-in regions where chemical rearrangements occur. For such regions one then has to add additional "molecular" configurations that vanish formally when the fragments separate. Their contributions to the energy lowering are usually called extra-molecular correlation energy [60] and require a different criterion of selecting the states with respect to the one to use for picking up configurations contributing to intra-molecular correlation energy that acts over the whole surface.

2.3 - Some more approximate treatments.

The methods discussed in the previous section have in common the fact that the full electrostatic (non relativistic) Hamiltonian is employed in the search for the eigenfunctions and their energy eigenvalues. This type of approach, albeit approximate in nature, is usually considered to be ab initio in the sense that no further approximations are introduced beside the standard B.O. separation and the choice of the $\mathcal{H}$ operator as given in eq. (2.16).

Since we have shown above how quickly the computational difficulties get out of hand within the HF or the CI approach, especially for more realistic many-electron systems, it is very often necessary either to simplify the Hamiltonian of (2.16) or to correct its HF wavefunctions without invoking an extensive CI or MCHF treatment.

These further approximations for potential surface calculations have been usually discussed by starting with the Valence Bond (VB) method rather than the MO-LCAO approach [60]. The rigorous VB method can be presented in its more conventional way, whereby the component atoms are 'prepared' in valence states which are not eigenfunctions of the atoms themselves before the electron-pair bonds are formed, or can be given via molecular wavefunctions that are built up by atomic Hamiltonian eigenfunctions [61]. The former constitutes the basis for the numerous types of hybridization schemes that are frequently invoked by chemists, while the latter is also called the spin-valence method.

The so called London approach can be most easily seen for the standard H_3 problem. By centering a 1s orbital on each of the three H atoms one can in fact write down the two independent doublet VB structures:

$$^2\Phi_1 = (a\ b,\ c) \quad ; \quad ^2\Phi_2 = (a,\ b\ c) \tag{2.38}$$

where the main diagonals of the determinants involved (with separated spatial and spin functions) allow us to write for, say, the $^2\Phi_1$ structure:

$$^2\Phi_1 = \mathcal{N}\{|1S_a\alpha(1)\ 1S_b\beta(2)\ 1S_c\alpha(3)| - |1S_a\beta(1)\ 1S_b\alpha(2)\ 1S_c\alpha(3)|\} \tag{2.39}$$

with $\mathcal{N}$ being a normalization factor and the bond being between the a and b atoms.

The total wavefunction can then be given as:

$$\Psi(H_3) = c_1\ {}^2\Phi_1 + c_2\ {}^2\Phi_2 \qquad (2.40)$$

thus, the variation principle provides the coefficients c_1, c_2 and the energy expression:

$$E_{\pm} = \frac{1}{A}\{-B \pm (B^2 - BC)^{\frac{1}{2}}\} \qquad (2.41)$$

with A, B and C analytic expressions [62] that contain the following types of integrals:

$$Q = \text{Coulombic Integral} = \langle abc|\mathcal{H}|abc\rangle \qquad (2.42a)$$
$$J_i = \text{Exchange Integral} = \langle abc|\mathcal{H}|bac\rangle \qquad (2.42b)$$
$$S_k = \text{Overlap Integral} = \langle a\,|\,b\rangle \qquad (2.42c)$$

with all the permutations of {abc} in the J_i's and S_k's.

The solution of the secular equation then yields two energy values, of which $E^{(-)}$ is the surface of lower energy except for the energy degeneracy at the equilateral triangular geometry[62]. Although no approximations have been introduced thus far, improvements over the simple basis expansion of (2.39) are obviously necessary even for the H_3 case, let alone for the more complex, many-electron atoms composing a realistic supermolecule. The most elementary change is to start off with orthogonal orbitals, thus recovering the original London formula [63] that only contains one and two-electrons integrals:

$$E = Q \pm\{\tfrac{1}{2}(J_1 - J_2)^2 + \tfrac{1}{2}(J_2 - J_3)^2 + \tfrac{1}{2}(J_1 - J_3)^2\}^{\frac{1}{2}} \qquad (2.42)$$

where:

$$Q = \sum_i Q_i = Q_1 + Q_2 + Q_3 \qquad (2.43)$$

with:

$$Q_1 = \int a(1)\, b(2)\, \mathcal{H}_{ab}(1,2)\, a(1)\, b(2)\, d\tau_1\, d\tau_2 \qquad (2.44a$$

$$J_1 = \int a(1)\, b(2)\, \mathcal{H}_{ab}(1,2)\, a(2)\, b(1)\, d\tau_1\, d\tau_2 \qquad (2.44b)$$

and:

$$\mathcal{H}_{ab}(1,2) = -\frac{1}{r_{1b}} - \frac{1}{r_{2a}} + \frac{1}{r_{12}} + \frac{1}{R_{ab}} \qquad (2.45)$$

with the familiar meaning of the one- and two-electron operators and the nuclear repulsion term.

We have now obtained the integrals in eq. (2.42) as depending only on two atoms , hence one can make use of the information that is given by experimental or theoretical studies of diatomic molecules energy curves. One of the earlier suggestions made [63] involved, in fact, the VB formula with zero overlap for the singlet state of the H_2 molecule, plus a proportionality constant that corrected the Q_i and J_i values according to the internuclear distance. Obviously, if the experimental triplet curves are available, they can also be used to obtain values for the two-electron integrals of the above [64]. When overlap is not neglected, however, the exchange integrals J_i's of (2.42b) acquire further contributions from triatomic terms and the following energy expressions become much more complicated than (2.42), particularly if one wants to introduce higher-ℓ orbitals and more ionic and atomic excited states besides the basic structure of (2.38) [65].

An interesting approximate treatment was also introduced by Moffitt [66] by starting off with the <u>ab initio</u> VB approach and then applying some experimental data to correct the asymptotic behaviour of the complete system. The first step then was the formation of approximate eigenfunctions for the states of the isolated atoms, ϕ_i^A, ϕ_k^B , ϕ_ν^C ..... which are antisymmetric with respect to the exchange of electrons and are eigenfunctions of the various angular momentum operators. They can provide a basis set for expanding the total molecular eigenfunction, provided one antisymmetrizes each of their direct products to allow for the interchange of the various electrons originally assigned to the separate atoms. These basis functions are called Composite Functions (CF) and are of course not eigenfunctions of the angular momentum operators for the whole molecule. The corresponding eigenfunctions can be formed by taking simple linear combinations of the CF's and thus obtaining the spin eigenfunctions which correspond to the rigorous VB approach described before [61].

The total molecular wavefunctions can in turn be expressed as linear combination of the spin eigenfunctions:

$$\Psi_{tot}^{|SM>} (ABC \; ....) = \sum_k C_k \, \Phi_k^{|SM>}(ABC \; ....) \qquad (2.46)$$

The coefficients C_k are determined by using the variational principle through the usual set of secular equations and determinant, as in the CI treatment of eq. (2.34). As one easily sees, from the way the CF functions have been obtained, the total molecular wavefunction dissociated correctly for each of its $|SM>$ states, since the spin eigenfunctions provide a good description of the specific states of the final fragments. Despite this advantage, however, the rigorous VB approach has been used much more rarely than its MO counterpart, the reason for this lying in the so called non-orthogonality problem, which deals with the difficulty of calculating the Hamiltonian matrix elements between VB structures that can be represented as linear combinations of Slater determinants built up from non-orthogonal orbitals [67].

Thus, the Moffitt suggestions, often called the Atom-in-Molecule (AIM) method, was directed at choosing those CF functions that contribute the most to the total molecular energy, and then correcting the approximate atomic energies given by the chosen contributions with known experimental atomic energies. This correction also counteracts the necessary limitation in the number of CF's that can be used in the expansion of the approximate total wavefunction. In the case of a diatomic molecule, a typical Hamiltonian matrix element between the CF $|\alpha>$ and $|\beta>$ is given by:

$$H_{\alpha\beta} = H_{k\ell,mn} = \overline{H}_{k\ell,mn} + \frac{\overline{S}_{k\ell,mn}}{2} \{\Delta E_k^A + \Delta E_\ell^B + \Delta E_m^A + \Delta E_n^B\} \qquad (2.47)$$

where the $\overline{H}$ and $\overline{S}$ are the theoretical elements of the Hamiltonian and the overlap matrices between the CF $|\alpha\rangle$, with atomic eigenfunctions $|k\rangle$ and $|\ell\rangle$ on atoms A and B, and the CF $|\beta\rangle$, with the atomic eigenfunctions $|m\rangle$ and $|n\rangle$ on the same atoms. The energy corrections in brackets indicate the various differences between computed and experimental values for the atomic configurations under examination.

Any improvement in the description of the atomic states will progressively reduce the value of the Moffitt correction, which will always have the important property that as the atoms or molecular fragments are pulled apart, and the CF become orthogonal, the computed energies tend toward the experimental energies of the separated components, both for their ground and excited states. On the other hand, when the atoms approach each other the various wavefunctions representing the separated states become increasingly less independent, hence the Moffitt expansion is overcomplete for finite and small $\underline{R}$ because of the non-zero overlap between the different atomic wavefunctions.

To correct for this, a new modification has been suggested and has been called the Orthogonalized **M**offitt (O**M**) method [68]. Essentially it modifies the AIM method in the region of small distances by first performing a Schmidt orthogonalization procedure between CF basis and then applying the correction (2.47) to the Hamiltonian matrix elements in the new basis. In all the cases where it has been applied [68,69], the O**M** method improves the calculated dissociation energies by improving the description of the composite system, while leaving unsolved the question of its accuracy for the overall quality of the potential surface.

Another approximate approach that is worth mentioning for its great wealth of future possibilities for dealing with more complex molecular systems regards the various attempts at casting a convincing form of a Pseudopotential [70] for part of the electron density. The rigorous ab initio approach encounters in fact several difficulties when dealing with systems with heavy atoms, arising from the large number of electrons with which they are associated. Since most of the involved electrons are occupying inner "core" orbitals, according to the usual chemical understanding of the bonding in molecules they should play only a minor role in determining inter-atomic forces , especially at distances different from the equilibrium geometries. This immediately suggests a way of simplifying the computational problem by dealing only with outer electronic orbitals and thus reproducing the effect of the "core" orbitals interactions by some effective, local or non-local, potential that manages to exclude the valence electrons form the inner region and from a variational collapse into "core" orbitals.

Some of the earlier forms have been local and spherical, like the Hellmann potential [71], that includes some parameters which allow the ground and first excited atomic states to yield the correct experimental binding energies. Other attempts have allowed for the fact that the effective Fock operator should behave differently for different angular momenta of the valence orbitals, thus producing a Pseudopotential non-local in the angular variables [72] and very effective in matching the quality of all-electrons calculations on similar systems. Recently [73] extensions have been presented for a Pseudopotential approach within a multi-structure VB approach that allows for the correct inclusion of valence electron correlation effects at the intramolecular and extramolecular level.

2.4 - The Electron Gas Model

In the previous sections we have shown that the calculations of potential hypersurfaces, even for simple systems, require a great deal of effort and computational skill in order to produce reliable results over the whole region of interactions that is involved in inelastic collision processes. The various *a priori* methods reviewed before are in general rather accurate in the strong-overlapping regions of the near equilibrium geometries, but become rapidly more uncertain in the intermediate and long-range distances of the dissociating species, hence requiring an even greater computational labour both for reacting and non-reactive interactions of the partners.

It is however true to say that the interest in scattering calculations of recent years has been focussed on many atom-atom, atom-molecule and molecule-molecule systems for which chemical interactions took place only at very high energies, and were therefore concerned with computing the whole potential curve or surface for near-thermal energy scattering along the ground electronic states, where one species or both were given by closed-shell configurations. This means that recent years have also seen rapid progress and many new methods for calculating the long-range attractive van der Waals potentials [74], while the calculations of reliable short-range repulsive interactions still require extensive SCF or CI computations.

In particular, it is worth discussing in detail a recently developped statistical methods which, in spite of its remarkable simplicity, appears to give reliable, quantitative, information on both the repulsive and attractive regions of the spherical atom-atom interactions and the anisotropic molecular interactions as well. This so called Electron Gas Model (EGM) approach, was presented first for atom-atom interactions [75,76] and so we will start with a brief review for spherical systems. It has however quickly received a great deal of attention and has been extended to several systems of interest to molecular scattering.

It is well known that the calculation of the interaction between pairs of closed-shell atoms in the region of the van der Waals minimum presents a great many problems. Conventional SCF calculations usually fail to produce any minimum at all in their computed potentials and the corresponding CI approach is of such complexity as to be practical only in the simpler cases such as He-He [77]. Further efforts have thus been made to apply perturbation theory, which very successfully predicts the long-range (almost zero overlap) dispersion energy, up to the region of the potential well and again rather good agreement has been obtained in the He-He case [78]. The different perturbation approaches do not lead, however, to consistent results in second and higher orders of perturbation while their practical application still requires very considerable computational effort.

On the other hand, the principal assumption of the EGM method provides a direct way of simplifying the treatment at the outstart: if no chemical bonds are formed during the approach then the electron density distribution of the two atoms (or molecules) are not distorted as they interpenetrate, and hence the electron density of the combined supermolecule is given at any point along the relative C.M. $\underline{R}$ distance by the sum of the densities of the separate atoms:

$$\rho_{tot}(R) = \rho_A(\underline{r}_A) + \rho_B(\underline{r}_B) \tag{2.48}$$

where the electron density ρ_A pertains to atom A with atomic number Z_A and position $\underline{r}_A$ in a SF frame of reference. The electron density ρ_B means the same for the other atomic partner B. The vector R specifies the relative distance of the two atoms.

The idea of approximating the behaviour of various electronic systems by means of the uniform electron gas is certainly not new in quantum mechanics

[79,80]. Essentially one decides to describe a system of many (N) electrons only by their probability distribution in three-dimensional space, rather than by a complete wavefunction in 3N-dimensional space. One further assumes that, at any point in the distribution, the system behaves locally the same way as the uniform electron gas with a density equal to that of the system at that point. The whole problem is then reduced to that of finding the total density distribution of the supermolecule and of solving the uniform electron gas total energy expression, problems which are usually both easier to solve than the original one.

Once the prescription for answering the first question is given by eq. (2.48), one then moves on to the idea that the energies of the real atomic systems interacting at various $\underline{R}$ values may be approximated by the simple energy expression for the electron gas, thus providing a direct way of obtaining the interaction energy difference, i.e. the potential $V(\underline{R})$. In the usual B.O. approximation this is in fact given by:

$$
\begin{aligned}
V(\underline{R}) &= E(\underline{R}) - E(\infty) \\
&= V_{HF}(\underline{R}) + V_{corr}(\underline{R})
\end{aligned}
\qquad (2.49)
$$

where the total energy is given by the usual sum of Hartree-Fock energy contributions, E_{HF}, plus correlation energy corrections, E_{corr}.

The total energy at infinite separation is given by the sum of the two energies of the atoms within the electron gas approximation:

$$
\begin{aligned}
E_A = {} & C_k \int \{\rho_A(\underline{r}_A)\}^{5/3} d\underline{r}_A + C_e \int \{\rho_A(\underline{r}_A)\}^{4/3} d\underline{r}_A \\
& + \int \varepsilon_c(\rho_A) \times \rho_A(\underline{r}_A)\, d\underline{r}_A - Z_A \int \{\rho_A(\underline{r}_A)/r_A\} d\underline{r}_A \\
& + \tfrac{1}{2} \iint \{\rho_A(\underline{r}_{A1})\, \rho_A(\underline{r}_{A2})/r_{12}\}\, d\underline{r}_{A1}\, d\underline{r}_{A2}
\end{aligned}
\qquad (2.50)
$$

where, in atomic units:

$$C_k = \frac{3}{10}\,(3\pi^2)^{2/3} \quad ; \quad C_e = -\frac{3}{4}\,(3/\pi)^{1/3} \tag{2.51}$$

Here the first term in eq. (2.50) represents the kinetic energy and the second term the exchange energy of atom A, with $\varepsilon_c(\rho_A)$ the correlation energy for density of the electron gas:

$$\varepsilon_c = 0.0311 \cdot \ln r_s - 0.048 + 0.009\, r_s \cdot \ln r_s - 0.01\, r_s;\ r_s \leq 0.7 \tag{2.52a}$$

$$= 0.06156 + 0.01898 \cdot \ln r_s;\quad 0.7 \leq r_s \leq 10, \tag{2.52b}$$

$$= -0.438\, r_s^{-1} + 1.325\, r_s^{-3/2} - 1.47\, r_s^{-2} - 0.4\, r_s^{-5/2};\ 10 \leq r_s. \tag{2.52c}$$

where the radius r_s is related to the electron density by the normalization condition:

$$\frac{4\pi}{3}\,\rho(\underline{r}_s) \times r_s^3 = 1 \tag{2.53}$$

and the three equations (2.52a-c) describe the high electron density expansion form [81], the interpolation formula [76,82] and the low-density expansion form [83]. The last two terms of eq. (2.50) are the Coulombic energy parts and in the HF approximation they can be evaluated from the computed atomic electron densities, after combining them to give a single term:

$$\iint \rho_A(\underline{r}_{A1})\, \rho_A(\underline{r}_{A2}) \{(2r_{12})^{-1} - (r_A)^{-1}\}\, d\,\underline{r}_{A1}\, d\,\underline{r}_{A2} \tag{2.54}$$

where use has been made of the relation:

$$Z_A = \int \rho_A\,(\underline{r}_A)\, d\,\underline{r}_A \tag{2.55}$$

while for an ion of charge m eq. (2.55) becomes:

$$\int \rho_A\,(\underline{r}_A)\, d\,\underline{r}_A = Z_A - m \tag{2.56}$$

which then provides an alternative form for the integral of (2.54):

$$\iint \rho_A(\underline{r}_{A1})\, \rho_A(\underline{r}_{A2}) \left\{ \frac{1}{2r_{12}} - \frac{Z}{Z-m}\,\frac{1}{r_A} \right\} d\,\underline{r}_{A1}\, d\,\underline{r}_{A2} \tag{2.57}$$

After all the angular integrations are carried out, the integrals (2.54) and (2.57) can be reduced to two-dimensional integrals over the atomic radii r_{A1} and r_{A2} and can be carried out numerically [76,82].

When comparing total energies obtained via the EGM calculations with those given by HF calculations, it has been found that the accuracy is generally of about 10% including He atom and Li^+ ion which have only two electrons [84]. The accuracy for the kinetic energy and the exchange energy is a little worse for the lighter atoms, while it becomes better for the heavier atoms. This general trend of getting better accuracy when the number of electrons increases is in keeping with the increased validity of the uniform gas model for atomic electrons and useful to treat those systems that are out of reach for the HF-SCF treatment.

For the correlation energy one has, however, quite a different situation since the EGM method predicts much larger correlation energies than the "experimentally" known values [85]. It is in fact overestimated by a factor of 3 for the lighter atoms while decreasing to a factor of 2 for the systems with higher Z. This has been related to the fact that the EGM approach treats each electron as a continuous electron density and calculates the correlation between different parts of this density. Hence, by treating all the electrons in an atom as with continuous electron densities one includes a sort of correlation within an electron together with the correlation energy among different electrons, thus overestimating the correct value of this energy contribution.

By using the energy expression (2.50) and the approximation (2.48), it is now possible to write down an explicit expression for the atom-atom potential curve of (2.49):

$$\begin{aligned} V(R) &= V_{HF}(R) + V_{corr}(R) \\ &= V_k(R) + V_c(R) + V_e(R) + V_{corr}(R) \end{aligned} \tag{2.58}$$

where:

$$V_C(R) = \iint \rho_A(\underline{r}_1)\, \rho_B(\underline{r}_2) \left\{\frac{1}{R} + \frac{1}{r_{12}} - \frac{1}{r_{1B}} - \frac{1}{r_{2A}}\right\} d\underline{r}_1\, d\underline{r}_2 \tag{2.59a}$$

and:

$$V_i(R) = \int d\underline{r}\Big\{[\rho_A(\underline{r}_A) + \rho_B(\underline{r}_B)]\, \varepsilon_i(\rho_A + \rho_B) - \rho_A(\underline{r}_A)\, \varepsilon_i(\rho_A) - \rho_B(\underline{r}_B)\, \varepsilon_i(\rho_B)\Big\} \tag{2.59b}$$

with i = K, e and corr. and the ε_i's given by the expressions (2.51) and (2.52) for both the isolated atoms and the interacting system. By using a spheroidal coordinates system [82] the volume element $d\,\underline{r}$ is $= \frac{R^3}{8}(\lambda^2 - \mu^2)\; d\mu\, d\lambda\, d\phi$, with:

$$\lambda = \frac{r_A + r_B}{R} \quad ; \quad \mu = \frac{r_A - r_B}{R} \tag{2.60}$$

the angular integration over ϕ being trivial for linear symmetry. The λ and μ integrations can then be performed numerically using Gauss-Laguerre and Gauss-Legendre quadratures [76,82].

The potential curves computed in this way for rare-gas pairs of the A_2 and AB types work surprisingly well for all R less than R_m (the distance of the minimum) for large atoms, although it gives potential wells of the order of 3 or 4 times the known depths for lighter systems like He_2 [76,82,85]. Moreover, it provides the long-range part of the interaction with the wrong R-behaviour, since we have seen above that the EGM approach greatly overestimates atomic correlation energies and strongly reduces the interatomic correlation contributions that depend on the overlap among the partner atoms.

It was then suggested [86,87] that the exchange energy contributions for a uniform electron gas should be modified when the total number of electrons in the system is finite and small, as for the lighter atoms, in order to correct for the self-energy exchange contributions that do not cancel out as the coulombic self-energy terms but become instead very important. The long-range dispersive terms could also be included by simply adding the corresponding perturbation series contributions to the EGM calculations, to correct for the wrong asymptotic behaviour [87]. The V_e (R) terms in eq. (2.58) can then be replaced by the following expression:

$$V_e(R) = V_e(R) \{1 - 8\delta/_3 + 2\delta^2 - \delta^4/_3\} \qquad (2.61)$$

where δ is the solution of:

$$(4N)^{-1} = \delta^3 \{1 - 9\delta/_8 + \delta^3/_4\} \qquad (2.62)$$

Such a correction turns out to be substantial even for large atoms, reducing the V_e (R) contributions for Xe_2 by about 2/3 [85]. It is also very effective in reducing well depths for lighter pair interactions. It results, however, that the direct inclusion of van der Waals energy still gives potential wells that are much too deep for nearly all systems [87]. Another possible correction can then be obtained if one notes that for sufficiently large values of R the van der Waals dispersion energy is the interatomic correlation energy and provides the only interaction energy. For decreasing R values, however, the exchange and overlap contributions begin to count and the asymptotic series expansion begins to diverge [87]. Since for closed-shell systems this effect appears at R much smaller than R_m, one could try the empirical prescrition:

$$V_{corr}(R) = V_{disp}(R) \quad ; \quad R \geqslant R_c \qquad (2.63a)$$

$$= \frac{V_{disp}(R_c)}{V_{corr}(R_c)} V_{corr}(R); \quad R < R_c \qquad (2.63b)$$

where R_c is obtained by computing the relative distance for which the ratio on the r.h.s. of eq. (2.36b) goes through a minimum. The value of the computed ratio is usually 2 to 3 and R_c is less than R_m and close to σ, the point at which V(R) crosses zero. Such a procedure obviously requires a knowledge of the van der Waals coefficients C_6, C_8 and C_{10}, together with combining rules for the AB systems for the rare gases [87]. The general result is one of rather good agreement with the available experimental data, as is shown by a few selected examples in Table 1. The table also reports the results obtained via a further modification of the EGM approach [88] that treats short-range and long-range forces via a generalized form of the Drude's model [89] . This latter model deals with the atoms as three-dimensional harmonic oscillators in order to provide particularly simple expressions for dispersion forces and for short-range interactions that correct the various shortcomings of the EGM approach discussed before. It is worth pointing out that for the first time a simple computational model provides quantitative agreement with molecular beams data without having to resort to some adjustable parameters. The required computer times are typically of 1-2 second per point with last-generation machines, to be compared with the hours-long demands of CI calculations.

It should be mentioned, however, that part of the success of the present approximation can be related to the opposite effects of the prescritions (2.48) and (2.50). The total energy calculated within the EGM approach yields an average error of $\simeq +9\%$ with respect to experimental values, while the error introduced by the undistorted densities summed up to represent the total density of the pseudomolecule is of the order of $\simeq -11\%$ with respect to the correctly computed united atom density [90] . This holds at the small R values representing the repulsive region of the potentials, while it is not true any more in the asymptotic region, where the van der Waals energies need to be included since they are not given by the non-overlapping electronic densities that appear in the correlation energy term of (2.59b).

Finally the increasing demand for a more accurate understanding of the adiabatic potential curves governing the dynamics in a wide variety of scattering experiments which involve ions of alkali atoms, both in their ground-states and excited states, has provided a further area of application for the EGM approach. In the case of ionic systems, however, the additional distorsions introduced by the strong charge effects play a different role according to what region of interaction one is examining with the experiments. At large distances, in fact, the calculated potential curves of eq. (2.49) have a rather large error because the EGM model described so far fails to account for the induction and dispersion contributions, thus causing an even larger discrepancy in the ion-neutral atom systems because of the larger induction contribution of these partners. Therefore we do not expect that the EGM potentials discussed

TABLE 1

Calculated properties of interatomic potential curves for the noble gas pairs via different forms of the EGM approach.

		He-He	He-Ar	Ar-Ar	Ar-Xe
R_m (Å)	a	2.50	3.15	3.60	3.92
	b	2.85	3.26	3.71	4.00
	c	3.22	3.51	3.82	4.11
	d	2.96	3.37-3.56	3.70	4.01
σ (Å)	a	2.30	2.80	3.29	3.53
	b	2.55	3.02	3.32	3.56
	c	2.87	3.13	3.40	3.65
	d	2.65	2.98	3.32	3.60-3.70
ε (10^{-16} erg)	a	60.0	90.0	180.0	245.0
	b	36.3	74.0	235.0	318.0
	c	8.89	41.84	178.3	250.2
	d	15.2	34.54	195.0	2.36-2.62

a) ref. [82]; b) ref. [87]; c) ref. [88]; d) experiments.

so far would properly represent the attractive well regions of the interaction.

At short distances, however, these long-range contributions are small compared with the large repulsive energy and one can therefore expect that EGM curves provide reasonable estimates of true potentials from very short R values until somewhere just inside σ, the distance at which V(R) crosses zero. Reliable comparisons and checking of the results can then be made with experimental molecular beams data, whenever available for more than one group [90,91,82]. The method fails however in providing the attractive tails, the well positions and depths for all the systems examined.

To compensate for this failure to account for the induction forces, one might attempt simply to add the known long range asymptotic form of the induction [74], $-\alpha/2R^4$, where α is the dipole polarizability of the atom. This term gives the interaction of the ionic charge and the induced dipole on the atom, treating them as a point charge and a point dipole. This should hold well at very large distances where the overlap of the charge distributions are negligible. However, as one goes to shorter distances, this simple term starts to fail to describe the induction properly owing to the large overlap and also owing to the effect of the stronger repulsion, which will tend to oppose the polarization for the positive ion cases and help it for the negative ions. It therefore appears that the asymptotic induction term remains comparable in magnitude with the interactions themselves, until one comes to rather shorth distances where the simple addition of this term would bring the potential curves too low in these regions when compared with the experimental curves. At even shorter distances this term becomes small compared with the total interactions and its addition would be unnecessary and not accurate.

To correct the potential behaviour for these systems, two suggestions have recently been put forward in order to mantain the overall simplicity of the treatment while improving the ion-atom curves in the well region and at long distances. The first [88] is based on the previously discussed modification of the Drude model [89], whereby the dispersion and induction forces are included in the asymptotic region and then modified in the short -range region which contains the EGM repulsive part without correlation correction. One empirical parameter is required to describe a sort of effective number of electrons, N_f, which forces the model to have the correct long-range form of the dispersion and induction forces. In the latter case the N_f value for the ions was taken as the one given by the isoelectronic rare gas |88|.

Another approach [92] directly corrects the leading induction term at short distances by defining an effective induction potential $V_{ind}^{eff}(R)$:

$$V_{ind}^{eff}(R) = -\frac{\alpha_n}{2R} \times f(R) \qquad (2.64)$$

with f(R) being a cut-off function that correctly goes to zero as R^2 within the overlap region that still includes the correlation energy correction of the EGM approach:

$$f(R) = \{ 1 - \exp(-\beta^6) \} \qquad (2.65a)$$

or:

$$f(R) = \{ 1 - \exp(-\beta) \}^6 \qquad (2.65b)$$

Both forms contain a "volume" factor β which is a function of R:

$$\beta = {}^{R}/_{\gamma} \qquad (2.66)$$

where γ is related to the expectation values $\langle r^2 \rangle$ of the outer-shell valence orbitals of the relevant target atom:

$$\gamma = [\langle r^2 \rangle]^{\frac{1}{2}} \qquad (2.67)$$

In other words, the penetration into the inner region by the incoming ions and its effect on the interaction potential are described by the function f(R) for those distances for which the overlap begins to cause the divergent behaviour of the perturbation expansion containing the polarization term.

A comparison of the potential curves obtained via eqs. (2.65a) and (2.65b) with the semiclassical treatment of the Drude model is shown in Figure 1 for the $K^{\pm}$-Ne system, while a set of potential parameters computed for various ion-rare gas pairs is listed for comparison in Table 2. One sees that the EGM approach, modified as above, provides reasonable agreement with the available experimental data and its results can be used to yield other meaningful properties of gaseous systems.

Once the potential is known, one can in fact obtain the transport properties of a diluite gas by expressing them in term of a set of reduced collision integrals, obtained in a classical scheme after three successive integrations [93]. First one determines the classical deflection function $\vartheta(b, E)$ as a function of the impact parameter b and relative energy E:

$$\vartheta(b, E) = \pi - 2b \int_{r_c}^{\infty} \frac{dr}{r^2 [F(r,b,E)]^{\frac{1}{2}}} \tag{2.68}$$

where the classical turning point r_e is the outermost root of the F function. A further averaging over b yields for each value of E the relevant transport cross section:

$$Q^{(\ell)}(E) = 2\pi \left[1 - \frac{1+(-)^{\ell}}{2(\ \ +1)}\right]^{-1} \int_0^{\infty} (1 - \cos^{\ell} \theta)\, b\, d b \tag{2.69}$$

The above quantities can in turn be employed for a further energy averaging that produces the collision integrals as a function of temperature:

$$\bar{\Omega}^{(\ell,s)}(T) = \left[(s+1)!\ (KT)^{s+2}\right]^{-1} \int_0^{\infty} Q^{(\ell)}(E)\ E^{s+1} e^{-\frac{E}{KT}}\ d E \tag{2.70}$$

It therefore becomes important to obtain the $\bar{\Omega}^{(\ell,s)}$ at various temperatures rapidly via a simple analytic form of the interaction V(R). This has been done for the alkali ion-rare gas pairs and it has been found that the usually

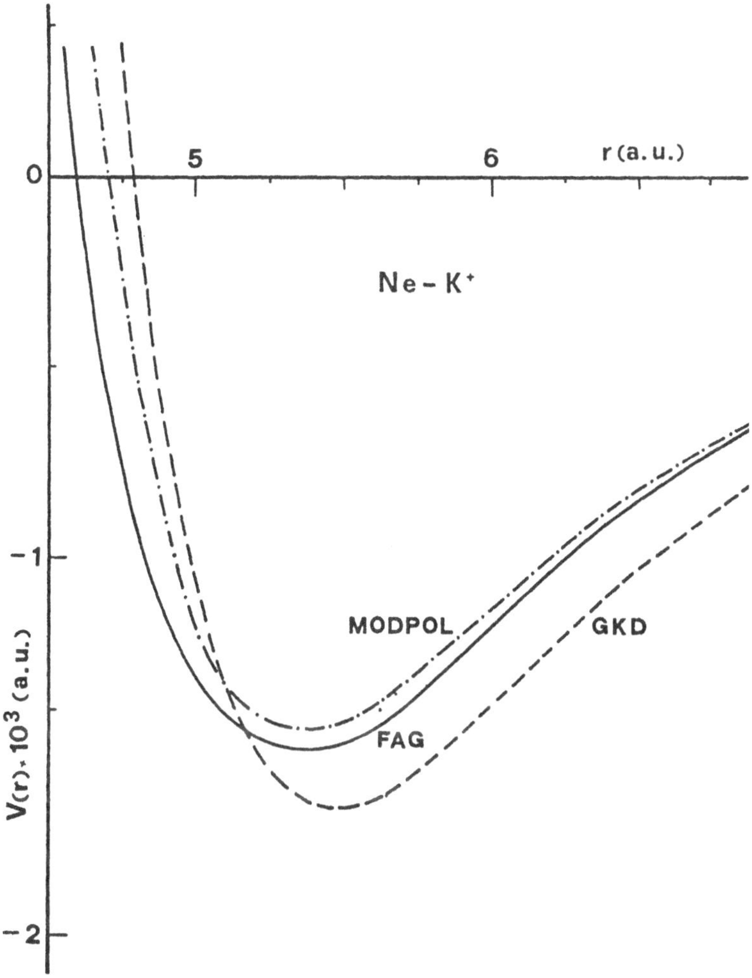

Figure 2.1

Ion-atom interaction potentials for the K^+- Ne pair around the well region as given by the EGM potential:

GKD = Drude model of ref. [88].

FAG = Polarized EGM treatment via eq. (2.65a)

MODPOL = Polarized EGM treatment via eq. (2.65b)

TABLE 2

Calculated properties of interaction potentials for some ion-rare gas pairs with different modifications of the EGM approach.

		Li^+-Ne	Na^+-Ne	K^+-Ar	Li^+-Xe
R_m (Å)	a	2.71	2.89	3.56	—
	b	1.99	2.41	3.09	—
	c	1.93	2.43	2.91	2.41
	d	—	—	3.00	—
σ (Å)	a	2.44	2.66	3.24	—
	b	1.70	2.10	2.66	—
	c	1.67	2.01	2.54	1.84
	d	—	—	2.63	—
ε (10^{-16} ergs)	a	8.3	24.0	92.9	—
	b	1986.5	1153.4	1938.0	—
	c	1980.0	1160.4	1938.0	9290.0
	d	—	—	1938.0	8170.2

a) ref. [91]; b) ref. [88]; c) ref. [92]; d) experiments.

suggested simple forms of V(R) only hold within limited ranges of temperature while better results are yielded by optimized coefficients which fit the potential curves given by the EGM approach [94,95,96].

The method has also been extended in recent years to treat the interaction between atoms and molecules, between ions and molecules and between molecules. Unfortunately, the amount of quality control or success assessment of the results becomes correspondingly more complicated. Although, in fact, it is fairly straighforward to compute observable properties from spherical potentials, this is by no means obvious for non-spherical potential surfaces. Moreover, it is neither simple nor, in general, unique to obtain potential information from scattering observables so that our experimental knowledge about anisotropic potentials is still quite limited.

The EGM approach has therefore been primarily directed towards interacting pairs that either were already studied by other theories or examined in molecular beam experiments, thus providing possible tests for its results. It has also dealt mainly with partners which have closed-shell electronic structures, in order to avoid tha mayor distortions caused by chemical bonding or reactive interactions. Its surprisingly good predictions [97] of three-body interactions [98], atom-surface interactions [99] and the spherical part of molecular interactions [100], has however encouraged its extension [101-103] to more complicated cases and molecular partners.

The interaction of a closed-shell linear molecule **M** with a neutral closed-shell atom A can be written in terms of the internal molecular coordinate R, the centre-of-mass distance r and the angle θ between them:

$$V(r,\theta;R) = E(r,\theta;R) - E(\infty,\text{any}\,\theta;R) \tag{2.71}$$

$$= V_{HF}(r,\theta;R) + V_{corr}(r,\theta;R)$$

where V_{HF} contains the Coulomb, kinetic and exchange contributions of eq. (2.58), while V_{corr} contains the correlation already discussed before. The Coulomb term is computed from exact electrostatic relations and consists of four terms which describe nuclear-nuclear and electron-electron repulsions and the nuclear-electron attraction. In atomic units this is simply given by:

$$V_c = \iint \rho_M(\underline{r}_1)\, \rho_A(\underline{r}_2)\, \{\frac{1}{r_{12}} - \frac{1}{r_{1A}} + \sum_{\alpha \in M} f_\alpha (\frac{1}{r_{\alpha A}} - \frac{1}{r_{2\alpha}})\}\, d\underline{r}_1\, d\underline{r}_2 \qquad (2.72)$$

whre the sum is over all the nuclei α of M and $f_\alpha = Z_\alpha/N_M$, with Z_α being the αth nucleus charge and N_M the number of electrons in molecule **M**.

Since A is a spherical atom the electrostatic potential due to it can be evaluated analytically after expanding its wavefunction over some analytic basis set (STO's or GTO's):

$$\Phi_A(\underline{r}_1) = \int \rho_A(\underline{r}_2)\, r_{12}^{-1}\, d\underline{r}_2 \qquad (2.73)$$

which then reduces eq. (2.72) to the following form:

$$V_c = \int d\underline{r}_1\, \rho_M(\underline{r}_1)\, \{\Phi_A(\underline{r}_1) - \frac{Z_A}{r_{1A}} + \sum_\alpha f_\alpha [\frac{Z_A}{r_{\alpha A}} - \Phi_A(\underline{r}_\alpha)]\} \qquad (2.74)$$

For molecule-molecule interactions the evaluation of the electron-electron contribution in (2.72) will of course require a 6D integration over the known HF density functions ρ_{M_1} and ρ_{M_2}. The electrostatic potential of (2.73) will be in fact given, in that case, by a 3D integral over ρ_{M_1} or ρ_{M_2} and the final expression of (2.74) can thus be formally put down again in terms of a three-dimensional quadrature as in the atom-molecule case. The greatest saving in computing time for a spherical Φ_A, however, comes not so much from the ease of calculating the electrostatic potential but from the possibility of tabulating it once at a suitable radial grid and then obtaining values as necessary by interpolation. Moreover, it has been shown [103] that it is possible for homonuclear diatomic molecules to perform a 2D tabulation and interpolation, while the general case requires at least two separate 3D quadratures. This has been in fact done by using spherical polar coordinates centered on the centre of mass of

M_1 (and M_2) [111] or by using prolate spheroidal coordinates [101].

Since the total interactions between non-reacting systems is dominated by the van der Waals potentials, the quadrature errors in the Coulomb terms are usually less than 5% over the whole range of overlapping densities which is within the expected accuracy of the EGM approach. For molecule-molecule interactions, however, a faster numerical treatment has been put forward [102], in which the electron density of the molecule 1, ρ_{M_1}, is expanded not in terms of products of MO's which are in turn expanded in basis functions, but directly in a basis of STO's, which turned out to be faster and more efficient than GTO's in fitting the tail-end distribution of ρ_{M_1}:

$$\rho_{M_1}(\underline{r}_1) = \bar{\rho}_{M_1}(\underline{r}_1) = \sum_i^{n_{M_1}} a_i \cdot \chi_i(\underline{r}_i) \tag{2.75}$$

the a_i are determined by linear least squares method adapted to yield better fits at large intermolecular distances. If both ρ_{M_1} and ρ_{M_2} are expanded in this way one can rapidly evaluate the coulomb energy by doing only ($n_{M_1} \times n_{M_2}$) integrals which are at the most two-centre integrals that can be done using standard programs [113]. The 3D quadratures become therefore the largest ones needed for all the other terms and for the full electrostatic contributions.

Some of the systems that have recently appeared in the litterature are listed in Table 3. One sees from it that the progress with the use of this model has been very rapid in providing the necessary $V(\underline{r})$ form to start inelastic collision calculations involving many molecules of experimental interest. One finds again that the short-range repulsive regions of interactions are remarkably well described via the EGM approach, and so are the qualitative features of weakly anisotropic potentials. On the other hand, the well depths and positions are usually inaccurate because of the difficulty in obtaining the correct asymptotic form of $V(\underline{r})$, unless modifications are introduced in the model [111]. Even then, however, care should be taken in properly joining in the various van der Waals terms. Comparison with HF or MCHF calculations in fact invariably shows constants departures in the tail-end of the well regions [109].

Finally, attempts have been made to extend the approach to closed shell-open shell interactions of atoms [114,115] by treating the noble gas halides

TABLE 3

Potential surface calculations via the EGM approach for molecular partners

SYSTEMS	CHARACTERISTICS	REFS. (year)
$Ar-N_2$	distance and angle dependence at N_2 eq. geometry	[103] (1973)
Ar-HCℓ	reasonable accuracy of short range repulsive forces, position (but not depth) of potential well.	[104] (1974)
He-HCN	linear rigid rotor in its ground vibrational level. 10-term Legendre expansion of potential. Rate constants for rotational excitation at low T.	[105] (1974)
He-CO	deeper well predicted and at a shorter distances than expectations. 5-term Legendre expansion. Transport properties calculations.	[106] (1976) [107] (1975)
He-HCℓ	rigid rotor in its lowest vibrational state. 9-term Legendre expansion.	[108] (1975)
$He-H_2CO$	rigid asymmetric top at H_2CO equilibrium geometry. Collinear and perpendicular approaches of projectile	[109] (1975)
Li^+-H_2	Several H bond distances. Collinear and perpendicular approaches of Li^+-. Comparison with HF calculations	[109] (1975)
$He-NH_3$	Equilibrium geometry of NH_3 and distance, angle dependence. 12-term Legendre expansion.	[110] (1976)

TABLE 3 cont.

He-CO_2	highly anisotropic. 8-term Legendre fitting. Reasonably good virial coeffs. Greatest inaccuracy in the well region.	[111] (1976)
Ar-CO_2	highly anisotropic. 10-term Legendre fitting. Reasonably good virial coeff.s. Greatest inaccuracy in the well region.	[111] (1976)
Ar-NO	NO($X^2\pi$) in its equilibrium geometry. A' and A" potentials given. 6-term Legendre expansions for average and difference potentials. Very anisotropic system.	[112] (1977)
HF-HF	Collinear and parallel geometries with HF at eq. distances. Comparison with SCF results. Good repulsive region and failure to get H-bonding.	[102] (1974)

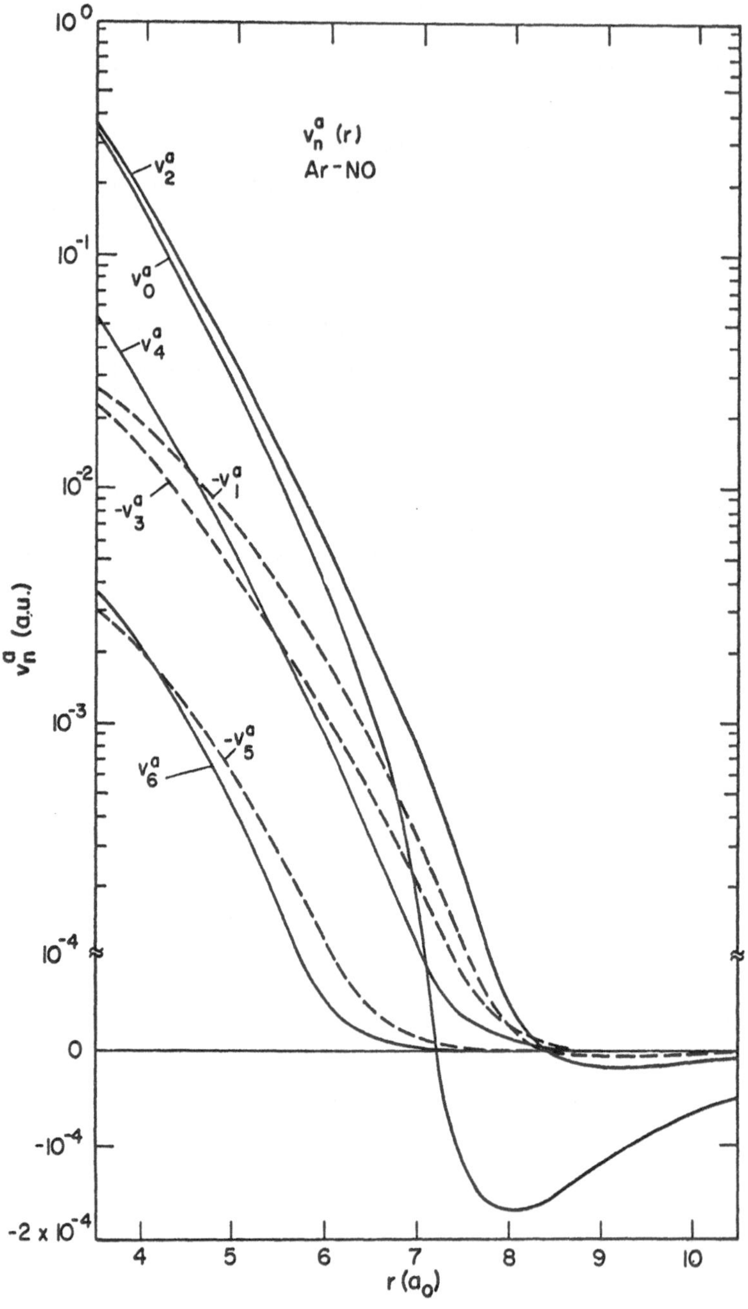

Figure 2.2

Average interaction potential expansion coefficients, V_i^a, for Ar atom and NO($X\,^2\pi$) molecule. The V_i^a's with odd values of i are plotted as negative [from G.C. Nielson et al., J. Chem. Phys. 66, 1396 (1977)].

and looking at the interpretation of their emission spectra. Essentially this extension involves some empirical prescription for obtaining the electron density of the alkali atoms and the comparison with ab-initio calculations for KrF shows that the prescription is successful for the Π state but poor for the Σ state. The latter involves the empty F orbital pointing at the Kr atoms, while the former has the empty F orbital perpendicular to Kr, hence the amount of charge transfer and configuration mixing caused in the Σ state by ionic bond formation is not correctly accounted for via the EGM approach. A further attempt has been made for atom-molecule interactions involving an open shell molecule, this time choosing the Ar-NO pair, where all the ground state interaction is of the non-bonding type with very little charge-transfer processes, hence possibly more suitable for an EGM treatment [116]. The asymptotic C_6 coefficients computed for the same system were also added into the long-range interaction [117] and the extra π orbital for NO was included with a different angular part for the $A'(V_+)$ and $A''(V_-)$ potentials. The results are shown in Figure 2, where the average of the two potentials, $V_a = \frac{1}{2}(V_+ + V_-)$, is plotted in terms of its radial coefficients in a 6-terms Legendre polynomial expansion [116]. Although the various approximations suggest only a qualitative accuracy for their shapes, it is worth pointing out that the great reduction in computational work afforded by this method makes it a very useful tool for mapping out the comparative behaviour of surfaces among similar partners and for readily provinding scattering-orientated forms of $V(\underline{r})$. In spite of its weaknesses and limitations the EGM approach has therefore yielded a good deal of surprisingly reliable results that have substantially extended our general knowledge of non-reacting systems interaction potential surfaces both in the repulsive and the weakly attractive regions.

2.5 - A survey of recent applications

In the last two sections we have examined the possibilities offered by some reductive treatments of the full electrostatic, spin-free non-relativistic Hamiltonian that controls the spectrum of the electronic bound states of simple closed-shell interacting partners within the B.O. approximation. The search for such reductions is obviously very important in order to minimize computational efforts without too drastic a loss of accuracy for the obtained potentials, but it still remains essential to also progressively upgrade the complexity of the attainable ab-initio results that naturally provide, when experiments lack or are uncertain, the benchmark calculations against which simplified treatments are tested. The progress in calculating complete potential surfaces can thus be assessed by considering in detail the work done on simple systems.

The first and oldest of such systems has been the simplest of exchange reactions, namely $H+H_2 \rightarrow H_2+H$. The history of the calculations of this reaction have been recently reviewed [46] and will not be repeated here. Suffice it to say that one of the recent attempts uses an approximate natural orbitals treatment [118] that was then fitted with a slightly modified analytic formula which also uses Morse functions to fit the H_2 potential [119]. Reaction probabilities were in turn computed via classical trajectory calculations and their results appear to confirm the validity of the saddle-point properties yielded by the ANO surface of this system [118].

A more strongly interacting pair of partners is provided by the H^+-H_2 system and by various other diatomic molecules interacting with protons, which have recently acquired a great deal of interest [120]. For the H_3^+ extensive SCF-CI calculations were carried out [121] and found an equilibrium geometry of an equilateral triangle with bond lengths of 1.66 a.u. . Further studies [122] with more extended basis sets have esentially confirmed the above results, while the detailed [123] study of the avoided crossing of the two potential surfaces from the charge-exchange has indicated that the lowest surface is attractive with a binding energy of 444 KJ/mol for H_3^+ with respect to the separate species. The second surface for $H + H_2^+$ was instead reported to be repulsive. Other proton-diatomic systems have also been studied for a few geometries and with a limited amount of CI included They were essentially meant to treat the Hydrogen bond regions but still show the feasibility of ab-initio calculations at the SCF level within the HF scheme to treat potential surface evaluations where no new bonds are formed and no electron pairs are broken. For slightly polarizable molecules, in fact, one might hope to even find simple electrostatic models for treating proton-molecule inelastic scattering [124]. Table 4 collects some of the recent results on surface calculations that only employ ab-initio

TABLE 4

Some *ab initio* computed potential energy surfaces from atom-molecule and molecule-molecule systems.

SYSTEMS	General Comments	References
1) atom-diatom		
$He-H_2$	SCF-CI treatment with van der Waals corrections. IEPA treatment for the long-range forces. Analytic fittings of repulsive and attractive regions. Dependence on H_2 distance	[125 - 127]
$Li-H_2$	HF calculations. 4-Term Legendre expansion. Analytic fits of the radial coefficients.	[128 , 129]
$Ne-H_2$	Collinear and perpendicular geometries. 2-conf. MCSCF and 2-electron CI. Analytic fits for a 2-term Legendre expansion. Dependence on the H_2 distance	[130]
Li-HF	HF equilibrium separation. HF calculations. 7-term Legendre expansion fitting.	[131 , 132]
2) ion-diatom		
Li^+-H_2	variable H_2 internuclear distance. SCF results and CI (IEPA) results. Analytic fittings to a 3-term Legendre expansion.	[133 - 135]
Li^+-HD	SCF calculations at various HD geometries. Isotope-corrected analytic fitting.	[136]

TABLE 4 cont.

Li^+-N_2	SCF calculations and CEPA correlation correction. N_2 geometry varies. Analytic fit to a 4-term Legendre expansion.	[137]
Li^+-CO	SCF calculations with CO eq. geometry. Surface dependence on CO bond length. Correlation effects on potential well-fit of the repulsive part of the surface.	[138]
H^+-CO	Protonation regions for HCO^+ and COH^+ linear geometries. Effects of CI and of CO bond length on potential wells. Analytic fit and Legendre expansion.	[139 - 142]
H^+-CO_2	SCF and CI Calculations of equilibrium structures of HCO_2^+. Non-linear geometry for protonated species. SCF potential surface and analytic fitting.	[143 , 144]
H^+-N_2	Protonation region of H^+N_2. SCF and limited CI calculations. Effect of proton approach on N_2 bond length.	[140 , 145]
<u>3) diatom-diatom</u> H_2- H_2	SCF calculations at various H_4 geometries. Double-zeta basis set and CI treatments.	[146]
HF-HF	SCF calculations with molecules at eq. geometry. 294 points of a 4D surface attempts at analytic fittings. Fitting of a triple series of Legendre polynomials. 30 (SF) terms are transformed into 20(BF) terms; the 6 largest are fit-	[147 , 148]

TABLE 4 cont.	ted analytically.	
$CO-H_2$	Several geometries of eq. CO and H_2 distances in the non-reactive region (> 4.0. a.u.). Dispersion energy corrections and C_6 coefficients. SCF calculations for the whole surface.	[149]

methods and gives some salient features of their findings.

It clearly appears from this Table that a great deal of progress has been made in the last few years to gather from quantum chemistry models the necessary information to start evaluating energy transfer by collision in molecules. It also indicates the lengthy procedure needed to fit the calculated points with the necessary accuracy to make scattering calculations somehow stabilized against insufficient expansions. It further suggests that bound state calculations are used at their best when directly linked with scattering calculations involving pure rotational inelasticity or rotovibrational inelasticity, since the accuracy attained in the numerical solutions of the coupled equations (see next Chapter) is strongly related to the number of Legendre expansion terms or to the radial grids over which the potential is reliably known.

As a final summary, Figure 2.3 presents a typical interaction potential for non-reacting atom-atom pairs (i.e. the Ne-Ne interaction) for which the various regions where different theoretical techniques have been employed are presented. One clearly sees that for non-chemical systems the HF approach only allows to obtain the repulsive regions of the potential, while perturbative approaches can be fairly successfully used for isotropic long-range interaction. It is in the interplay of these terms, i.e. around the well regions, where both experiments and theory are put to the greatest test in reliably yielding the correct potential behaviour.

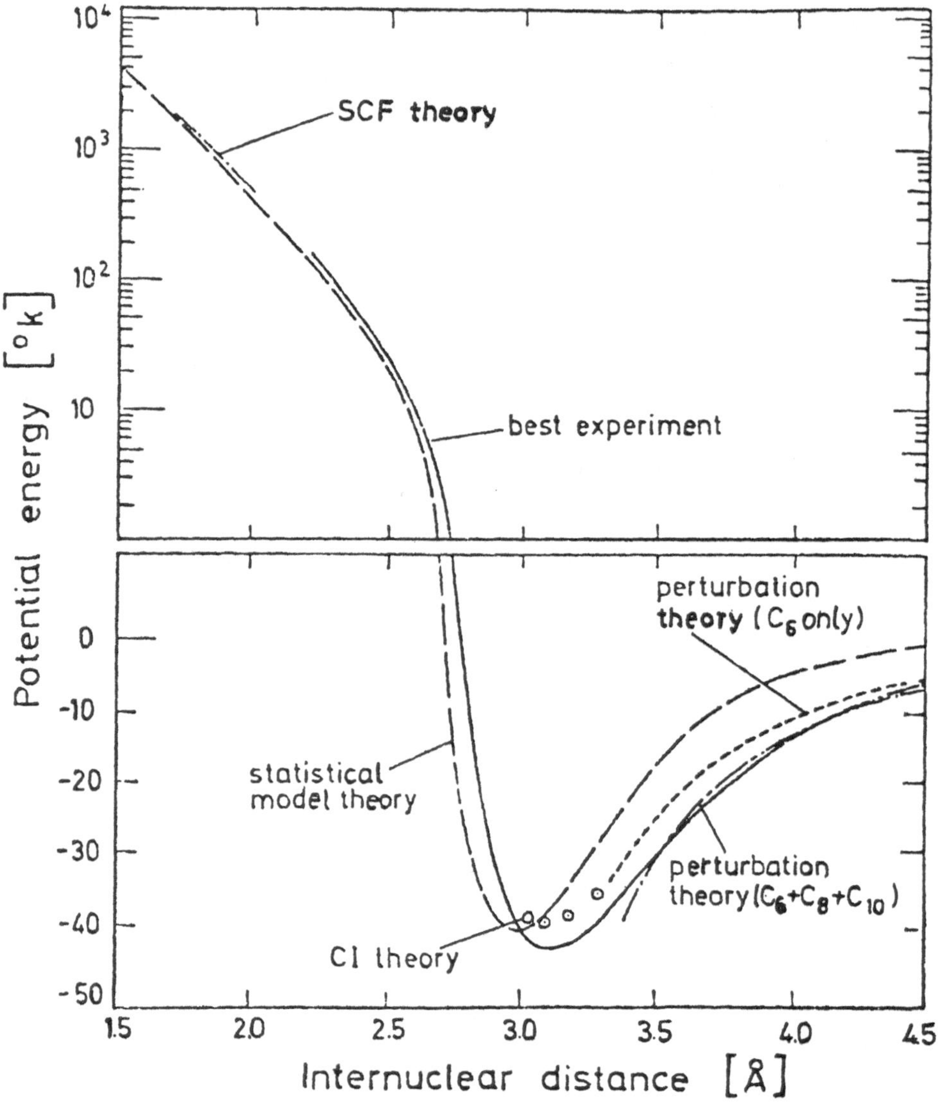

Figure 2.3

Comparison of best experimental results on Ne-Ne interaction potential and various theoretical methods that can be used in different radial domains [from J.P.Toennies, in: IX Int. Symposium on rarefied gas dynamics, Göttingen, July 1974].

REFERENCES

[1] A. Dalgarno and R. Mc Carroll, Proc. Roy. Soc. A237,(1956) 383.

[2] M. Born and R. Oppenheimer, Annln. Phys. 84,(1927) 457.

[3] M. Born and K. Huang, Dynamical Theory of Crystal Lattices, O.U.P., (1954).

[4] M. Born, Nachr. Akad. Wiss., Göttingen, Mat. Phys. Kl., 1.

[5] R.B. Gerber, Int. J. Quant. Chem., 1, (1967) 781.

[6] R.B. Gerber, Proc. Roy. Soc. A309, (1969) 221.

[7] C.A. Coulson and R.B. Gerber, Int. J. Quant. Chem. 2, (1968) 607.

[8] J.P. Toennies, Chem. Soc. Rev., 3, 407 (1974).

[9] J.P. Toennies, Ann. Rev. Phys. Chem., 27, 225 (1976).

[10] M.P. Kelly, Advan. Chem. Phys., 14, 129 (1969).

[11] R.K. Nesbet, Advan. Chem. Phys., 14, 1 (1969).

[12] R.K. Nesbet, Int. J. Quantum Chem., 43, 117 (1971).

[13] C. Eckart, Phys. Rev.,36, 878 (1930).

[14] E.A. Hylleraas and B. Undheim, Z. Physik, 65, 759 (1930).

[15] e.g., see: E.B. Wilson Jr., J. Chem. Phys., 43, S172 (1965).

[16] H. Conroy, J. Chem. Phys., 47, 930, 5307 (1967).

[17] S.F. Boys and N.C. Handy, Proc. Roy. Soc. Sez. A309, 209 (1969).

[18] S.F. Boys and N.C. Handy, Proc. Roy. Soc. Ser. A310,43 (1969).

[19] N.C. Handy and S.T. Epstein, J. Chem. Phys.,53, 1392 (1970).

[20] S.F. Boys and N.C. Handy, Proc. Rpy. Soc. Ser. A311, 309 (1969).

[21] E. Clementi and D.R. Davis, J. Comp. Phys., 1, 223 (1966).

[22] E.F. Hayes and R.G. Parr, Progr. Theor. Phys. Suppl. 40, 78 (1967).

[23] F.A. Gianturco and D.G. Thompson, Chem. Phys. 14, 111 (1976).

[24] F.A. Gianturco and N. Chandra, Chem. Phys. Lett., 24, 326 (1974).

[25] F.A. Gianturco and N.K. Rahman, J. Phys. B., 10, L219 (1977).

[26] J.C. Slater, Quantum Theory of Atomic Structure, vol.2 (McGraw-Hill, N.Y.,1960).

[27] C.C.J. Roothaan, Rev. Mod. Phys., 23, 69 (1951).

[28] C.C.J. Roothaan, and P. Bagus, Methods Comput. Phys., 2, 47 (1963).

[29] M. Krauss, Tech. note N. 438, N.B.S., Washington D.C. (1967).

[30] R.K. Nesbet, Advan. Quantum Chem., 3, 1 (1967).

[31] R.K. Nesbet, Advan. Chem. Phys., 14, 1 (1969).

[32] C.A. Coulson, Valence, O.U.P., Oxford, (1961).

[33] A.C. Wahl, J. Chem. Phys., 41, 2600 (1964).

[34] A.D. Mc Lean, J. Chem. Phys., 39, 2653 (1963).

[35] I.H. Hillier, V.R. Saunders and M.H. Wood, Chem. Phys.Lett., I, 323 (1970).

[36] R.K. Nesbet, Proc. Roy. Soc., A230, 312 (1955).

[37] P.O. Löwdin, Adv. Chem. Phys., 2, 207 (1959).

[38] R. Mc Weeny and B.T. Sutcliffe, Methods of Molecular Quantum Mechanics (Academic Press, N.Y., 1969).

[39] E. Clementi, J. Chem. Phys., 38, 2248 (1963).

[40] T. Amos and L.C. Snyder, J. Chem. Phys., 41, 1773 (1964).

[41] P.O. Löwdin, Rev. Mod. Phys., 36, 966 (1964).

[42] J.C. Slater, Quantum Theory of Molecules and Solids (McGraw-Hill, N.Y.,1963).

[43] L.C. Snyder, J. Chem. Phys., 46, 3602 (1967).

[44] L.C. Snyder and H. Basch, J. Amer. Chem. Soc. 91, 2189 (1969).

[45] J.A. Pople et al., J. Amer Chem. Soc., 92, 4796 (1970).

[46] R.F. Bader and R.A. Gangi, Theoretical Chemistry Vol. II, (A Specialist Periodical Report, The Chemical Soc., 1975).

[47] T.L. Gilbert and A.C. Wahl, J. Chem. Phys., 47, 3425 (1967).

[48] E.A. Hylleraas, Z. Physik, 48, 469 (1928).

[49] G. Das and A.C. Wahl, Adv. Quantum Chem., 5, 261 (1970).

[50] R.P. Hosteny, A.R. Hinds and A.C. Wahl, Chem. Phys. Lett., 23, 9 (1973).

[51] K.J. Miller and K. Ruedemberg, J. Chem. Phys. 48, 3450 (1968).

[52] M. Jungen and R. Ahlrichs, Theor. Chim. Acta, 17, 339 (1970).

[53] O.Sinanoglu, Advan. Chem. Phys., 14, 237 (1969).

[54] C. Moeller and M.S. Plesset, Phys. Rev., 46, 618 (1934).

[55] P.O. Löwdin, Phys. Rev., 97, 1474 (1955).

[56] S. Hagstrom and H. Shull, Rev. Mod. Phys. 35, 624 (1963).

[57] C. Edmiston and M. Krauss, J. Chem. Phys., 45, 1833 (1966).

[58] F. Driessler et al., Theoret. Chim. Acta, 30, 315 (1973).

[59] V. Staemmler, Chem. Phys., 17, 187 (1976).

[60] A.C. Whal, in Theoretical Chemistry, W. Byers-Brown Ed., MTP Int. Review of Science, pg. 41 (Oxford, U.K., 1972).

[61] W. Moffitt, Rep. Progr. Phys., 17, 173 (1954).

[62] R.N. Porter, R.M. Stevens and M. Karplus, J. Chem. Phys., 49, 5163 (1968).

[63] S. Sato, J. Chem. Phys., 23, 592 (1965).

[64] M. Karplus, in Molecular Beams and Reaction Kinetics, pg. 320 (Academic Press, N.Y., 1970).

[65] L. Pedersen and R.N. Porter, J. Chem. Phys., 47, 4751 (1967).

[66] W. Moffitt, Proc. Roy. Soc. (London), A210, 245 (1951).

[67] P.O. Löwdin, Phys. Rev., 97, 1474 (1955).

[68] G.G. Balint-Kurti and M. Karplus, J. Chem. Phys. 50, 478 (1969).

[69] G.G. Balint-Kurti, Mol. Phys., 25, 393 (1973).

[70] J.D. Weeks, A. Hazi and S.A. Rice, Adv. Chem. Phys., 16, 283 (1969).

[71] e.g. see: W. Kutzelnigg, Chem. Phys. Lett., 4, 435 (1969).

[72] R.N. Dixon and J.M.V. Hugo,Mol. Phys. 29, 953 (1975).

[73] R.N. Dixon , P.W. Tasker and G.G. Balint-Kurti, Mol. Phys. 34, 1455 (1977).

[74] For reviews see: J.O. Hirschfelder (Ed.), Adv. Chem. Phys., 12, (1967).

[75] V.I. Gaydaenks and V.K. Nikulin, Chem. Phys. Lett., 7, 360 (1970).

[76] R.G. Gordon and S. Kim, J. Chem. Phys., 56, 3122 (1972).

[77] H.F. Schaefer, D.R. Mc Laughlin, F.E. Harris and B.J. Selder, Phys. Rev. Lett., 255, 991 (1970).

[78] J.N. Murrell and G. Shaw, Molec. Phys., 15, 325 (1968).

[79] e.g. see: P. Gambas, Die Statistiche Theorie des Atoms und ihre Auwendungen (Springer, Vienna, 1969).

[80] N.H. March, Adv. Phys., 6, 1 (1957).

[81] W.J. Carr Jr. and A.A. Manedunin, Phys. Rev., 133, 371 (1964).

[82] F.A. Gianturco and M. Dilonardo; J. Chim. Phys., 72, 315 (1975).

[83] W.J. Carr Jr. et al., Phys. Rev., 124, 747 (1961).

[84] Y.S. Kim and R.G. Gordon, J. Chem. Phys., 60, 1842 (1974).

[85] J.S. Cohen and R.T. Pack, J. Chem. Phys., 61, 2372 (1974).

[86] A.I.M. Rae, Chem. Phys. Lett., 18, 574 (1973).

[87] A.I.M. Rae, Mol. Phys., 29, 467 (1975).

[88] Y.S. KIm and R.G. Gordon, J. Chem. Phys., 61, 1 (1974).

[89] P.K.L. Drude,The Theory of Optics (Longmans Green, London 1933).

[90] V.K. Nikulin and Y.N. Tsarev, Chem. Phys., 10, 433 (1975).

[91] Y.S. Kim and R.G. Gordon, J. Chem. Phys., 60, 4323 (1974).

[92] F.A. Gianturco, J. Chem. Phys., 64, 1973 (1976).

[93] E.W. Mc Daniel and E.A. Mason, The mobility and diffusion of ions in gases (Wiley-Interscience, N.Y. 1973).

[94] F.A. Gianturco et al., J. Chim. Phys. 74, 66) (1977).

[95] F.A. Gianturco and U.T. Lamanna, Physics 223C, 1 (1977).

[96] I.R. Gatland et al., J. Chem. Phys., 66, 537 (1977).

[97] D.F. Heller at al., J. Chem. Phys., 62, 1949, 3601, 3854 (1974).

[98] Y.S. Kim, Phys. Rev., A11, 796, 804 (1975).

[99] D.L. Freeman, J. Chem. Phys., 62, 941 (1975).

[100] B. Schneider et al., Chem. Phys. Lett., 27, 577 (1974).

[101] S. Green and R.G. Gordon, Q.C.P.E., University of Indiana, Bloomington, Indiana 47401. Program. n. 251 (1974).

[102] G.A. Parker et al., Chem. Phys. Lett., 33, 599 (1975).

[103] Y.S. Kim, Thesis, Harvard U., 1973.

[104] S. Green, J. Chem. Phys., 60, 2654 (1974).

[105] S. Green and P. Thaddeus, Astrophysical J., 191, 653 (1974).

[106] S. Green and P. Thaddeus, Astrophysical J., 205, 766 (1976).

[107] L. Monchick and S. Green, J. Chem. Phys., 63, 2000 (1975).

[108] L. Monchick and S. Green, J. Chem. Phys., 63, 4198 (1975).

[109] S. Green et al., J. Chem. Phys., 633, 1154 (1975).

[110] S. Green, J. Chem. Phys., 64, 3463 (1976).

[111] G.A. Parker et al., J. Chem. Phys., 64, 1668 (1976).

[112] G.C. Nielson et al., J. Chem. Phys., 66, 1396 (1977).

[113] A.D. Mclean and M. Yoshimine, IBM J. Res. Develop. Suppl. (1967).

[114] M.J. Clugston and R.G. Gordon, J. Chem. Phys., 66, 239 (1977).

[115] M.J. Clugston ans R.G. Gordon, J. Chem. Phys., 66, 244 (1977).

[116] G.C. Nielson et al., J. Chem. Phys., 66, 1396 (1977).

[117] G.C. Nielson et al., J. Chem. Phys., 64, 2055 (1977).

[118] B. Lin, J. Chem. Phys., 58, 1925 (1973).

[119] A.C. Yates and W.A. Lester Jr., Chem. Phys. Lett., 24, 305 (1974).

[120] C.F. Giese and W.R. Gentry, Phys. Rev., A10, 2156 (1974).

[121] I.G. Csizmadia et al., J. Chem. Phys., 52, 6205 (1970).

[122] G.D. Carney and R.N. Porter, J. Chem. Phys., 60, 4251 (1974).

[123] C.W. Bauschlicher et al., J. Chem. Phys., 59, 1286 (1973).

[124] F.A. Gianturco, Int. J. Quantum Chem., 10, 37 (1976).

[125] M.D. Gordon and D. Secrest, J. Chem. Phys., 52, 120 (1973).

[126] B. Tsalpine and W. Kutzelnigg, Chem. Phys. Lett., 23, 173 (1973).

[127] P.J. Geurts et al., Chem. Phys. Lett., 35, 444 (1975).

[128] A. Karo et al., Status Report, Argonne Nat. Lab., (1974).

[129] B.H. Choi et al., Chem. Phys. Lett., 48, 237 (1977).

[130] J.W. Birks et al., J. Chem. Phys., 63, 1741 (1975).

[131] W.A. Lester Jr. and M. Krauss, J. Chem. Phys., 52, 4775 (1970).

[132] W.A. Lester Jr., J. Chem. Phys., 53, 1611 (1970).

[133] W.A. Lester Jr., J. Chem. PHys., 53, 1511 (1970).

[134] W.A. Lester Jr., J. Chem. Phys., 54, 3171 (1971),

[135] W. Kutzelnigg et al., Chem. Phys., 1, 27 (1973).

[136] W.A. Lester Jr., IBM J. Res. Dev., 15, 222 (1971).

[137] V. Staemmler, Chem. Phys., 7, 17 (1975).

[138] V. Staemmler, Chem. Phys., 17, 187 (1976).

[139] H.B. Jansen and P. Ros, Chem. Phys. Lett., 3, 140 (1969).

[140] S. Forsen and B. Roos, Chem. Phys. Letters, 6, 128 (1970).

[141] P.J. Bruna et al., Chem. Phys. 10, 323 (1975).

[142] F.A. Gianturco and U.T. Lamanna (unpublished results).

[143] S. Green et al., Chem. Phys., 17, 479 (1976).

[144] F.A. Gianturco and U.T. Lamanna (in preparation).

[145] K. Vasudevan et al., Chem. Phys., 5, 149 (1974).

[146] D.M. Silver and R.M. Stevens, J. Chem. Phys., 59, 3378 (1973).

[147] D.R. Yarkony et al., J. Chem. Phys., 60, 855 (1974).

[148] M.H. Alexander and A.E. De Pristo, J. Chem. Phys., 65, 5009 (1976).

[149] J. Prissette et al., Chem. Phys., to be published (1978).

3. ROTATIONAL AND VIBRATIONAL INELASTICITY IN MOLECULAR ENCOUNTERS

3.1 - Introduction

The theory of energy transfer in molecular collisions is of importance for the understanding of a number of physical problems. Moreover, the precise study of chemical reactivity and gas-phase reaction activation energies is strongly related to the reagents "preparation", that is to the distribution, storage and disposal of available molecular energies among internal degrees of freedom.

In several of these problems one is chiefly interested in determining microscopic and macroscopic probabilities of energy transfer between the translational motion and one or more of the internal motions of the molecules. As a consequence, a great deal of work in recent years has been done on the calculation of vibrational transition probabilities and of the inelastic cross sections in which essentially vibro-rotational states are excited in the molecules. Such collisions are in fact a typical process in simple molecular gases under ordinary room temperature conditions or in the T range from 100° to 5000° K. They are also the elementary collision processes mainly responsible for relaxation in fast flows that occur, for instance, during the rapid gas compression in shock and acoustic waves or during expasion in nozzle flow.

Moreover, both unimolecular and bimolecular reactions have their rates determined by early vibrational triggering or by the amount of energy stored in the vibrational degrees of freedom of the reactants. The ensuing redistribution among the internal energy degrees of freedom of the reactants is also determined by similar collision processes. One of the most recently studied processes, i.e. the competition of collision energy transfer with radiative decay in chemical lasers, is also strongly linked with the understanding of the individual encounter flux redistribution among allowed channels, since such an understanding is obviously required for further optimization of the chemical laser itself.

Electronic transitions during collision become significant only at higher temperatures and will not be discussed in this Chapter. The prime objective of an energy tranfer study thus the determination of whether any specific change in the internal states of the molecule has taken place after an individual encounter, since the treatment of a multiple scattering approach is still out of reach while, under many experimental conditions, the single-collision hypothesis seems to hold reasonably well.

In the present Chapter we will therefore outline the fully quantum approach to the three-dimensional treatment of vibro-rotational inelastic encounters between atoms and molecules. This turns out to be quite a formidable problem due to the many channels that are open even at low collision energies and for simple diatomic targets. Moreover, the corresponding knowledge of an accurate potential surface becomes rapidly prohibitive when increasing the atomic numbers of the component atoms and one is then forced to search for correspondingly simpler approaches to the quantum dynamics of the process.

The reduction in size of the number of coupled equations that have to be solved whill thus provide the subject for the following Chapter, while we will be discussing in the last section of the present one the computational results and the formulation of some general conclusion about the projected extensions of a rigorous treatment to more complex systems or to higher collisional energies. The link of the microscopic results on cross sections with individual transition rates and macroscopic relaxation times will be further discussed and examined in a separate Chapter.

3.2 - Quantum treatments of inelastic collisions

The quantum approach to the scattering between structureless particles interading via a spherical potential has altready been outlined (see Chapter 1), while the more general form of realistic interaction potentials has been examined in Chapter 2 where the computational techniques for evaluating such potentials have also been described.

In the case of scattering between particles with structure, it was pointed out that it is often convenient to separate explicitely the relative position vector, $\underline{R}$, of the collision partners and the remaining (3N-3) degrees of freedom for the N component nuclei, $\underline{X}$. The next step then is to extend the potential scattering theory to take into account a possible change in the internal state $\varphi_i^\Gamma(\underline{X})$ of the system, where $\varphi_i^\Gamma(\underline{X})$ typically represents a product of rotational-vibrational functions for the two colliding molecule in a given electronic state Γ and is evaluated at infinite separation. Each of these states is said to define a different channel of the system.

The first step here is to set up the equations of relative motion once we assume to have a known set of internal states $\varphi_i^\Gamma(\underline{X})$ and a known form of the hypersurface $V(\underline{R}\ ,\ \underline{X})$. The choice of a specific form of the boundary conditions and their relationship with the collision cross sections leads then to the introduction of a scattering matrix $\underline{\underline{S}}$ that relates the outgoing probability amplitudes in the different channels to their possible incoming values. Such a definition of the S-matrix is a direct generalization for the many-channel case of the potential scattering S-matrix defined in Chapter 1. Finally, one can obtain expressions for the scattering amplitudes $f_{ij}(\Omega)$ in terms of the matrix elements of $\underline{\underline{S}}$ and according to whether one is in the presence of a spherically symmetric field or of a more general interaction.

The time-indipendent Schrödinger equation:

$$\mathcal{H}\ \Phi\ (\underline{R}\ ,\ \underline{X}) = E_{TOT}\ \Phi\ (\underline{R}\ ,\ \underline{X}) \tag{3.1}$$

where the Hamiltonian $\mathcal{H}$ is the one defined in eq. (2.16), may be cast into two limiting forms according to whether the perturbation of the initial system appears via the potential itself or via a variation in the form of the internal states of the whole system. Other possible coupling schemes have also been suggested [1,2].

In the coupling scheme via the potential the full wavefunction is expanded in terms of the unperturbed (diabatic) orthonormal eigenstates $\varphi_j(\underline{X})$ of $\mathcal{H}_{int}(\underline{X})$ as given in eq. (2.17). In other words:

$$\Phi(\underline{R}\ ,\ \underline{X}) = \sum_j \psi_j(\underline{R})\ \varphi_j\ (\underline{X}) \tag{3.2}$$

so that after substitution into (3.1), multiplication by $\varphi_i(\underline{X})$ and interegration over the internal variables one can write that:

$$\{\nabla^2_{\underline{R}} + k^2_i\} \Psi_i(\underline{R}) = \sum_j U_{ij}(\underline{R}) \Psi_j(\underline{R}) \tag{3.3}$$

where:

$$k^2_i = 2\mathcal{M}(E_{tot} - E_i)$$

$$U_{ij}(\underline{R}) = 2\mathcal{M}\int \varphi_i(\underline{X}) \; V(\underline{R}, \underline{X}) \; \varphi_j(\underline{X}) d\underline{X} \tag{3.4}$$

Equation (3.3) represents the basic equation for the scattering problems we will be interested in and it is written here in its general form. It includes the important, simple models of the scattering of an atom by an harmonic oscillator or by a rigid rotor (the numerical and analytic approaches to its solution are described in a following Chapter). The different forms of V obviously directly affect the possible reduction in dimensionality of the coupled equations implied by the infinite sums of (3.3).

From a physical viewpoint the diagonal terms $U_{ii}(\underline{R})$ contribute directly to the elastic scattering. By distorting the $\Psi_j(\underline{R})$ from their outgoing plane wave forms they also affect the inelastic cross sections. The off-diagonal terms $U_{ij}(\underline{R})$, which couple together the different channels, are responsible for any inelastic process which occurs and for polarization contributions to the elastic scattering.

This formulation represents the diabatic picture of the problem. In actual cases that are concerned with electromic excitations or with processes where the relaxation of internal motion may be assumed as fast when compared with relative velocity, another limiting picture can be used. This means that eq. (3.2) is replaced by an expasion over a set of adiabatic internal states $\chi_k(\underline{R}, \underline{X})$ which are eigenfunctions (orthonormal) of the full internal hamiltonian at a given relative distance $\underline{R}$:

$$\{\mathcal{H}_{int}(\underline{X}) + V(\underline{R}, \underline{X})\} \; \chi_k(\underline{R}, \underline{X}) = E_K(\underline{R})\chi_K(\underline{R}, X) \tag{3.5}$$

both the eigenenergies $E_k(\underline{R})$ and the eigenfunctions therefore depending parametrically on $\underline{R}$. Since the interaction potential is defined as vanishing at infinity, they go over at long range to their couterparts E_j and $\varphi_j(\underline{X})$.

The expansion (3.2) then becomes:

$$\Phi(\underline{R}, \underline{X}) = \sum_K \tilde{\Psi}_k(\underline{R}) \; \chi_k(\underline{R}, \underline{X}) \tag{3.6}$$

which, substitued into (3.1) leads to the following equations for the $\tilde{\Psi}_k$'s:

$$\{\nabla^2_{\underline{R}} + k^2_k(\underline{R})\}\ \tilde{\Psi}_k(\underline{R}) = \sum_{\ell} \{A_{k\ell}(\underline{R}) \cdot \nabla_{\underline{R}} + B_{k\ell}(\underline{R})\}\ \tilde{\Psi}_{\ell}\ (\underline{R}) \tag{3.7}$$

where:

$$k^2_k(\underline{R}) = 2\mathcal{M}\{E_{tot} - E_k(\underline{R})\} = \tag{3.8}$$

and:

$$A_{k\ell}(\underline{R}) = -\,2 \int \chi^*_k(\underline{R}\ ,\ \underline{X})\ \nabla_{\underline{R}}\ \chi_{\ell}(\underline{R}\ ,\ \underline{X})\ d\underline{X}$$
$$= -\,2 < \chi_k | \nabla_{\underline{R}} |\ \chi_{\ell} > \tag{3.9a}$$

$$B_{k\ell}(\underline{R}) = -\int \chi^*_k(\underline{R}\ ,\ \underline{X})\ \nabla^2_{\underline{R}}\ \chi_{\ell}(\underline{R}\ ,\ \underline{X})\ d\underline{X}$$
$$= -\,< \chi_k\ | \nabla^2_{\underline{R}} |\ \chi_{\ell} > \tag{3.9b}$$

This is termed the adiabatic, or kinetic, coupling scheme. The non-adiabatic effects in (3.7) may arise from both the radial and angular terms in the A's and B's, although the radial contributions are normally the most important for molecular systems [3,4].

The adiabatic approach, in full analogy with the treatment outlined in Chapter 2 of the electronic bound states in the B-O approximation, requires for the perturbation treatment to be valid that the energy difference, $\Delta E_{k\ell}$, between initial state $|k>$ and final state $|\ell>$ should dominate the radial part of the interaction term. This means that, under semiclassical conditions, one can write:

$$\frac{\{E_k(\underline{R}) - E_{\ell}(\underline{R})\}}{<\chi_k|^d/dR|\chi_{\ell}> \cdot \underline{v}} >> 1 \tag{3.10}$$

where $\underline{v}$ is the relative velocity and the integral over the radial gradient is the inverse of the characteristic lengh. Such a description is therefore favoured by large adiabatic energy differences and low collision velocity.

The boundary conditions on the two forms of the equations of motion (3.3) and (3.7) may all be expressed in the same form since the physical nature of the problem should not be affected by the choice of representation. In a laboratory, space fixed reference system these conditions therefore require, by extension of the equation (1.16), that the solution will have the asymptotic form of a plane wave together with an outgoing wave in the entrance channel, and outgoing waves in all the other channels:

$$\Psi_i(\underline{R}) \underset{R\to\infty}{\sim} e^{ik_i z} + f_{ii}(\Omega)\,\frac{e^{ik_i R}}{R} \tag{3.11a}$$

$$\Psi_i(\underline{R}) \underset{R\to\infty}{\sim} f_{ij}(\Omega)\,\frac{e^{ik_i R}}{R} \tag{3.11b}$$

$f_{ij}(\Omega)$ can now be interpreted as the probability amplitude for scattering flux into channel j from the initial channel i.

Hence, again in analogy with eq. (1.24), and allowing for a corresponding change in the asymptotic velocity after an inelastic collision, the differential cross section becomes:

$$\sigma_{i\to j}(\theta,\varphi) = {}^{k_i}/_{k_j}\; |\, f_{ij}(\theta,\varphi)\,|^2\; d\Omega \tag{3.12}$$

and the total inelastic cross section:

$$\sigma_{i\to j} = {}^{k_j}/_{k_i} \int_0^{2\pi}\!\!\int_0^{\pi} |f_{ij}(\theta,\varphi)|^2 \sin\theta\, d\theta\, d\varphi \tag{3.13}$$

the relation betweeen a particular $f_{ij}(\Omega)$ and the S-matrix follows by the generalization of the partial wave analysis of Section 3 in Chapter 1. One must however take into account the fact that, due to the internal structures of either of the partner or of both, several quantum numbers arise from the description of their internal states. Thus, they have to be coupled with the orbital angular momentum, $\bar{\ell}$, arising from the relative motion of the collision partners since only the magnitude of the total angular momentum component along a chosen axis of quantization is strictly conserved.

Each of the above quantum numbers becomes however an independent constant of the motion whenever: $V(\underline{R}\,,\,\underline{X}) \to V(R\,,\,\underline{X})$ hence no coupling exists for such a spherical potential.

For such a case a simple partial wave decomposition of the functions describing the translation in eq. (3.3) becomes therefore possible:

$$\Psi_j(\underline{R}) = \frac{1}{R}\cdot\sum_{\ell=0}^{\infty} C_\ell\, \psi_{j\ell}(R)\cdot P_\ell(\hat{R}\cdot\hat{Z}) \tag{3.14}$$

and this leads to the following radial equations for the unknown $\Psi_{j\ell}(R)$ functions:

$$\left\{ \frac{d^2}{dR^2} + k_i^2 - \frac{\ell(\ell+1)}{R^2} \right\} \psi_{i\ell}(R) = \sum_j U_{ij}(R)\, \psi_{j\ell}(R) \tag{3.15}$$

Each orbital angular momentum quantum number ℓ contributing to the expansion (3.14) gives rise then to a different set of coupled equations like (3.15) and to a different S-matrix $\underline{\underline{S}}(\ell)$. Since the boundary conditions (3.11) require a plane wave component in the incoming channel, i, the corresponding coefficient of (3.14) can be found as given by:

$$C_\ell^{(i)} = \frac{(2\ell + 1)\, i^\ell}{k_i} \tag{3.16}$$

hence the general asymptotic form of each solution $\Psi_j(R)$ to be used in eq. (3.2) is given by:

$$\Psi_j(\underline{R}) = \delta_{ij}\, e^{ik_j z} + f_{ij}(\theta) \frac{e^{ik_j z}}{R} \tag{3.17}$$

which is the multichannel version of eq. (1.50) and where the scattering amplitude is written in a form that is in turn an extension of eq. (1.52):

$$f_{ij}(\theta) = \frac{1}{2(k_i k_j)^{\frac{1}{2}}} \sum_\ell (2\ell + 1) \{S_{ij}^{(\ell)} - \delta_{ij}\} P_\ell (\cos\theta) \tag{3.18}$$

This central field amplitude is again cylindrically symmetric and the terms in curly brackets on the r.h.s. of (3.18) now contain the elements of a generalized $S(\ell)$-matrix. In fact, for the diagonal elements of such a matrix one can write a complex expression of the type:

$$S_{ii}^{(\ell)} = \exp\,[\,2i(\alpha_\ell + i\beta_\ell)] \tag{3.19}$$

its imaginary part, in this case, has to account for the depletion of the initial channel into the inelastic channels. Hence: $\beta_\ell > o$ and $|S_{ii}^{(\ell)}|^2 < 1$. The scattering amplitude $f_{ii}(\theta)$ therefore acquires a form more transparently analogous to (1.52):

$$f_{ii}(\theta) = \frac{1}{2ik_i} \sum_\ell (2\ell + 1)\{\exp[\,2i(\alpha_\ell + i\beta_\ell)] - 1\}\, P_\ell(\cos\theta) \tag{3.20}$$

and the corresponding eleastic cross section becomes:

$$\sigma_{ii}(k^2) = \frac{\pi}{k_i^2} \sum_\ell (2\ell + 1) \left\{ |S_{ii}^{(\ell)}|^2 + 1 - 2 \, \mathrm{Re} \, S_{ii}^{(\ell)} \right\}$$

while the inelestic, differential and total cross sections are obtained via eq. (3.18):

$$\sigma_{ij}(\theta) = \frac{1}{4k_i^2} \left| \sum_\ell (2\ell + 1)(\delta_{ij} - S_{ij}^{(\ell)}) \, P_\ell \, (\cos\theta) \right|^2 \qquad (3.22)$$

$$\sigma_{ij}(k^2) = \frac{\pi}{K_i^2} \sum_\ell (2\ell + 1) \, | \, (\delta_{ij} - S_{ij}^{(\ell)}) |^2 \qquad (3.23)$$

Now, because of the structure of the S-matrix: $S_{ij}^{(\ell)} = S_{ji}^{(\ell)}$ and: $\sum_j |S_{ij}^{(\ell)}|^2 = 1$. Hence:

$$k_i^2 \, \sigma_{ij}(\theta) \; = \; k_j^2 \, \sigma_{ji}(\theta) \qquad (3.24a)$$

and:

$$k_i^2 \, \sigma_{ij}(k^2) = k_j^2 \, \sigma_{ji}(k^2) \qquad (3.24b)$$

The above equalities are called the equations of detailed balance, a result which will frequently be employed in the following Chapters.

Moreover, if one writes:

$$\sum_{j \neq i} \sigma_{ij}(k^2) = \frac{\pi}{k_i^2} \sum_\ell (2\ell + 1) \sum_{j \neq i} | \, S_{ij}^{(\ell)} \, |^2$$

$$= \frac{\pi}{k_i^2} \sum_\ell (2\ell + 1) \, \{ \, 1 - | \, S_{ii}^{(\ell)} \, |^2 \} \qquad (3.25)$$

$$= \frac{\pi}{k_i^2} \sum_\ell (2\ell + 1) \, \{ \, 1 - \exp(-4\beta_\ell) \}$$

the total cross section becomes:

$$\sigma_{tot}(k^2) = \sigma_{ii}(k^2) + \sum_{j\neq i} \sigma_{ij}(k^2)$$

$$= \frac{2\pi}{k_i^2}\sum_{\ell}(2\ell + 1)(1 - \mathrm{Re}\, S_{ii}^{(\ell)}) = \frac{4\pi}{k_i^2}\,\mathrm{Im}\, f_{ii}(o) \tag{3.26}$$

which is the formulation of the optical theorem of eq. (1.63) for a multichannel scattering situation and for a spherical potential.

3.3 The rotational behaviour of molecules

Before examining the actual ways in which the expansions of the total wavefunction $\Phi(\underline{R}\ ,\ \underline{X})$ discussed in the previous ection can be carried out, let us examine the general form of the asymptotic wavefunctions describing the molecules as rigidly rotating in space.

This will obviously represent an important starting point for the initial, less complex, problem of the pure rotational excitations and de-excitations of molecular targets by collisions. The further extension to polyatomic molecules also vibrating in space requires a detailed discussion of the form of the B-O potential surface within the spatial asymptotic region where the (3N-3) coordinates correspond to near-equilibrium values of the internal coordinates for the separate molecular partners.

It constitutes a subject of its own [5] and will interest us here only as far as the simplest case of diatomic molecules is concerned, as these have thus far been the maintargets dealt with by the recent computational literature.

Let us begin by definig two right-handed coordinate systems (x, y, z) and (x', y', z') such that the former refers to the space-fixed (SF) frame while the latter refers to the molecular or body-fixed (BF) frame. A rotation of the frame (x, y, z) by the Euler angles (α, β, γ), as defined in Rose [6] will bring it to coincide with the frame (x', y', z), as indicated in Fig. 3.1 (page 121).

Let $\underline{j}$ denote the total rotational angular momentum of the molecule, whose magnitude and direction are fixed in both coordinate systems. Its components in the

SF and BF frames will be given, respectively, by (j_x, j_y, j_z) and (j'_x, j'_y, j'_z) while $\underline{j}$ is understood as measured in the SF-frame.

The free rotation of the molecule as a rigid body and treated within the B-O approximation is given by the classical Euler's equations of motion where the three principal moments of inertia $I_{x'}, I_{y'}, I_{z'}$, of the rigid body appear. They provide the conditions:

$$j^2_{x'} + j^2_{y'} + j^2_{z'} = j^2 = \text{const}$$

$$(\text{k.E.})_{tot} = \tfrac{1}{2}\left\{ \frac{j^2_{x'}}{I_{x'}} + \frac{j^2_{y'}}{I_{y'}} + \frac{j^2_{z'}}{I_{z'}} \right\} = \text{const} \qquad (3.27)$$

That is, the square of the total angular momentum and the kinetic energy of rotation both remain constant in a free rotation.

In momentum space the first of eq.s (3.27) represents a sphere, while the second a momental ellipsoid. From such equations one recovers the well-known definitions of asymmetric top molecules $(I_{x'} \neq I_{y'} \neq I_{z'})$ and symmetric top molecules $(I_{x'} = I_{y'} \neq I_{z'}$, with cyclic permutation of indeces) where the momental ellipsoid becomes a rotational ellipsoid; and finally of spherical top systems $(I_{x'} = I_{y'} = I_{z'})$ where the ellipsoid becomes a sphere.

For linear polyatomic molecules the third unequal moment may be extremely small in the symmetric top and the ellipsoid becomes a circular cylinder, while $j_{z'}$ = constant.

In the SF frame we have that:

$$[j_x, j_y] = i\, j_z \quad \text{with cyclic permutations} \qquad (3.28)$$

and

$$[j^2, j_x] = o \qquad (3.29)$$

while in the BF frame the commutation relations are [7]:

$$[j_{x'}, j_{y'}] = -\, i\, j_{z'} \qquad (3.30)$$

For symmetric top molecules, if z' is chosen to be the axis of symmetry of the molecule, then both j^2 and $j_{z'}$ are constants of motion, i.e. both can be simultaneously defined:

$$[j^2, j_{z'}] = o \tag{3.31}$$

Hence for any molecule it must be that:

$$j_x^2 + j_j^2 + j_z^2 = j_{x'}^2 + j_{y'}^2 + j_{z'}^2 = j^2 = \text{const.} \tag{3.32}$$

while for a symmetric top it also holds that:

$$[j^2, j_z] = [j^2, j_{z'}] = o \tag{3.33}$$

Let us now define an eigenfunction of the operaters j^2, j_z and $j_{z'}$ as $\chi^j_{mm'} = |j\ mm'\rangle$, then it can be shown that [8] :

$$\begin{aligned} j^2\ |j\ mm'\rangle &= j(j+1)\ |j\ mm'\rangle \\ j_z\ |j\ mm'\rangle &= m\ |j\ mm'\rangle \\ j_{z'}|j\ mm'\rangle &= m'|\ j\ mm'\rangle \end{aligned} \tag{3.34}$$

The rotational Hamiltonian $\mathcal{H}_{rot}$ associated with the rotational energy of (3.27) is given by:

$$\mathcal{H}_{rot} = \tfrac{1}{2}\left\{\frac{\mathcal{L}x'}{I_{x'}} + \frac{\mathcal{L}y'}{I_{y'}} + \frac{\mathcal{L}z'}{I_{z'}}\right\} \tag{3.35}$$

where the $\mathcal{L}$'s are now the components of the rotational angular momentum operator along the BF axis.

This hamiltomian operator yields the following matrix elements between the $\chi^j_{mm'}$ eigenfunctions:

$$\langle \chi^j_{mm'}|\ \mathcal{H}_{rot}|\ \chi^j_{mm'}\rangle = \frac{1}{4(I_{x'} + I_{y'})}\left\{j(j+i) - m'^2\right\} + \frac{m'^2}{2I_{z'}} \tag{3.36a}$$

and

$$\langle \chi^j_{mm'}|\mathcal{H}_{rot}|\chi^i_{mm'\pm 2}\rangle = \frac{1}{8(I_{x'} + I_{y'})}\cdot [j(j+1) - m'(m'\pm 1)]^{\frac{1}{2}}\cdot [j(j+1) - (m'\pm 1)(m'\pm 2)]^{\frac{1}{2}} \tag{3.36b}$$

for spherical-top molecules one then gets:

$$\langle \chi^j_{mm'}|\ \mathcal{H}_{rot}|\ \chi^j_{mm'}\rangle = j(j+1)/_{2I} \tag{3.37}$$

since in (3.35) $I_{x'} = I_{y'} = I_{z'} = I$.

While for symmetric-top, with z' defining the symmetry axis, one has that: $I_{x'} = I_{y'} = I_{(xy)'}$ and hence the energy eigenvalue for a given state is:

$$< \chi^{j}_{mm'} | \mathcal{H}_{rot} | \chi^{j}_{mm'} > = E_{jm'} = \tag{3.38}$$
$$= \frac{1}{2} \left\{ \frac{j(j+1)}{I_{(xy)'}} + \frac{m'^2}{I_{z'} + I_{(xy)'}} \right\}$$

For the physical situation where $I_{z'} < I_{(xy)'}$ one talks about a prolate symmetric-top, while for the cases where: $I_{z'} > I_{(xy)'}$ one has an oblate symmetric-top.

The orthonormal wavefunctions for the spherical top with the eigenvalues E_j's given by eq. (3.37) are the usual spherical harmonics . The Wigner D-functions [6] form a complete set of eigenfunctions for the symmetric-top. In fact, the rotation by an angle γ in Fig. 3.1 is carried out about the BF axis Oz' and the corresponding momentum operators are given by:

$$j_z = -i \frac{\partial}{\partial \alpha} \tag{3.39}$$

$$j_{z'} = -i \frac{\partial}{\partial \gamma} \tag{3.39}$$

now, the definition of the D-functions [7] gives:

$$\mathcal{D}^{j}_{mm'}(\alpha\beta\gamma) = e^{-im\alpha} d^{j}_{mm'}(\beta) e^{-im\gamma} \tag{3.40}$$

and eq.s (3.34) give the following results:

$$j_z \mathcal{D}^{j}_{-m,-m'} = m \mathcal{D}^{j}_{-m,-m'}$$
$$j_{z'} \mathcal{D}^{j}_{-m,-m'} = m' \mathcal{D}^{j}_{-m,-m'} \tag{3.41}$$

The symmetric-top eigenfunctions exhibit two different components of the angular momentum j according to whether one is considering the BF frame (m') or the SF frame (m). For m'= 0, one recovers the spherical-top forms of eigenfunctions, i.e. the spherical harmonics. Hence:

$$\mathcal{D}^{j}_{mo}(\alpha\beta\gamma) = \sqrt{\frac{4\pi}{2j+1}}\; Y^{*}_{jm}(\beta\alpha)$$

$$= (-)^{m}\sqrt{\frac{4\pi}{2j+1}}\; Y_{j-m}(\beta\alpha) \qquad (3.42)$$

therefore:

$$(-)^{m}\mathcal{D}^{j}_{-mo}(\alpha\beta\gamma) = \sqrt{\frac{4\pi}{2j+1}}\; Y_{jm}(\beta\alpha) \qquad (3.43)$$

which thus finally gives:

$$(-)^{m}\mathcal{D}^{j}_{-m,-m'}(\alpha\beta\gamma) = \mathcal{D}^{j*}_{mm'}(\alpha\beta\gamma) \qquad (3.44)$$

This is the definition of the normalized wavefunctions for the symmetric top:

$$S^{j}_{mm'} = \sqrt{\frac{2j+1}{8\pi^2}}\;\mathcal{D}^{j*}_{mm'}(\alpha\beta\gamma) \qquad (3.45)$$

The asymmetric-top wavefunctions are represented by a linear combination of $S^{j}_{mm'}$ functions:

$$\mathcal{A}^{j}_{m\tau}(\alpha,\beta,\gamma) = \sum_{m'} a_{jm'\tau}\, S^{j}_{mm'}(\alpha,\beta,\gamma) \qquad (3.46)$$

where τ is now a 'pseudo' quantum number that distinguishes the different rotational levels with the same j, while the expansion coefficients a's are determined by diagonalizing the asymmetric top Hamiltoniam used before. Moreover, since:

$$\langle j_1 m_1 m_1' \mid j_2 m_2 m_2' \rangle = \delta_{j_1 j_2}\cdot\delta_{m_1' m_2'}\cdot\delta_{m_1' m_2'} \qquad (3.47)$$

expansion coefficients can be chosen in such a way as to make the $\mathcal{A}$'s orthonormal, which gives:

$$\langle j'm'\tau' \mid j\; m\;\tau\rangle = \delta_{jj'}\cdot\delta_{mm'}\cdot\delta_{\tau\tau'} \qquad (3.48)$$

This therefore requires that:

$$\sum_{m'} a_{jm'\tau'}\cdot a_{jm\tau} = \delta_{\tau\tau'} \qquad (3.49)$$

In conclusion the generalized, time-independent Schödinger equation for the rotational motion of an asymmetric top molecule can be written as:

$$\mathcal{H}_{rot}\, \mathcal{A}^{j}_{m_j\tau} = E_{j\tau}\, \mathcal{A}^{j}_{m_j\tau} \qquad (3.50)$$

with the definition of $\mathcal{H}_{rot}$ from eq. (3.35) and with the eigenfunction expansion for the $\mathcal{A}$'s given by:

$$\mathcal{A}^{j}_{m_j\tau}(\alpha\beta\gamma) = \sum_{\lambda} a_{j\lambda\tau}\, S^{j}_{m_j\lambda}(\alpha\ \beta\ \gamma) \qquad (3.51)$$

where, according to (3.41), we define the symmetric-top eigenfunctions via the indeces given by the following equations:

$$\begin{aligned} \mathcal{L}_{z'}\, S^{j}_{m_j\lambda} &= \lambda\, S^{j}_{m_j\lambda} \\ \mathcal{L}_{z}\, S^{j}_{m_j\lambda} &= m_j S^{j}_{m_j\lambda} \end{aligned} \qquad (3.41b)$$

The definition of the final channel in an inelastic scattering process will therefore require, for pure rotational excitations, the establishing of a channel index α, corresponding to a given rotational state.

For spherical-top molecules in their electronic totally symmetric ground state $\alpha = j$. For symmetric top molecules $\alpha = j\Omega$, where Ω is now the projection of j along the molecular principal symmetric axis in the BF frame. Finally, $\alpha = j\tau$ for asymmetric-top molecules, with the definition of τ as already given before. The inclusion of vibrations will obviously increase the number of indeces needed for the channel definition.

3.4 Rotational excitation in atom-molecule collisions: the space-fixed reference frame

This is the least complex of the problems that deal with the collisional excitation of molecular internal degrees of freedom, since the form of the total Hamiltonian of eq. (2.16) now becomes:

$$\mathcal{H} = - \frac{1}{2\mathcal{M}} \nabla^2_{\underline{R}} + \mathcal{H}_{rot} (\Omega) + V (\underline{R} , \Omega) \qquad (3.42)$$

where the only molecular internal coordinate is now Ω that denotes collectively the three Euler angles of the molecule in a SF frame.

In order to solve the corresponding time-indipendent Schrödinger equation (3.1), the wavefunction has to be expanded in a basis set. The most common choice for low energy collisions has been provided by basis sets of the 'static' type in that they are based on a description of the non-interacting partners and do not change during the course of the collisions. They are the ones defines as 'diabatic' in eq. (3.2). Obviously it is possible to define an adiabatic set that changes contінously during the course of the collision and diagonalizes the Hamiltonian of eq. (3.5). In both cases the exact treatments should yield the same results, but we will see later how different truncations provide approximate schemes where the simpler physics involved becomes more or less transparent according to the choice of representation.

For the present case it is convenient to define two different coordinate systems and wes hall confine ourselves to molecules with a vanishing component of their electronic angular momentum along the main axis and to structureless atoms in a ('S) configuration. The theory, as we will mention later, can be extended to more complex cases. However this is still seldom worthwhile since we have altready seen how one has very little knowledge of the intermolecular potentials for polyatomic partners. Further, it would be markedly more difficult to compute the cross sections reliably.

As we have discussed in the first ection of the present hapter, the existence of a non-spherical potential poses the problem of angular momentum coupling. In fact, the total angular momentum and its component about a chosen axis of quantization, usually the incident direction, are of course strictly conserved, but angular momentum (torque) can be exchanged between internal and translational motions. The two choices for the coordinate system of reference originate from the two different descriptions proposed for this problem.

The first employs a fixed quantization axis along the direction of incident motion and hence chooses an SF coordinate system. It is conceptually simpler and had been extensively discussed in the recent literature. The second, due originally to Jacob and Wick [8], allows the quantization axis to rotate with the interparticle vector $\underline{R}$, leads to more tractable results for further approximations and employs a BF frame of reference. Both reference frames are reported in Fig. 3.2 for the case of atom-diatomic encounters: the unprimed coordinates are body-fixed (BF) while the prime coordinates are space-fixed (SF). The origin is chosen in the centre-of-mass of the whole system.

In performing the expansion of eq. (3.2) the functions $\varphi_j(\underline{X})$ describe the molecular rotational wavefunctions and the corresponding channel index j has to include the whole range of conserved quantum numbers in order to allow for a reduction of the number of equations (3.3) for each choice of constants of motion. One type of mathematical formulation of the representation problem, developed by Curtiss and collaborators over an extensive series of paper [9, 10, 11, 12, 13, 14], defines the orientation of the colliding system by the rotation that brings the whole system to a standard configuration, and expresses the rotational wavefunction in terms of elements of the irreducible representation matrix of the rotation group.

The second formulation, which we will be discussing in this section, has been studied by Arthurs and Dalgarno [15], Takayanagi [16], Davison [17] and Bernstein and collaborators [18]. Here a complete set of functions is chosen, each member of which represents a state with a definite value of the total angular momentum J and of its z component. The analysis therefore becomes similar to that described for nuclear reaction by Blatt and Biedenharn [19].

One defines at the start the operator $\underline{J}$:

$$\underline{\underline{J}} = \underline{\underline{L}} + \underline{\underline{j}} \tag{3.43}$$

where $\underline{\underline{L}}$ is the angular momentum operator for the kinetic enrgy term of $\mathcal{H}$ in polar coordinates and $\underline{\underline{j}}$ the rotational angular momentum operator for the molecule as a spherical top.

One can then verify that the total Hamiltonian of (3.42) satisfies the following relations:

$$[\mathcal{H}, \underline{\underline{L}}] \neq 0 \quad ; \quad [\mathcal{H}, \underline{\underline{j}}] \neq 0 \tag{3.44a}$$

while for an isolated system:

$$[\mathcal{H}, \underline{\underline{J}}] = 0 \tag{3.44b}$$

Hence a set of compatible observables is given by: $\mathcal{H}$, $\underline{\underline{J}}$, $\underline{\underline{J}}_z$, the eigenvectors of which appear in the following eigenvalue equation:

$$\mathcal{H} \mid E\,J\,M > = E \mid E\,J\,M > \qquad (3.45)$$

In the Schrödinger representation, the SF coordinate basis is given, from Fig. 3.2, by $|\underline{R}'\ \hat{r}'>$. Hence:

$$< \hat{r}'\ \underline{R}' |\mathcal{H}| E\,J\,M > = E < \hat{r}'\ \underline{R}' | E\,J\,M > \qquad (3.46)$$

Since:

$$\int d\underline{R}'\ d\hat{r}' \mid \underline{R}'\ \hat{r}' > < \hat{r}'\ \underline{R}' \mid = 1 \qquad (3.47)$$

One then obtains:

$$\int < \hat{r}'\ \underline{R}' |\mathcal{H}| \underline{R}''\ \hat{r}'' > d\,\underline{R}''\ d\,\hat{r}'' < \hat{r}''\ \underline{R}'' \mid E\,J\,M > = E < \hat{r}'\ \underline{R}' \mid E\,J\,M > \qquad (3.48)$$

If one defines:

$$< \underline{R}'\ \hat{r}' \mid E\,J\,M > = \varphi^{E}_{JM}\ (\underline{R}',\ \hat{r}') \qquad (3.49)$$

and since the potential in (3.42) is a local potential:

$$< \hat{r}'\ \underline{R}' \mid \underline{\underline{V}} \mid \underline{R}''\ \hat{r}'' > = \delta(\underline{R}' - \underline{R}'')\quad \delta(\hat{r}' - \hat{r}'')\ V\ (\underline{R}',\ \Omega) \qquad (3.50)$$

One can write the eigenvalue equation for each eigenvector that is going to be used for the coupled expansion (3.2) in the SF frame of reference:

$$< \hat{r}''\ \underline{R}' |\mathcal{H}| \underline{R}'\ \hat{r}' > \varphi^{E}_{JM}\ (\underline{R}',\ \hat{r}') = E\ \varphi^{E}_{JM}(\underline{R}'',\hat{r}') \qquad (3.51)$$

Since the potential is nonspherical the eigenvalues of eq.s (3.44a), ℓ and j, are not good quantum numbers. The possibility of separating radial and angular variables is therefore barred in the present case. One can however choose a set of functions where j and ℓ are good quantum numbers in the asymptotic situation, before the particles interact. Let us call such a set $|j\ell JM>$, hence its projection onto the SF basis of the Schrödinger equation (fig. 3.2) is given by:

$$< \hat{r}'\ \hat{R}' \mid j\ell\ JM > = \mathcal{Y}^{JM}_{j\ell}\ (\hat{r}'\ \hat{R}') \qquad (3.52)$$

where we have dropped the index E, that describes now a constant for the isolated

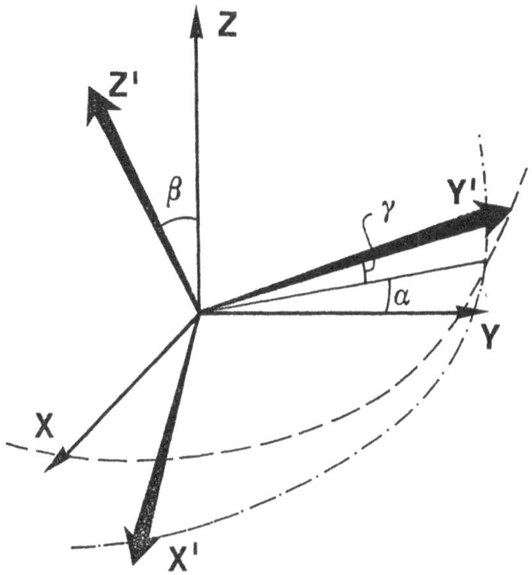

Fig. 3.1

Space-fixed (SF) coordinate system (x,y,z) and Body-fixed (BF) coordinate system (x',y',z') for molecular rotations. The Euler angles (α,β,γ) follow the notation used by M.E.Rose [6].

system. Once $|E\ J\ M\rangle$ is chosen, however, there is a degeneracy over new channel labels (j, ℓ) since: $|j-\ell| \leqslant J \leqslant j+\ell$. The general form or the expansion of φ^{E}_{JM} is then given by:

$$\varphi^{E}_{JM}(\underline{R}', \hat{r}') = \sum_{j,\ell} A^{JM}_{j\ell}\, \psi^{EJM}_{j\ell}(\underline{R}', \hat{r}') \tag{3.53}$$

where j and ℓ are simply labels of asymptotic eigenvalues.

Now by using the basis set of (3.52) one gets:

$$\psi^{EJM}_{j\ell}(\underline{R}', \hat{r}') = \sum_{j'\ell'} u^{Jj\ell}_{j'\ell'}(R)\cdot\frac{1}{k_j\, R}\cdot\mathcal{Y}^{JM}_{j'\ell'}(\hat{r}', \hat{R}') \tag{3.54}$$

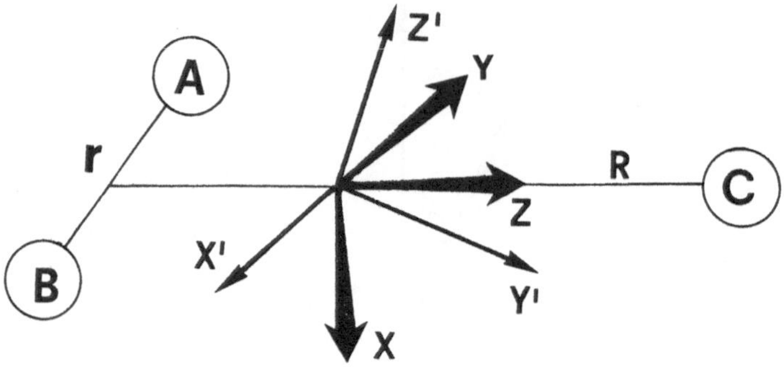

Fig. 3.2

Atom-plus-diatomic-molecule frames of reference used in text to describe inelastic collisions. The unprimed coordinates are defined as Body-fixed (BF) and their z-axis follows the incoming projectile C. The primed coordinates define a Space-fixed (SF) system of axis and their z'-axis usually contains the molecular axis of quantization.

It next becomes possible to try to obtain a purely radial equation for the u's, as discussed for the simpler case of potential scattering. The choice of the $\mathcal{J}^{JM}_{j\ell}$ basis of (3.52) introduces in (3.54) a further summation over j' and ℓ', thus preparing a representation for the effect of the anisotropic potential, which allows transitions from the initial asymptotic state $|j\ell\rangle$ to the final asymptotic state $|j'\ell'\rangle$. The functions of (3.54) are a complete set in the Hilbert space, since one can define a corresponding set of operators in that space which all commute with one another: $\mathcal{H}$, $\underline{\underline{J}}$, $\underline{\underline{j}}_o$, $\underline{\underline{\ell}}_o$. The last two operators only act upon the incoming part of the wavefunction and yield the initial states of the partners in the region of no interaction. They obviously commute with each other and with $\mathcal{H}$ and $\underline{\underline{J}}$, since the system is prepared in that particular state $|j_o\ell_o\rangle$ and the potential V does not act on it. Hence, only one eigenvector is associated with the set of quantum numbers: (E J M j ℓ).

For the more general case of an asymmetric top molecule, we have already seen in section 3.2 that: $j \rightarrow \alpha$ and thus the coefficients of (3.52) can more specifically be written as:

$$\mathcal{J}^{JM}_{\alpha\ell}(\hat{\underline{R}}', \hat{r}') = \sum_{\substack{m_j \\ m_\ell}} C(j\ell J;\, m_j m_\ell\, M)\, \mathcal{A}^{j}_{m_j\tau}(\hat{r}')\, Y^{m_\ell}_{\ell}(\hat{\underline{R}}') \tag{3.55}$$

where the C's are Clebsh-Gordan coefficients [6] .

The unitarity of the C's can be used to write:

$$\mathcal{A}^{j}_{m_j\tau}(\hat{r}')\, Y^{m_\ell}_{\ell}(\hat{\underline{R}}') = \sum_{JM} C(j\ \ell\ J;\, m_j m_\ell\ M)\, \mathcal{J}^{JM}_{\alpha\ell}(\hat{r}', \hat{\underline{R}}') \tag{3.56}$$

and similarly, making use of the orthonormality relation between asymmetric top wavefunctions, one can further write:

$$< \mathcal{J}^{J'M'}_{\alpha'\ell'} \mid \mathcal{J}^{JM}_{\alpha\ell} > = \delta_{JJ'}\delta_{MM'}\delta_{\alpha\alpha'}\delta_{\ell\ell'} \tag{3.57}$$

therefore, by using atomic units throughout, the radial Schrödinger equation via the Hamiltonian (3.42) becomes:

$$\left\{ \frac{d^2}{dR'^2} + k^2_{\alpha\alpha'} - \frac{\ell'(\ell'+1)}{R'^2} \right\} u^{J\alpha\ell}_{\alpha'\ell'}(R') = \sum_{\alpha''\ell''} < \alpha'\ell'|U|\alpha''\ell''>^J u^{J\alpha\ell}_{\alpha''\ell''}(R') \tag{3.58}$$

where:

$$k^2_{\alpha\alpha'} = 2\mathcal{M}\left\{ E_\alpha - E_{j'\tau'} \right\} \tag{3.59a}$$

and:

$$E = K.E. + E_{j\tau} = E_\alpha \tag{3.59b}$$

Moreover the coupling matrix elements are given by the following angular integrals:

$$<\alpha'\ell'|U|\alpha''\ell''>^J = 2\mathcal{M}\int d\hat{\underline{R}}'\, d\hat{r}'\, \mathcal{J}^{JM}_{\alpha'\ell'}\, V(\underline{R}', \hat{r}')\, \mathcal{J}^{JM}_{\alpha''\ell''} \tag{3.60}$$

The physics of the problem requires the radial functions to satisfy the following boundary conditions:

$$u^{J\alpha\ell}_{\alpha'\ell'}(R') \underset{R'\to 0}{\sim} 0$$

$$u^{J\alpha\ell}_{\alpha'\ell'}(R') \underset{R'\to\infty}{\sim} k^{-\frac{1}{2}}_{\alpha\alpha'} \Big\{ \delta_{\alpha\alpha'} \cdot \delta_{\ell\ell'} \exp[-i(k_{\alpha\alpha}R'-\tfrac{1}{2}\ell\pi)] - - S^{J}_{\alpha'\ell',\alpha\ell} \exp[i(k_{\alpha\alpha'}R'-\tfrac{1}{2}\ell'\pi)] \Big\} \tag{3.61}$$

Hence the use of expansion (3.54) into (3.53) allows one to write the total wavefunction of eq. (32) in the following form:

$$\Phi(\underline{R}', \hat{r}') = \sum_{J,M} \sum_{\alpha,\ell} A^{JM}_{\alpha\ell} \psi^{JM}_{\alpha\ell}(\underline{R}', \hat{r}') \tag{3.62}$$

which will have the asymptotic form:

$$\Phi \underset{R'\to\infty}{\sim} R'^{-1} \sum_{J,M} \sum_{\alpha,\alpha'} \sum_{\ell,\ell'} k^{-\frac{1}{2}}_{\alpha\alpha'} A^{JM}_{\alpha\ell} \Big\{ \delta_{\alpha\alpha'} \delta_{\ell\ell'} \exp[-i(k_{\alpha\alpha}R'-\tfrac{1}{2}\ell\pi)] - - S^{J}_{\alpha'\ell',\alpha\ell} \exp[i(k_{\alpha\alpha}R'-\tfrac{1}{2}\ell'\pi)] \Big\} \mathcal{Y}^{JM}_{\alpha\ell} \tag{3.63}$$

the incident plane wave in the α channel is: $\exp[i\,\underline{k}_{\alpha\alpha'}\underline{R}']\mathcal{A}^{j}_{m_j\tau}(\hat{r}')$ with propagation vector, $\underline{k}_{\alpha\alpha}$. Now, the propagation vector can be chosen so as to define the z-axis in the SF frame, hence:

$$\exp[i\,k_{\alpha\alpha}z']\mathcal{A}^{j}_{m_j\tau}(\hat{r}') = \sqrt{4\pi} \sum_{\ell} i^{\ell} \sqrt{2\ell+1}\, j_{\ell}(k_{\alpha\alpha}R')\, Y^{0}_{\ell}(\hat{R}')\mathcal{A}^{j}_{m_j\tau}(\hat{r}') \tag{3.64}$$

which, using (3.56)), can be rewritten as follows in the asymptotic region:

$$\exp[i\,k_{\alpha\alpha}z']\mathcal{A}^{j}_{m_j\tau} \underset{R'\to\infty}{\sim} -\frac{\sqrt{\pi}}{k_{\alpha\alpha}R'} \sum_{J,M,\ell} i^{\ell+1} \sqrt{2\ell+1}\, C(j\ell J; m_j o M) \cdot \cdot \mathcal{Y}^{JM}_{\alpha\ell}(\Omega', \hat{r}') \Big\{ \exp[i(k_{\alpha\alpha}R-\tfrac{1}{2}\ell\pi) - \exp[-i(k_{\alpha\alpha}R-\tfrac{1}{2}\ell\pi)] \Big\} \tag{3.65}$$

Now, since the total wavefunction (3.62) should have the following asymptotic form after scattering:

$$\Phi \underset{R' \to \infty}{\sim} \Phi_{\text{incoming}} + \Phi_{\text{scattered}} \tag{3.66}$$

to determine the part of the wavefunction that is due to the scattering the incoming part should be substracted from Φ becouse, prior to scattering, the incoming part of Φ and the incoming part of the incident wavefunction should be the same. Hence, by equating the coefficients ot $e^{-ik_{\alpha\alpha}R'}$ from eq.'s (3.63) and (3.65), we get:

$$\sum_{\substack{J,M\\ \alpha,\alpha'\\ \ell,\ell'}} k_{\alpha\alpha'}^{-\frac{1}{2}} A_{\alpha\ell}^{JM} \cdot \delta_{\alpha\alpha'} \cdot \delta_{\ell\ell'} \cdot \mathcal{Y}_{\alpha'\ell'}^{JM} \exp(\tfrac{1}{2} i \ \ell \ \pi) = \frac{\sqrt{\pi}}{k_{\alpha\alpha}} \sum_{J,M,\ell} i^{\ell+1} \sqrt{2\ell+1} \ C(j\ell J;\ m_j o M) \mathcal{Y}_{\alpha\ell}^{JM} \exp(\tfrac{1}{2} i \ \ell\pi) \tag{3.67}$$

This expression can be further simplified via eq. (3.57), and we finally get:

$$\begin{aligned} A_{\alpha\ell}^{JM} &= i^{\ell+1} \ C(j\ell J;\ m_j o M) \ \frac{\sqrt{(2\ell+1)\pi}}{k_{\alpha\alpha}} \quad ; \quad \text{in the } (\alpha\ell) \text{ channel.} \\ &= o \quad \text{in the other channels.} \end{aligned} \tag{3.68}$$

We thus have, from eq.s (3.63), (3.65) and (3.66), that:

$$\Phi_{\text{scattered}} \underset{R' \to \infty}{\sim} \frac{\sqrt{\pi}}{R' k_{\alpha\alpha}} \sum_{JM\ell} i^{\ell+1} \sqrt{2\ell+1} \ C(j\ell J;\ m_j o M) \exp[\, i(k_{\alpha\alpha} R' - \tfrac{1}{2}\ell\pi)] \cdot \mathcal{Y}_{\alpha\ell}^{JM} - R'^{-1} \sum_{\substack{J,M\\ \alpha,\alpha'\\ \ell,\ell'}} k_{\alpha\alpha'}^{-\frac{1}{2}} A_{\alpha\ell}^{JM} S_{\alpha'\ell',\alpha\ell}^{J} \exp[\, i(k_{\alpha\alpha'} R' - \tfrac{1}{2}\ell'\pi)] \cdot \mathcal{Y}_{\alpha'\ell'}^{JM} \tag{3.69}$$

which can be finally simplified, with the help of eq.(3.68), to yield:

$$\Phi_{\text{scattered}} \underset{R \to \infty}{\sim} \sum_{\alpha'} \frac{i}{k_{\alpha\alpha}} \left(\frac{k_{\alpha\alpha}}{k_{\alpha\alpha'}} \right)^{\frac{1}{2}} \cdot R'^{-1} \exp(i \, k_{\alpha\alpha'} R') \sum_{m_{j'}} q(\alpha' m_{j'};\ \alpha m_j | \hat{R}') \cdot \mathcal{A}_{m_{j'}\tau}^{j'} \tag{3.70}$$

where the reaction amplitude q is given by:

$$\mathfrak{q}(\alpha' m_{j'};\ \alpha m_j |\ \hat{R}') = \sum_{\substack{J,M \\ \ell,\ell' \\ m_{\ell'}}} \sqrt{\pi}\ i^{\ell-\ell'} \sqrt{2\ell+1}\ C(j\ell J;\ m_j o M) \cdot C(j'\ell' J;\ m_{j'} m_{\ell'} M) \cdot \\ \cdot\ T_{\alpha'\ell',\alpha\ell} \cdot Y_{\ell}^{m_{\ell'}}(\hat{R}') \tag{3.71}$$

where the reactance matrix $\underset{\approx}{T}$ is defined as follows as already discussed in Chapter 1:

$$\underset{\approx}{T} = 1 - \underset{\approx}{S} \tag{3.72}$$

The final expression for the reaction amplitude $\mathfrak{q}$ is the same as the one given by Arthurs and Dalgarno [15] for rigid rotors, the only difference here be the more general extension to asymmetric tops with a different set of quantum numbers. The differential cross-section for excitation from the $|\alpha\ m_j>$ state of the asymmetric molecular rotor to the $|\alpha' m_{j'}>$ state is given by:

$$\frac{d\sigma}{d\hat{R}'}(\alpha m_j \rightarrow \alpha' m_{j'}) = k_{\alpha\alpha}^{-2} \cdot |\mathfrak{q}(\alpha' m_{j'}, \alpha m_j | \hat{R}')|^2 \tag{3.73}$$

Averaging over the m_j components and summing up the final $m_{j'}$ components one can write down the final expression for the $(\alpha \rightarrow \alpha')$ transition:

$$\frac{d\sigma}{d\hat{R}'}(\alpha \rightarrow \alpha') = \frac{(-)^{j'-j'}}{4(2j+1)k_{\alpha\alpha}^2} \sum_L A_L \cdot P_L(\cos\theta) \tag{3.74}$$

where:

$$A_L = \sum_{\substack{J_1,J_2 \\ \ell_1,\ell_2 \\ \ell_1',\ell_2'}} Z(\ell_1 J_1 \ell_2 J_2;\ jL) \cdot Z(\ell_1' J_1 \ell_2' J_2;\ j'L) \cdot T^{J_1 *}_{\alpha'\ell_1',\alpha\ell_1} \cdot T^{J_2}_{\alpha'\ell_2',\alpha\ell_2} \tag{3.75}$$

and

$$Z(a\ b\ c\ d;\ e\ f) = (-)^{\frac{1}{2}(f-a+c)} \cdot [(2a+1)\ (2b+1)\ (2c+1)\ (2d+1)]^{\frac{1}{2}} \cdot \\ \cdot\ C(a\ c\ f;\ ooo) \cdot W(a\ b\ c\ d;\ e\ f) \tag{3.76}$$

In eq.(3.76) the W's are Racah coefficients [19].

The corresponding inelastic integral cross section in then given by:

$$\sigma(\alpha \to \alpha') = \int \frac{d\sigma}{d\hat{R}'} (\alpha \to \alpha')\, d\hat{R}' =$$

$$= \frac{(-)^{j-j}}{4(2j+1)\, k^2_{\alpha\alpha}} A_0 \qquad (3.77)$$

$$= \frac{\pi}{(2j+1)\, k^2_{\alpha\alpha}} \sum_{J\ell\ell'} (2J+1)\, |T^J_{\alpha'\ell',\alpha\ell}|^2$$

and the momentum transfer cross section:

$$\sigma_m(\alpha) = \sum_{\alpha'} \sigma_m (\alpha\to\alpha') = \int \sum_{\alpha'} \frac{d\sigma}{d\hat{R}'} (\alpha\to\alpha')\, (1 - \cos\theta)\, d\hat{R}'$$

$$= \frac{\pi\, (-)^{j-j'}}{(2j+1)\, k^2_{\alpha\alpha}} \sum_{\alpha'} (A_0 - \frac{1}{3} A_1) \qquad (3.78)$$

A consequence of the symmetry of the S-matrix is the reciprocity relation:

$$\sigma(\alpha \to \alpha') = k^{-2}_{\alpha\alpha}\, k^2_{\alpha'\alpha'}\, (2j+1)^{-1}\, (2j'+1) \cdot \sigma(\alpha'\to\alpha) \qquad (3.79)$$

Whenever the interaction potential $V(\underline{R}', \hat{r}')$ can be written in the following multipolar expansion:

$$V(\underline{R}', \hat{r}') = \sum_\lambda V_\lambda (R')\, P_\lambda (\hat{R}' \cdot \hat{r}') \qquad (3.80)$$

The corresponding coupling matrix elements for the rigid rotor, spherical-top molecule can be rewritten from eq. (3.60) as follows:

$$< j\ell|U^J|j'\ell' > = 2\mathcal{M} \sum_\lambda f_\lambda (j\ell, j'\ell'; J)\, V_\lambda (R') \qquad (3.81)$$

where the f coefficients [20] are given by:

$$f_\lambda (j\ell, j'\ell'; J) = \int \mathcal{Y}^{JM*}_{j\ell} (\hat{R}', \hat{r}')\, P_\lambda (\hat{R}' \cdot \hat{r}')\, \mathcal{Y}^{JM}_{j'\ell'} (\hat{R}', \hat{r}')\, d\hat{R}'\, d\hat{r}'$$

$$= (-)^{j+j'-J} [(2j+1)\, (2\ell+1)\, (2j'+1)\, (2\ell'+1)]^{\frac{1}{2}} \cdot \qquad (3.82)$$

$$\cdot \begin{pmatrix} j' & \ell & j \\ 0 & 0 & 0 \end{pmatrix} \cdot \begin{pmatrix} \ell' & \lambda & \ell \\ 0 & 0 & 0 \end{pmatrix} \cdot \begin{Bmatrix} J & \ell & j \\ \lambda & j' & \ell' \end{Bmatrix}$$

where the (:::) are 3-j coefficients and the {:::} are 6-j coefficients [6].

On the other hand, for a non-linear target, either symmetric or asymmetric-top molecule, the multipolar expansion of eq. (3.80) requires the inclusion of the relative orientation of the projectile with respect to a reference frame that corresponds to the principal moments of inertia used to define the rotor wavefunctions in the previous section. Thus:

$$V(\underline{R}', \hat{r}') = V(R, \theta_R, \phi_R) = V(\underline{R}) \tag{3.81}$$

where $\underline{R}$ is now the position vector of the projectile with respect to a set of BF axes taken along the principal moments of inertia of the rotor. The expansion (3.80) therefore becomes:

$$V(R, \theta_R, \phi_R) = \sum_{\lambda,\mu} V_{\lambda\mu}(R)\, Y_\lambda^\mu(\theta_R, \phi_R) \tag{3.82}$$

Hence, to transform back to the SF frame [21], one has to write:

$$\begin{aligned} V(\underline{R}', \hat{r}') &= \mathcal{D}^{-1}(\theta_{r'}, \varphi_{r'})\, V(\underline{R}) \\ &= \sum_{\lambda\mu} V_{\lambda\mu}(R) \sum_{\rho} \{\mathcal{D}^{-1}(\theta_{r'}, \varphi_{r'})\}^\lambda_{\rho\mu}\, Y_\lambda^\rho(\theta_R, \phi_R) \\ &= \sum_{\lambda,\mu,\rho} V_{\lambda\mu}(R)\, \mathcal{D}^{\lambda *}_{\rho\mu}\, Y_\lambda^\rho(\theta_R, \phi_R) \end{aligned} \tag{3.83}$$

The spherical harmonics phase convertion (3.42) plus the use of a real and local interaction potential as defined in (3.50) then require that:

$$V_{\lambda,-\mu}(R') = (-)^\mu V^*_{\lambda,\mu}(R') \tag{3.84}$$

In might be convenient to choose the principal moment of inertia axes that also make the coefficients of (3.82) to be real. This requires the local, body-fixed plane (XZ) to be a plane of reflection symmetry , hence:

$$V(R, \theta_R, \phi_R) = V(R, \theta_R, -\phi_R) \tag{3.85}$$

This is always possible for symmetric tops with a n-fold symmetric axis along Z. Thus, one can rewrite eq.(3.82) in the following way:

$$V(R, \theta_R, \phi_R) = \sum_{\lambda \geq \mu} V_{\lambda\mu} (R) \cdot [Y_\lambda^\mu (\hat{R}) + (-)^\mu Y_\lambda^{-\mu} (\hat{R})] \cdot$$
$$\cdot (1+\delta_{o\mu})^{-1} = \sum_{\lambda,\mu}{}' (2\pi)^{-\frac{1}{2}} V_{\lambda,\mu} (R) (2 - \delta_{\mu o}) \cdot P_{\lambda\mu} (\cos\theta_R) \cos (\mu\phi_R) \quad (3.86)$$

where the $P_{\lambda\mu}$ are unnormalized associated Legendre polynamials.

Moreover, from eq.(3.36a), one sees that symmetric-top eigenfunctions are energy degenerate over the m'quantum number, since E_α depends on $(m')^2$. Hence, in order to obtain proper eigenfunctions of the inversion operator one has to define the following linear combinations:

$$|jm'm>_{s,a} = [2(1 + \delta_{om'})]^{-\frac{1}{2}} \cdot [|jm'm> \pm |j-m'm>] =$$
$$= |jm'm\eta> \quad (3.87)$$

where the l.h.s. subscriptions s and a are respectively associated to the upper and lower signs within the r.h.s. bracket and the additional η index refers to both. This will in turn affect the form of the coupling matrix elements (3.60) and (3.81) which are now given by the following expressions:

$$< \alpha\ell|V| \alpha'\ell' >^J = \sum_{\lambda\mu} V_{\lambda\mu} (R) <\mathcal{J}_{\alpha\ell}^{JM}| Y_\lambda^\mu |\mathcal{J}_{\alpha'\ell'}^{JM}> =$$
$$= \sum_{\lambda\mu} V_{\lambda\mu} (R) (-)^{j+j'-k-J} + [(2j+1)(2j'+1)(2\ell+1)(2\ell'+1)\cdot \quad (3.88)$$
$$\cdot (2\lambda+1)]^{\frac{1}{2}} \cdot (4\pi)^{-\frac{1}{2}} \begin{pmatrix} \ell' & \lambda & \ell \\ 0 & 0 & 0 \end{pmatrix} \begin{pmatrix} j & \lambda & j' \\ k & \mu & -k' \end{pmatrix} \begin{Bmatrix} J & \ell & j \\ \lambda & j' & \ell' \end{Bmatrix}$$

where the (m') quantum number has bees called k and the terms in the brackets are the same as those defined before. When properly symmetrized eigenfunctions are used, however, each of the above terms in the summation becomes:

$$V_{\lambda\mu} (R) < j\, k\eta\, \ell| [Y_\lambda^\mu + (-)^\mu Y_\lambda^{-\mu}] (1 + \delta_{o\mu})^{-1} | j'k'\eta'\ell' >_J =$$
$$= V_{\lambda\mu} (R) \cdot \underline{\text{Norm.}} \cdot \underline{\text{Fact.}} \quad (3.89)$$

where Norm. and Fact. are given by the following expressions:

$$\underline{\text{Norm.}} = \{2(1+\delta_{\mu o})\}^{-1} \cdot \left\{(1+\delta_{ko})\ (1+\delta_{k'o})\right\}^{-\frac{1}{2}} \tag{3.90a}$$

$$\begin{aligned}\underline{\text{Fact.}} = \Big\{ & < \mathcal{J}^{J,M}_{jk,\ell}\ Y^{\mu}_{\lambda} + (-)^{\mu}\ Y^{-\mu}_{\lambda}\ | \mathcal{J}^{J,M}_{j'k',\ell'} > + \\ & + \eta < \mathcal{J}^{J,M}_{j-k,\ell} |\ Y^{\mu}_{\lambda} + (-)^{\mu}\ Y^{-\mu}_{\lambda} | \mathcal{J}^{J,M}_{j'k',\ell'} > + \\ & + \eta' < \mathcal{J}^{J,M}_{jk,\ell} | \dots + \dots | \mathcal{J}^{J,M}_{j'-k',\ell'} > + \\ & + \eta \cdot \eta' < \mathcal{J}^{J,M}_{j-k,\ell} | \dots + \dots | \mathcal{J}^{J,M}_{j'-k',\ell'} > \Big\}\end{aligned} \tag{3.90b}$$

By using the triangular relations between μ, k and k' plus the coupling transformations of spherical harmonics, one obtains [22] :

$$\begin{aligned} & V_{\lambda\mu}(R) \left\{1+\eta\eta'(-)^{j+j'+\lambda+\mu}\right\} \cdot \frac{1}{2} \cdot \left\{(1+\delta_{ko})\ (1+\delta_{k'o})\right\}^{-\frac{1}{2}} \cdot \\ & \cdot \Big\{ \omega < \mathcal{J}^{J,M}_{jk,\ell} |\ Y^{\pm\mu}_{\lambda}\ | \mathcal{J}^{J,M}_{j'k',\ell'} > + \\ & + \eta < \mathcal{J}^{J,M}_{j-k,\ell} |\ Y^{\mu}_{\lambda}\ | \mathcal{J}^{J,M}_{j'k',\ell'} > \Big\} \end{aligned} \tag{3.91}$$

where: $\omega = 1$ if $(k'-k) \geq 0$ and $\omega = (-)^{\mu}$ if $(k'-k) < 0$. There will be a contribution from the first term in the curly bracket on the r.h.s. of (3.90b) only if $\pm\mu = k'-k$, while the second term will contribute if $\mu = k'+k$. If k and/or k' vanish, only one of these terms will contribute to the coupling.

The conservation of parity requires that the coupling is non-zero only if:

$$\eta \cdot (-)^{j+k+\ell} = \eta' + (-)^{j'+k'+\ell'} \tag{3.92}$$

The coupled equations can now be split up into two blocks of non-interacting parity fo each J value.

In conclusion the final form obtained for individual function of the basis set required in eq. (3.53) is therefore given by the general expression for a rigid, spherical rotor:

$$\psi^{EJM}_{j\ell}(\underline{R}', \hat{r}') = \sum_{j'\ell'} \frac{1}{k_j R'} \left\{ -\frac{1}{2i} \exp[-i(k_j R' - \ell'\pi/_2)] \cdot \delta_{jj'} \cdot \delta_{\ell\ell'} + \frac{1}{2i}\left(\frac{k_j}{k_j}\right)^{\frac{1}{2}} S^J_{j'\ell',j\ell} \exp[i(k_j R' - \ell'\pi/_2)] \right\} \mathcal{Y}^{JM}_{j'\ell'}(\hat{R}', \hat{r}') \tag{3.93}$$

3.5 - Rotational excitation in atom-molecule collisions: the helicity representation.

As we saw in the previous Section, the initial relative velocity vector was used to define the z'-axis in the SF frame (eq. (3.64)). This resulted in a simplification of the problem in that one could have an initial situation without any angular momentum of relative motion about the z'-axis, which in quantum mechanical terms means: $m_\ell = 0$. The final relative velocity vector however lies along the $\hat{R}$ direction and the corresponding z-components of the angular moments for the scattered partial waves have to be referred to the original, fixed, coordinate system. As a consequence, there is a resulting lack of symmetry in the treatment of the total wavefunction before and after collision.

In the helicity representation the problem is treated in a symmetric manner, in that the scattered wave in described in a rotating coordinate system whose z-axis lies along the $\hat{R}$ direction and follows the target during interaction (BF frame of fig. 3.2). This last approach is often referred to as the **R**-helicity representation to distinguish is from another frame called the P-helicity representation. In the latter the molecular angular momenta for the scattered wave are referred to the direction of linear momentum $\hat{P}$[23, 24, 25].

In both references, however, the orbital angular momentum, $\underline{\ell} = \underline{R} \times \underline{p}$ has a vanishing projection along the z-axis.

Let us now define as $\mathcal{R}$ the rotation that brings the coordinate system $\underline{x}'$ into the $\underline{x}$ coordinate system through the matrix $\underline{\underline{R}}$:

$$\underline{x} = \underline{\underline{R}}\, \underline{x}' \tag{3.94}$$

Its effect on any function defined with respect to $\underline{x}'$, say $\psi(\underline{x}')$, is obtained by transforming the basis functions of the functional space $\underline{x}'$, $\{\psi(\underline{x}', JM)\}$ according to the following prescription [26] :

$$\psi(\underline{x},\ JM) = \psi(\underline{\underline{R}}\underline{x}',\ JM) = \mathcal{R}^{-1}\ \psi(\underline{x}',\ JM) \tag{3.95}$$

This becomes, in terms of the coefficients of the irreducible representations of rotation group [20] :

$$\psi(\underline{x}',\ JM) = \sum_{\Omega} \mathcal{D}^{J}_{\Omega M}(\mathcal{R})\ \psi(\underline{x},\ J\ \Omega) \tag{3.96}$$

where Ω is now the projection of J along the z-axis.

The corresponding eigenvalue equation in the SF frame, i.e.:

$$\mathcal{H}'\ \psi(\underline{x}') = E\ \psi(\underline{x}') \tag{3.96a}$$

becomes now:

$$\mathcal{H}'\mathcal{R}\ \psi(\underline{x}) = E\ \mathcal{R}\ \psi(\underline{x}) \tag{3.97b}$$

and projecting it onto one basis function defined as follows:

$$\varphi(\underline{x}') = \mathcal{R}\ \varphi(\underline{x}) \tag{3.97}$$

one can further write:

$$\begin{aligned} &< \mathcal{R}\ \varphi(\underline{x})\ |\ E - \mathcal{H}'|\ \mathcal{R}\ \psi(\underline{x}) > = \\ &= <\ \varphi(\underline{x})|\ \mathcal{R}'\ \mathcal{H}'\ \mathcal{R}\ |\ \psi(\underline{x})\ > \end{aligned} \tag{3.98}$$

One thus sees that new representation of $\mathcal{H}'$ in the BF frame is given by:

$$\mathcal{H} = \mathcal{R}^{-1}\ \mathcal{H}'\ \mathcal{R} \tag{3.99}$$

The new BF-functions are related to the previous SF functions via the $\mathcal{D}$-coefficients, whose arguments are operated upon by the angular momentum operator in the SF frame $\mathcal{L}'$. Therefore they are no longer angular momentum eigenfunctions.

We then speak of scattering from a state of helicity Ω about the initial direction to a state of helicity Ω' about the final direction, rather than from a state $|jm_j>$ to a state $|j'm_{j'}>$ as in the previous SF representation.

The main mathematical consequence of this transformation is to remove terms with $m_{\ell'} \neq o$ in the final coupling terms, because the choice of the BF axes is such that they are coincident with the unit vectors of $\underline{R}$ in the spherical polar coordinates of the SF frame. Therefore there can be no orbital angular momentum about $\underline{R}$. Following the convention of ref. [20], the Euler angle rotation performed here is the one by which the polar coordinates of $\underline{r}$ and $\underline{R}$ in the new BF system can be written as: $\hat{r} = (\theta, o)$ and: $\hat{R} = (0, 0)$, with θ being the angle that the molecular bond forms with the $\underline{R}$ vector (see Fig. 3.2). The helicity index Ω describes the projection of J on the BF z-axis.

In the case of a spherical top rotor the BF rotated wavefunctions can be obtained [27] from an expansion similar to the one shown in eq. (3.96) for the SF wavefunctions:

$$\Psi^{JM}_{j\Omega}(\underline{R}, \hat{r}) = \mathcal{R}^{-1} \varphi_{JM}(\underline{R}', \hat{r}') = \sum_{\Omega'=-J}^{J} \mathcal{D}^{J}_{\Omega' M}(\varphi', \theta', o)\, \psi^{Jj\Omega}_{\Omega'}(R, \hat{r}) \qquad (3.101)$$

Now the total orbital angular momentum operator $\mathcal{L}'$ can be rewritten in the following form:

$$\begin{aligned} \mathcal{L}'^2 &= (\underset{\sim}{J} - \underset{\sim}{j})^2 = \underset{\sim}{J}^2 + \underset{\sim}{j}^2 - 2\, \underset{\sim}{J} \cdot \underset{\sim}{j} = \\ &= \underset{\sim}{J}^2 + \underset{\sim}{j}^2 - 2\, \underset{\sim}{J}_z \cdot \underset{\sim}{j}_z - 2\, \{\underset{\sim}{J}_x \cdot \underset{\sim}{j}_x + \underset{\sim}{J}_y \cdot \underset{\sim}{j}_y\} \end{aligned} \qquad (3.101)$$

however:

$$\begin{aligned} \underset{\sim}{J}_x &= \frac{1}{2i} (\underset{\sim}{J}_+ + \underset{\sim}{J}_-) \\ \underset{\sim}{J}_y &= \frac{1}{2i} (\underset{\sim}{J}_+ + \underset{\sim}{J}_-) \\ \underset{\sim}{j}_x &= \frac{1}{2i} (\underset{\sim}{j}_+ + \underset{\sim}{j}_-) \\ \underset{\sim}{j}_y &= \frac{1}{2i} (\underset{\sim}{j}_+ - \underset{\sim}{j}_-) \end{aligned} \qquad (3.102)$$

Hence, by substituting the results of (3.102) into eq. (3.101), one gets:

$$-2\, \{\underset{\sim}{J}_x \underset{\sim}{j}_x + \underset{\sim}{J}_y \underset{\sim}{j}_y\} = -(\underset{\sim}{J}_+ \underset{\sim}{j}_-) - (\underset{\sim}{J}_- \underset{\sim}{j}_+) \qquad (3.103)$$

The Hamiltonian operator in the SF frame now becomes:

$$\begin{aligned} \mathcal{H}' = \frac{1}{2\mathcal{M}} R'^{-1} \left(\frac{\partial^2}{\partial R'^2}\right) \cdot R' + \left(\frac{1}{2\mathcal{M}R'^2}\right) \cdot \underset{\sim}{J}^2 + \left(\frac{1}{2\mathcal{M}R'^2}\right) \cdot \underset{\sim}{j}^2 + \left(\frac{1}{2\mathcal{M}R'^2}\right) \underset{\sim}{J}_z \underset{\sim}{j}_z + \\ + \left(\frac{1}{2\mathcal{M}R'^2}\right) \cdot \underset{\sim}{J}_+ \underset{\sim}{j}_- + \left(\frac{1}{2\mathcal{M}R'^2}\right) \underset{\sim}{J}_- \underset{\sim}{j}_+ + \mathcal{H}'_{rot}(\hat{r}') + V'(\underline{R}', \hat{r}') \end{aligned} \qquad (3.104)$$

which yields the following contributions after the expansion of (3.100) is operated on by each term of (3.104). More in detail, one can write in fact that from the first two terms on the r.h.s. of (3.104) one gets:

(i); $$\sum_{\Omega'} \mathcal{D}^{J}_{\Omega' M} \frac{1}{2\mathcal{M} R} \left(\frac{\partial^2}{\partial R^2} \right) R \ \psi^{Jj\Omega}_{\Omega'} (R, \hat{r}) \qquad (3.105a)$$

and:

(ii); $$\sum_{\Omega'} \mathcal{D}^{J}_{\Omega' M} \left(- \frac{1}{2\mathcal{M} R^2} \right) J(J+1) \ \psi^{Jj\Omega}_{\Omega} (R, \hat{r}) \qquad (3.105b)$$

since the $\mathcal{D}^{J}_{\Omega' M}$ are eigenfunctions of $\underset{\sim}{J}^2$ and $\underset{\sim}{J}'^2$ with eigenvalue J(J+1) and are also eigenfunctions of $\underset{\sim}{J}_z$ and $\underset{\sim}{J}'_z$ with eigenvalues Ω' and M respectively [26].

The third term contribution of (3.104) yields in turn:

(iii); $$\sum_{\Omega'} \mathcal{D}^{J}_{\Omega' M} \left(- \frac{1}{2\mathcal{M} R^2} \right) \underset{\sim}{j}^2 \ \psi^{Jj\Omega}_{\Omega'} (R, \hat{r}) \qquad (3.106$$

since the coefficients $\mathcal{D}$'s are independent of the transformation: $\hat{r} = \underline{\underline{R}}\ \hat{r}'$ and $\underset{\sim}{j}^2$ is not touched by the rotation. Moreover, the $\mathcal{D}$'s are eigenvectors of $\underset{\sim}{J}_z$ with eigenvalues Ω' and the $\underset{\sim}{j}_z$ operator does not act on the $\hat{R}'$ coordinates.

Because of the present choice for the coordinate transformation, the angular momentum component along the z-axis is equal to zero, hence: $\underset{\sim}{J}_z = \underset{\sim}{\mathcal{L}}_z + \underset{\sim}{j}_z = \underset{\sim}{j}_z$. The result of the fourth term in eq. (3.104) therefore becomes:

(iv); $$\sum_{\Omega'} \mathcal{D}^{J}_{\Omega' M} \cdot 2\Omega'^2 \ \psi^{Jj\Omega}_{\Omega'} (R, \hat{r}) \left(\frac{1}{2\mathcal{M} R^2} \right) \qquad (3.107)$$

The fifth and sixth terms of the $\mathcal{H}$ operator in (3.107) can be obtained from the well known properties of the raising and lowering operators (in atomic units):

$$\underset{\sim}{J}_+ \ |J\ \Omega' > = \{(J - \Omega')(J + \Omega' + 1)\}^{\frac{1}{2}} \ |J, \Omega' + 1 > \qquad (3.108a)$$

$$\underset{\sim}{J}_- \ |J\ \Omega' > = \{(J + \Omega')(J - \Omega' + 1)\}^{\frac{1}{2}} \ |J, \Omega' - 1 > \qquad (3/108b)$$

and from the relationship:

$$[\underset{\sim}{J}_+, \underset{\sim}{j}_-] = 0 \qquad (3.108c)$$

Hence:

$$(v); \quad \frac{1}{2\mathcal{M}R^2} \sum_{\Omega'} \mathcal{D}^J_{\Omega' M} \{(J - \Omega')(J + \Omega' + 1)\}^{\frac{1}{2}} \, j_- \, \psi^{Jj\Omega}_{\Omega'+1} (R, \hat{r}) \tag{3.109}$$

and:

$$(vi); \quad \frac{1}{2\mathcal{M}R^2} \sum_{\Omega'} \mathcal{D}^J_{\Omega' M} \{(J + \Omega')(J - \Omega' + 1)\}^{\frac{1}{2}} \, j_+ \, \psi^{Jj\Omega}_{\Omega'-1} (R, \hat{r}) \tag{3.110}$$

Finally, for the last two terms in (3.104) we can write:

$$\text{(vii) and (viii):} \quad \{\mathcal{H}'_{rot}(\hat{r}') - V'(R, \hat{R}', \hat{r}')\} \sum_{\Omega'} \mathcal{D}^J_{\Omega' M} (\varphi', \theta', 0) \, \psi^{\Omega}_{\Omega} (R, \hat{r}) =$$

$$= \sum_{\Omega'} \mathcal{D}^J_{\Omega' M} \{\mathcal{H}'_{rot}(\underline{\underline{R}}^{-1} \hat{r}) - V'(R, \underline{\underline{R}}^{-1} \hat{R} = 0, \underline{\underline{R}}^{-1} \hat{r})\} \, \psi^{J\Omega}_{\Omega'} (R, \hat{r}) = \tag{3.111}$$

$$= \sum_{\Omega'} \mathcal{D}^J_{\Omega' M} \{\mathcal{H}_{rot}(\hat{r}) - V(R, \hat{r})\} \, \psi^{J\Omega}_{\Omega'} (R, \hat{r})$$

Now one can sum all the contributions from (3.105) through (3.111). By also remembering eq.s (3.96), i.e. that: $\mathcal{H}' \mathcal{R} = \mathcal{R} \, \mathcal{H}$, one further has that:

$$\sum_{\Omega'} \mathcal{D}^J_{\Omega' M}(\varphi', \theta', 0) \, \{......\} \, \psi^{Jj}_{\Omega'} \, (R, \hat{r}) = E \sum_{\Omega'} \mathcal{D}^J_{\Omega' M} \, \psi^{Jj\Omega}_{\Omega'} (R, \hat{r}) \tag{3.112}$$

where the part within curly brackets is now the new Hamiltonian in the BF frame of reference.

If one now defines:

$$- \frac{1}{2\mathcal{M}R} \left(\frac{\partial^2}{\partial R^2} \right) R + \frac{1}{2\mathcal{M}R^2} \{ [J(J+1) - 2\,\Omega'^2] + j^2 \} +$$

$$+ \mathcal{H}_{rot}(\hat{r}) + V(R, r, \theta) = \mathcal{H}_{\Omega'\Omega'} \tag{3.113}$$

and:

$$\{[(J \pm \Omega' + 1)\,(J \pm \Omega')]^{\frac{1}{2}} \, j_\pm\} \frac{1}{2\mathcal{M}R^2}$$

$$= - \lambda_\pm \, (J, \Omega') \, j_\pm / 2\mathcal{M}R^2 = \mathcal{H}_{\Omega', \Omega' \pm 1} \tag{3.114}$$

The eigenvalue equation (3.112) becomes:

$$- \mathcal{H}_{\Omega',\Omega'+1} \psi^{Jj\Omega}_{\Omega'+1} + [E - \mathcal{H}_{\Omega',\Omega'}] \psi^{Jj\Omega}_{\Omega'} - \mathcal{H}_{\Omega',\Omega'-1} \psi^{Jj\Omega}_{\Omega'-1} = 0 \tag{3.115}$$

In order to recover more closely the coupled equations of the previous Section, we further expand the BF wavefunction in the following form:

$$\psi^{Jj\Omega}_{\Omega'} (R, \hat{r}) = \sum_{j''} R^{-1} u^{Jj\Omega}_{j''\Omega'} (R) Y^{\Omega'}_{j''} (\hat{r}) \tag{3.116}$$

where $\hat{r} = (\theta_r, \varphi_r)$ are the polar angles of $\underline{r}$ in the BF frame (see Fig. 3.2). The BF axes are the unit vectors for $\underline{R}(\hat{e}_R, \hat{e}_\theta, \hat{e}_\varphi)$. The angular expansion over $\hat{R}$, the $Y^{\Omega'}_J (\hat{R})$ coefficients, is obviously missing in this representation as apposed to the SF frame of reference, since here: $\hat{R} = (0, 0)$. The Hamiltonian of (3.115) contains a centrifugal potential $1/R^2$ which allows for the fact that, as we are no longer in an inertial system, one would observe the projectile classically incoming along the z-axis and outgoing along the same z-axis after collision.

By substituing the expansion (3.116) into the eigenvalue equation (3.115), premultiplying by: $2\mathcal{M} Y^{*}_{j'\Omega'}(\hat{r})$ and integrating over d $\hat{r}$, one finally obtains the close-coupling or coupled-channel (CC) equations in the helicity representation [29]:

$$h^{j}_{\Omega',\Omega'-1} u^{Jj\Omega}_{j'\Omega'-1} (R) + h^{Jj'}_{\Omega',\Omega'} u^{Jj}_{j'\Omega'} (R) + h^{Jj'}_{\Omega',\Omega'+1} u^{Jj\Omega}_{j'\Omega'+1} (R) = \tag{3.117}$$

$$= 2 \sum_{j''} < j'\Omega' |V| j''\Omega' > u^{Jj\Omega}_{j''\Omega'} (R)$$

where:

$$h^{Jj'}_{\Omega',\Omega'} = \frac{d^2}{dR^2} + k^2_{j'} - \frac{[J(J+1) - 2\Omega'^2 + j'(j'+1)]}{R^2} \tag{3.118a}$$

and:

$$h^{Jj'}_{\Omega',\Omega'\pm 1} = \underset{\sim}{\lambda}_{\pm} (J, \Omega') \underset{\sim}{\lambda}_{\pm} (j', \Omega' \pm 1)_{/R^2} \tag{3.18b}$$

$$k^2_{j'} = 2\mathcal{M} [E - \varepsilon_{j'}] \tag{3.18c}$$

The coupling matrix elements that appear on the r.h.s. of (3.117) are given by:

$$< j'\Omega' |V| j''\Omega' > = \int d\hat{r} \, Y^{*}_{j'\Omega'} (\hat{r}) \, V (\hat{r}, \underline{R}) \, Y_{j''\Omega'} (\hat{r}) \tag{3.119}$$

and, via the expansion (3.80), which does not depend on the frame chosen but only on internal coordinates, one can further write that:

$$< j'\Omega' |V| j''\Omega' > = \sum_{\lambda} V_{\lambda} (R) \int d\hat{r}\, Y_{j'\Omega'}(\hat{r})\, P_{\lambda} (\hat{r} \cdot \hat{R})\, Y_{j''\Omega'} (\hat{r}) \qquad (3.120)$$

In the BF frame, $\theta_R = 0$ and $\theta_r = \theta = \hat{r}.\hat{R}$. One can then directly compute the above integral and write it down as follows:

$$< j'\Omega' |V| j''\Omega' > = \sum_{\lambda} (-)^{\Omega'} [(2j'+1)(2j''+1)]^{\frac{1}{2}} \cdot \begin{pmatrix} j' & \lambda & j'' \\ 0 & 0 & 0 \end{pmatrix} \cdot \begin{pmatrix} j' & \lambda & j'' \\ \Omega' & 0 & \Omega \end{pmatrix} \cdot V_{\lambda}(R) \qquad (3.121)$$

We thus see that eq. (1.121) does not contain the 6-j coefficients which appeared in the corresponding expression of the SF frame in eq. (3.82). This is because the internal coordinate $\theta = \hat{r}.\hat{R}$, over which the integration is performed in the multipolar expansion, is now equal to θ_r, the corresponding unit vector of $\hat{r}$ in polar coordinates.

The eigensolutions of the CC equations (3.117) have to satisfy the following boundary conditions:

$$u^{Jj\Omega}_{j'\Omega'} (0) = 0$$

$$u^{Jj\Omega}_{j'\Omega'} (R) \underset{R\to\infty}{\sim} k_{j'}^{-\frac{1}{2}} \Big\{ \delta_{jj'} \cdot \delta_{\Omega\Omega'} \exp [-i(k_j R - (J + j)^{\pi}/_2] - \qquad (3.122)$$

$$- S^{J}_{j'\Omega',j-\Omega} \cdot \exp [-i(k_{j'} R - (J + j')^{\pi}/_2] \Big\}$$

The usual total wavefunctions of eq. (3.2), should then asymptoticaly behave like eq.s (3.11) and should be expanded over an SF complete set of eigenfunctions analogous to those of eq. (3.115) used in eq. (3.100):

$$\Phi^{j}_{m_j} (\underline{R}', \hat{r}') = \sum_{JM\Omega} \mathcal{A}(JMjm_j\Omega)\, \psi^{JM}_{j\Omega} (\underline{R}', \hat{r}') \qquad (3.123$$

where the $\mathcal{A}$'s coefficients are determined by making its incident part equal to the incoming plane wave (P. W.).

Now, in the SF reference frame we can write:

$$\text{P.W.} = \exp(i\, k_j\, z') = \sum_\ell (2\ell + 1)\, i^\ell\, j_\ell\, (k_j\, R)\, P_\ell\, (\cos\theta_{R'}) \tag{3.124}$$

and, since:

$$P_\ell\,(\cos\theta_{R'}) = \mathcal{D}^{\ell}_{oo}\,(\varphi_{R'}, \theta_{R'}, 0) \tag{3.125a}$$

and

$$j_\ell\,(k_j\, R) \underset{R\to\infty}{\sim} \sin\,[\,k_j R - \ell\pi/_2]\;/_{k_j R} \tag{3.125b}$$

then:

$$\text{P.W.} \underset{R\to\infty}{\sim} \sum_\ell (2\ell + 1)\, i^\ell\, \frac{\sin(k_j R - \ell\pi/2)}{kR}\, \mathcal{D}^{\ell}_{oo}\,(\varphi_{R'}, \theta_{R'}, 0) \tag{3.126}$$

In therefore follows that the incoming P.W. with the spherical rotor in the initial state can be written as:

$$\exp(i\, k_j\, z')\, Y_j^{m_j}(\hat{r}') = \left\{ \sum_\ell (2\ell + 1)\, i^\ell\, \sin(k_j R - \ell\pi/2)/_{k_j R} + \mathcal{D}^{\ell}_{oo} \right\} \cdot \\ \cdot \sum_{\Omega'} \mathcal{D}^{j}_{\Omega' m_j}\,(\varphi_{R'}, \theta_{R'}, 0)\, Y_j^{\Omega'}(\hat{r}) \tag{3.127}$$

a contraction of the $\mathcal{D}$ -coefficients yields [6] now the simpler expression:

$$\mathcal{D}^{\ell}_{oo}\, \mathcal{D}^{j}_{\Omega' m_j} = \sum_J (2J + 1) \begin{pmatrix} j & \ell & J \\ \Omega' & 0 & \Omega' \end{pmatrix} \cdot \begin{pmatrix} j & \ell & J \\ m_j & 0 & m_j \end{pmatrix} \mathcal{D}^{J}_{\Omega', m_j} \tag{3.128}$$

and since: $m_j = M$, because of $m_\ell = 0$, one can rewrite eq.(3.127) in the following form:

$$\text{P.W.} = \left\{ (2k_j R)^{-1} \sum_{J,\Omega'} i^{J+j+1} (2J+1)\, \mathcal{D}^{J}_{\Omega' M}\,(\hat{R}')\, Y_j^{\Omega'}(\hat{r}) \right\} \cdot \\ \cdot \left\{ \delta_{\Omega', -M} \exp\,[-i(k_j R - (J+j)\pi/2] - \delta_{\Omega', M} \exp\,[\,i(k_j R - (J+j)\pi/2] \right\} \tag{3.129}$$

where one sees that the incoming part of the wavefunction has: $\Omega' = - M$ since, after collision, the new rotating z-axis points along the direction $(-z')$ of the original orientation.

Let us now write down the incoming part of eq.(3.123). After its transformation into the BF frame via eq.(3.100) one can write:

$$\Phi^{j}_{m_j} = \sum_{J,M,\Omega} \mathcal{A}(JM\, j\, m_j\, \Omega) \sum_{\Omega'} \mathcal{D}^{J}_{\Omega' M}(\varphi', \theta', 0) \quad \psi^{Jj\Omega}_{\Omega'}(R, \hat{r}) \tag{3.130}$$

now, from eq. (3.116) one can also rewrite (3.130) as follows:

$$\Phi^{j}_{m_j} = \sum_{J,M,\Omega} \mathcal{A}(JM,\, jm_j\, \Omega) \sum_{\Omega'} \mathcal{D}^{J}_{\Omega' M} \sum_{j''} R^{-1}\, u^{Jj\Omega}_{j''\Omega'}(R)\, Y^{\Omega'}_{j''}(\hat{r}) \tag{3.131}$$

The radial functions have to satisfy the boundary conditions stated in eq. (3.122), hence their substitution into eq. (3.131) provides the incoming part of (3.130). The latter, in turn, has to be compared with the incoming part of the P.W. expression in the BF frame, as given in eq. (3.129). The $\mathcal{A}$ -coefficients are thus obtained by equating these two parts, i.e. they are given by the following simple relationship:

$$\mathcal{A}(JM\, jm_j \Omega) = \frac{1}{2}\, i^{J+j+1} (2J + 1)\, \delta_{M,m_j} \cdot \delta_{\Omega'-M} \cdot k_j^{-\frac{1}{2}} \tag{3.132}$$

Now, following eq.(3.66), one can write the total wavefunction for a single entrance channel $|jm_j\rangle$, in the SF frame, for the spherical-top case as having the following asymptotic form:

$$\Phi^{j}_{m_j} \underset{R\to\infty}{\sim} \exp(i\, k_j\, z'_R)\, Y^{m_j}_{j}(\hat{r}') \; \cdot$$

$$\cdot \sum_{j' m_{j'}} q(j' m_{j'}, jm_j \,|\hat{R}')\, Y^{m_{j'}}_{j'}(\hat{r}')\, R^{-1} \exp(i\, k_{j'}\, R) \tag{3.133}$$

From the second term on the r.h.s. of this equation one should now subtract the outgoing part of the P.W. of eq.(3.129) and equate it to the result of (3.134) when the radial functions are chosen as those which satisfy eq. (3.122). The final result is then:

$$\Phi_{\text{scattered}} = \sum_{J,M,\Omega',j} (2R)^{-1}\, i^{j-j'}\, (2J+1)\, \delta_{M,m_j}\, \mathcal{D}^{J}_{\Omega' M}(\varphi', \theta', 0) \cdot$$

$$\cdot\, Y^{\Omega'}_{j'}(\hat{r}) \cdot (k_j k_{j'})^{-\frac{1}{2}} \cdot \exp(k_{j'} R) \cdot T^{J}_{j'\Omega', jM} \tag{3.134}$$

where the T-matrix as given in definition (3.72) makes use of the S-matrix of (3.122) and of the equality: $\Omega' = M$.

In order to go back to the SF frame, where the measurements are actually performed, one has to remember that:

$$Y_{j'}^{\Omega'}(\hat{r}) = \sum_{m_{j'}} \mathscr{D}^{j'}_{\Omega' m_{j'}}(\hat{R}')\, Y_{j'}^{m_{j'}}(\hat{r}') \tag{3.135}$$

Hence its substitution in to (3.134) and a summation over $|J-j| \leq L \leq J+j$ produce an expression for $\Phi_{\text{scattered}}$ that can be compared with eq. (3.133) in order to obtain the final form of the scattering amplitude:

$$\begin{aligned} q(jm_{j'}, jm_j \mid \hat{R}') = \sum_{JM\Omega'L} & i^{j-j'+1} \cdot (-)^{m_j - \Omega'} \cdot \delta_{Mm_j} \cdot (2J+1) \cdot \pi^{\frac{1}{2}} \cdot \\ & \cdot [(2L+1)\, k_j k_{j'}]^{-\frac{1}{2}} \cdot (J\,\Omega' j' - \Omega' | Jj'L\,0) \cdot \\ & \cdot (JMj' - m_{j'} | Jj'\,L\,M - m_{j'}) \cdot T^J_{j'\Omega', jM} \cdot Y_L^{M-m_{j'}}(\hat{R}') \end{aligned} \tag{3.136}$$

The computational problem of searching for the solutions of the CC equations (3.117) is of the same order of complexity as the one produced in the SF frame. Each rotor state j', in fact, produces (2j' + 1) channels $|j'\Omega'\rangle$ (with: $-j' \leq \Omega' \leq j'$) as in the SF scheme. The biggest difficulty arises here from the off-diagonal terms (3.118) which persist even after the potential V in (3.117) has vanished. In the SF frame the overall R-dependence was instead controlled only by the asymptotic form of the B.O. potential surface. On the other hand, such off-diagonal terms become less important at smaller R-values, leaving only the j" summation of the coupling matrix elements on the r.h.s. of eq. (3.117). The SF-frame formulation produced insteads another summation over ℓ', as shown in eq. (3.58).

One can thus conclude that for small value of J and strong couplings over long-range potentials the BF treatment is more manageable, while the SF frame of reference becomes preferable at large J values and weak potential couplings that mainly act via short-range potentials.

As in the SF case one now obtains the differential and total cross sections for the inelastic transitions. The corresponding integral cross section of (3.77) now becomes (for the rigid-rotor target molecule):

$$\sigma(j \to j') = \pi\, k^{-2} \sum_{J} (2J + 1)\, \sigma^{J}_{j \to j'} \qquad (3.177a)$$

where:

$$\sigma^{J}_{j \to j'} = (2j + 1)^{-1} \sum_{M=-j}^{j} \sum_{\Omega'=-j'}^{j'} |T^{J}_{j'\Omega',\, jM}|^{2} \qquad (3.177b)$$

where the summations start from the smaller of j and J for M and the smaller of j' and J for Ω', respectively.

3.6 - The vibro-rotational extension

An extension of the procedures described in the previous sections to vibrating and rotating molecules requires a more detailed examination of the corresponding vibration-rotation Hamiltonian generally defined as $\mathcal{H}_{int}(\underline{X})$ in eq.(2.17). Obviously one would like to obtain a theretical treatment of its structure that could successfully be related to the observed spectra of in creasingly more complex collision partners, although the corresponding increase of complexity for the scattering equations or for the B.O. hypersurfaces might still prevent a full ab initio treatment for some time to come.

From the point of view of the target molecules, however, microwave rotational spectra and infrared Raman vibration-rotation spectra at high resolution allow us to observe differences between the vibration-rotation term values of a molecule. These terms may be expressed empirically as the sum of a vibrational term which is independent of the rotational quantum numbers and a rotational term which is largely indipendent of the vibrational quantum numbers:

$$T(v, j) = G(v) + F_{v}(j) \qquad (3.138)$$

The rotational term values $F_v(j)$ are interpreted as the eigenvalues of an effective rotational Hamiltonian which is slightly different for each vibrational state. For an asymmetric top, as already discussed in Section 2, the form of such an Hamiltonian can be written as follows:

$$\mathcal{H}_{rot} = A_v \cdot j_a^2 + B_v \cdot j_b^2 + C_v \cdot j_c^2 + \\ + \frac{1}{4} \sum_{\alpha,\beta} (\tau_{\alpha\alpha\beta\beta}) \cdot j_\alpha^2 \cdot j_\beta^2 + \ldots\ldots \tag{3.139}$$

where j_a, j_b, j_c are components of the total angular momentum in h units; A_v, B_v and C_v the effective rotational constants and the $\tau_{\alpha\alpha\beta\beta}$'s are the centrifugal distorsion constants [31] .

The indeces α and β are summed over a, b and c. The centrifugal distorsion constants are generally smaller that the rotational constants by factor of the order of 10^{-4}. In principle even higher power terms in the angular momenta are also possible in (3.139), but the series converges rapidly so that these higher-order terms are too small to observe. The invariance of the Hamiltonian to time reversal also causes the occurrence of only even powers of the j's.

The relationship of the rotational constants to the energy levels and hence to the observed spectra is complicated by the fact that it is not possible to write down explicit expression for the eigenvalues of $\mathcal{H}_{rot}$ as altready discussed in Section 2 of this Chapter [32] . Moreover, the various constants to be determined in (3.139) show a vibrational dependence that is usually given by the general form:

$$B_v = B_{eff} - \sum_r \alpha_r^{(B)} (v_r + \tfrac{1}{2}) + \sum_{r>s} \gamma_{rs}^{(B)} (v_r + \tfrac{1}{2})(v_s + \tfrac{1}{2}) + \ldots . \tag{3.140}$$

where B_{eff} is an effective value determined by rotational levels within one vibrational state and the sums run over all the normal modes of the molecule under study. In general, the α's have a magnitude of the order 10^{-2} of the rotational constants and the γ's are usually 10^{-2} smaller still; thus the expansion in vibrational quantum numbers converges fairly rapidly. The vibrational dependence of the τ's is also given by a similar equation but in this case even the first power dependence on v is generally too small to observe, thus they are usually treated as v-independent.

For polyatomic molecules the complete rotation-vibrational Hamiltonian has been discussed in the literature for a long time [33, 34] and has more recently [35] been shown to be given by:

$$\mathcal{H}_{tot} = \sum_{\alpha,\beta} \frac{1}{2} \mu_{\alpha\beta} (\underset{\sim}{j}_{\alpha} - \underset{\sim}{\pi}_{\alpha})(\underset{\sim}{j}_{\beta} - \underset{\sim}{\pi}_{\beta}) + \frac{1}{2} \sum_{r} P_r^2 + V(\sum_{r} Q_r) + \cup \tag{3.141}$$

where the $\mu_{\alpha,\beta}$ are elements of the matrix $\underset{=}{\mu}$ of the effective reciprocal inertial tensors:

$$\mu_{\alpha,\beta} = (I'^{-1})_{\alpha,\beta} \tag{3.142}$$

where $I'_{\alpha\beta}$ is in turn given by:

$$I'_{\alpha\beta} = I_{\alpha\beta} \sum_{r,s,q} \xi^{\alpha}_{rq} \cdot \xi^{\beta}_{sq} \cdot Q_r \cdot Q_s \tag{3.143}$$

where the $I_{\alpha\beta}$'s are the instantaneous inertia tensors evaluated at the molecular reference configuration [35]. The ξ^{α}_{rq} are Coriolis coupling coefficients that couples the normal coordinate Q_r to the Q_q through rotation about the α-axis. P_r is the conjugate momentum of the corresponding Q_r normal coordinate:

$$P_r = - i h \frac{\partial}{\partial Q_r} \tag{3.144}$$

and $\cup$ is a very small mass-independent correction to the vibrational potential energy that has been compactly written as $V(\sum_r Q_r)$.

The $\underset{\sim}{\pi}_{\alpha}$'s are the components of the vibrational angular momentum describing another vibration-rotation coefficient containing Coriolis coupling:

$$\underset{\sim}{\pi}_{\alpha} = \sum_{r,s} \xi^{\alpha}_{rs} \cdot Q_r \cdot P_s \tag{3.145}$$

Finally, the potential expression can generally be written as:

$$V(\sum_r Q_r) = \frac{1}{2} \sum_r \omega_r q_r^2 + \frac{1}{6} \sum_{r,s,t} \phi_{rst} \cdot q_r \cdot q_s \cdot q_t + \frac{1}{24} \sum_{r,s,t,u} \phi_{rstu}\, q_r\, q_s\, q_t\, q_u + \ldots\ldots \tag{3.146}$$

where:

$$q_r = (P_r)^{\frac{1}{2}} \cdot Q_r \tag{3.147a}$$

and:

$$\gamma_r = 2\pi c \omega_{r/h} \tag{3.147b}$$

with ω_r being the harmonic wavenumber of the rth normal mode and the ϕ_{rst} and ϕ_{rstu} are the cubic quartic anharmonic force constants.

One possible way of solving the formidable-looking eigenvalue equations generated by the $\mathcal{H}_{tot}$ of (3.141) is to apply a transformation Φ that generates a new Hamiltonian:

$$\tilde{\mathcal{H}} = \Phi^{-1} \mathcal{H}_{tot} \Phi \tag{3.141b}$$

such that the harmonic oscillator basis functions are the eigenfunctions of $\tilde{\mathcal{H}}$ to the desired degree of accuracy. The averaging of $\tilde{\mathcal{H}}$ over the appropriate, now known, vibrational eigenfunctions allows one to obtain the effective rotational Hamiltonian within each vibrational state. An alternative approach is to set up the zeroth order approximation wavefunctions as products of harmonic oscillator functions in the normal coordinates as given by the zeroth order vibrational Hamiltonian:

$$\mathcal{H}^{(o)}_{vib} = \sum_r \frac{1}{2} (P_r^2 + \lambda_r Q_r^2) \tag{3.148}$$

with only the quadratic terms of eq.(3.146).

One then computes the matrix elements H_{ij} over this basis and analyses the rotational angular momentum operators that are now given explicitely. The appropriate use of perturbation theory, to a desired order, then allows the removal of those terms in the matrix which are off-diagonal in the vibrational basis functions. The diagonal matris elements in the ensuing modified matrix give the desired effective rotational in the different vibrational states. To apply such a procedure is therefore necessary, in order to expand the terms of matrix $\underline{\underline{H}}$ is powers of P_r and Q_r to be able to assess the relative orders of the various terms. This has been discussed extensively in ref.s [35] and [36].

A similar perturbative treatment can finally also be applied to the vibrational constants contained in the G term of eq.(3.138) because of the structure of the potential (3.146):

$$G(v) = \sum_r \omega_r(v_r + \frac{1}{2}) + \sum_{r>s} x_{rs}(v_r + \frac{1}{2})(v_s + \frac{1}{2}) + \dots \tag{3.149}$$

where the x_{rs} contains second-order contributions from the cubic force constants ϕ_{rst} and first-order contributions from the quartic force constants ϕ_{rstu}[35].

This formulation will then allow us to include various energy corrections to the term values originally obtained over the harmonic oscillator basis functions.

As should be clear from the brief résumé of above one can conclude that the correct treatment of the vibrational-rotational target Hamiltonian for general molecules of arbitrary complexity is a rather difficult task that is usually circumverted by a perturbation expansion over harmonic oscillator w. f's and a further averaging of the complete, simplified Hamiltonian over all vibrational coordinates. One is thus left with an Hamiltonian involving only the rotational angular momentum operators, in a procedure completely analogous to the separation of vibrational from electronic degrees of freedom in the conventional B.O. approximantion discussed in Chapter 2.

For the simplest case of a diatomic molecule without electronic angular momentum component along its axis the above considerations lead to the following form of the eigenvalue equation (2.17):

$$\mathcal{H}_{int}(\underline{r})\,\varphi_j^v(\underline{r}) = \varepsilon_{vj}\,\varphi_j^v(\underline{r}) \tag{3.150}$$

where (in atomic units):

$$\mathcal{H}_{int}(\underline{r}) = \left\{ -\frac{1}{2\mu}\left[\left(\frac{\partial}{\partial r}+\frac{1}{r}\right)^2 - \frac{j_{\underline{r}}^2}{r^2}\right] + V_o(r)\right\} \tag{3.151}$$

where the square angular momentum operator $j_{\underline{r}}^2$ has its usual orthonormal eigenfunctions provided by spherical harmonics as in eq. (3.38) and μ is the molecular reduced mass. One way of writing down the $\varphi_j^v(\underline{r})$ wavefunctions is thus readily provided by the expression.:

$$\varphi_j^v(\underline{r}) = r^{-1}\,\chi_j^v(r)\,Y_j^{m_j}(\hat{r}) \tag{3.152}$$

where the χ's are now eigensolutions of the radial equation:

$$\left\{ -\frac{1}{2\mu}\frac{d^2}{dr^2} + \frac{j(j+1)}{2\mu r^2} + V_o(r)\right\}\chi_j^v(r) = \varepsilon_{v_j}\,\chi_j^v(r) \tag{3.153}$$

Once the intermolecular potential $V_o(r)$ (indipendent of $\hat{r}$ for the case of a diatomic molecule in the absence of external fields) is known for the system under study, one can follow up the previous outline by expanding the functions χ_s' in an harmonic oscillator basis as:

$$\chi_j^v(r) = \sum_m C_{vj,m} N_m H_m (\alpha x) \exp \left(- \frac{1}{2} \alpha^2 x^2\right) \tag{3.154}$$

where: $x = (r - r_{eq})$ and the $H_m(\alpha x)$ is an Hermite polynominal with normalization constant N_m.

One can then determine the expansion coefficients $C_{vj,m}$ variationally by solving the corresponding set of homogeneous equations:

$$\sum_{n=o}^{Max} \left\{ H_{mn} - \varepsilon_{vj} \delta_{mn} \right\} = 0 \qquad m = 0, 1, \ldots . \text{ Max} . \tag{3.155}$$

where the matrix elements are given by:

$$\begin{aligned} H_{mn} = N_m N_n \Big\{ & \int_{-\infty}^{\infty} H_m(\alpha x) V_0(r) H_n(\alpha x) \exp (-\alpha^2 x^2) dx + \\ & + \frac{1}{2\mu} \int_{-\infty}^{\infty} H_m(\alpha x) [j(j + 1) r^{-2}] H_n(\alpha x) \exp(-\alpha^2 x^2) dx \Big\} \end{aligned} \tag{3.156}$$

and which can be obtained in a closed form for analytical V_0 potentials.

A variety of numerical techniques therefore can be employed to obtain the target eigenfunctions. For instance, the above variational procedure has been used for H_2 [37] and CO [38] targets with Morse potentials as $V_0(r)$, while a direct integration of (3.153) using Numerov's numerical method has been used for H_2 targets with several choices of V_0 potentials [39].

The scattering wavefunctions expansions for the atom-molecule cases discussed in the previous sections can now be rewritten with the inclusion of the above eigenfunctions $\chi_j^v(r)$. For instance, the SF expansion of eq.(3.53) now becomes:

$$\psi_{j\ell}^{EJM} (\underline{R}', \hat{r}') \Rightarrow \psi_{j\ell v}^{EJM} (\underline{R}', \underline{r}') = \psi_{\gamma\ell}^{EJM} (\underline{R}', \underline{r}') =$$

$$= \sum_{j',\ell' v'} u_{j'\ell' v'}^{Jj\ell v} (R') (k_{jv})^{-1} \cdot R'^{-1} \cdot \mathcal{J}_{j'\ell'}^{JM} (\hat{r}', \hat{R}') \cdot \chi_{j'}^{v'}(r') \tag{3.157}$$

$$= \sum_{\ell',\gamma'} R'^{-1} \cdot u^{J\ell\gamma}_{\ell'\gamma'}(R')(k_\gamma)^{-1} \cdot \chi_{\gamma'}(r') \mathcal{Y}^{JM}_{j'\ell'} (\hat{r}', \hat{R}') \tag{3.157}$$

where: $\gamma = (j, v)$ is the subscript for the channel index denoting the initial state of the diatomic target.

The coupling matrix elements of eq. (3.60) are correspondingly modified in the following way:

$$\begin{aligned} &< \alpha'\ell' \,|V|\, \alpha''\ell'' >^J \Rightarrow < \gamma'\ell' \,|V|\, \gamma''\ell'' >^J = \\ &= 2\mathcal{M} \iiint d\hat{R}'\, d\hat{r}'\, dr\, \mathcal{Y}^{JM\,*}_{\ell'\chi'} \cdot \chi_{\gamma'}(r)\, V(\underline{r}', \underline{R}') \mathcal{Y}^{JM}_{\ell''\gamma''}\, \chi_{\gamma''}(r) \end{aligned} \tag{3.158}$$

where one now has to solve a triple integral, with a further integration over the internal molecular variable r. Such an integration has to be performed for each R' value needed in the radial coupled differential equations (3.58), whose radial solutions will provide the A_L elements of (3.74), which now also depend on γ.

Entirely similar considerations can be carried out when rewriting the BF expansion of eq.(3.117), where the treatment remains unaltered except for the substitution of the internal rotational state index j with the vibrational-rotational channel index γ defined above. The computational complexity is however greatly increased, for the number of rotational channels that have to be included when vibrational levels are also energetically accessible goes up very rapidly even for the widely spaced levels of the H_2 molecule and one is soon confronted with a very large number of coupled equations to be solved. It therefore becomes necessary to resort to simplyfying treatments of the vector coupling algebra, which hopefully would also reflect the simpler behaviour of certain systems as suggested by more general physical considerations, thus striving to provide useful links betweeen theoretical modelling and actual outcomes of realistic experiments. This aspect will however be discussed in greater detail in the following Chapter.

3.7 - Molecule-molecule inelastic encounters

The collisional transfer of vibrational and rotational energy between pairs of diatomic molecules plays an important role in most gas laser systems [40] . Thus, it has been the goal of theory for many years to completely understand these processes [41, 42]. Unfortunately, even with present day computer facilities, it is impossible to calculate accurately tha quantum mechanical transition probabilities between all the initial and final vibration-rotation states of the two colliding partners. Consequently, the usual approach is forced to resort to some kind of simplifying approximation. This could be done by retaining the multidimensional charecter of the collision while at the same time reducing the complexity of both the dynamics and the potential surface of interaction. Several interesting studies have appeared along these lines [43, 44] and will be examined later.

Alternatively, one can choose a more rigorous treatment of the dynamics and a farily accurate form of the potentials while at the same time rendering the collisional problem tractable by eliminating the rotational degrees of freedom. Traditionally this second approach has formed the basis for a great deal of work on atom-molecule vibrationally inelastic collisions [45, 46].

The initial treatment of the full problem requires the definition of a total Hamiltonian that closely follows eq.(2.16):

$$\mathcal{H} = -\frac{1}{2M}\nabla^2_{\underline{R}'} + \mathcal{H}_{int}(\underline{X}_1) + \mathcal{H}_{int}(\underline{X}_2) + V(\underline{R}', \underline{X}_1 \underline{X}_2) \tag{3.159}$$

where $\underline{X}_1$ and $\underline{X}_2$ define the internal coordinates of the colliding molecules. For diatomic systems, which will be those discussed here in some detail, $\underline{X}_1 = \underline{r}'_1$ and $\underline{X}_2 = \underline{r}'_2$. It will further be assumed that there is no electromic angular momentum component for each molecule along its internuclear axis (Σ states). The extension to polyatomic systems and to asymmetric tops can easily be done, although the notation would become more complicated. The unit vectors $\hat{r}_i$, i = 1, 2 describe each molecular bond orientation in an SF frame of reference (x', y', z'), as discussed earlier, while the vector $\underline{R}'$ connects the molecular centres of mass and is also defined with respect to the same SF-frame.

Each $\mathcal{H}_{int}(\underline{r}_i)$ admits orthonormal eigensolutions according to eq. (3.153). They are here written in the form:

$$\varphi_i(v_i, j_i; \underline{r}'_i) = \chi_i(\gamma_i; r'_i)\, Y_{j_i}^{m_{j_i}}(\hat{r}'_i) \qquad r = 1, 2 \tag{3.160}$$

with: $|\gamma_i\rangle = |v_i j_i\rangle$.

The total wavefunction could be written as a brute force expansion over the (3.160) basis set with functions of $\underline{R}$ as coefficients, the latter being also expanded over the corresponding spherical harmonics $Y_\nu^{m_\nu}(\hat{R}')$:

$$\Psi_{tot} = \sum_{\substack{\gamma_1,\gamma_2 \\ m_{s_1},m_{s_2} \\ \nu,m_\nu}} (k_\alpha R')^{-1} \cdot f(\gamma_1, \gamma_2, m_{s_1}, m_{s_2}, \nu, m_\nu |R') \cdot \tag{3.161}$$

$$\cdot\, \chi_1(\gamma_1; r_1') \cdot \chi_2(\gamma_2; r_2')\, Y_{s_1}^{m_{s_1}}(\hat{r}_1')\, Y_{s_2}^{m_{s_2}}(\hat{r}_1')\, Y_\nu^{m_\nu}(\hat{R}')$$

where the k_α wave vector of the relative motion is defined as:

$$k_\alpha^2 = 2\mathcal{M}\{E_{tot} - E_1 - E_2\} \quad \text{and} \quad \alpha = |\gamma_1\, \gamma_2> \tag{3.162}$$

A more rapidly converging basis set could be defined, still within the unperturbed expansion limits discussed in earlier parts of this Chapter, by choosing a coupled representation for the target molecular states as already done in eq.s (3.53), (3.54) and (3.100), (3.115) for the atom-molecule cases of the previous Sections:

$$\Psi_{tot}(\underline{r}_1', \underline{r}_2', \underline{R}') = \sum (k_\alpha R')^{-1}\, C^{JM}_{\substack{j,j_2j \\ v,v_2\ell}}\, u^{JMj_1j_2j\, v_1v_2\ell}_{j_1'j_2'j'v_1'v_2'\ell'}(R') \cdot \tag{3.163}$$

$$\cdot\, \chi_1(\gamma_1'; r_1') \cdot \chi_2(\gamma_2'; r_2')\, \mathcal{Y}^{JM}_{j_1'j_2'j'\ell'}(\hat{r}_1', \hat{r}_2', \hat{R}')$$

where the vector-coupling provides the following additional vectors:

$$\begin{aligned} \underline{\ell} &= \underline{\ell}_1 + \underline{\ell}_2 \\ \underline{j} &= \underline{j}_1 + \underline{j}_2 \\ J &= \underline{\ell} + \underline{j} \end{aligned} \tag{3.164}$$

and the sum in (3.163) is over all the indeces on the r.h.s. and M is the J projection along the z'-axis (in $\hbar$ units).

The $\mathcal{Y}$'s are given by the couplings of the angular spherical harmonics included in (3.161) original expansion:

$$\mathcal{J}^{JM}_{j_1 j_2 j \ell}(\hat{r}'_1, \hat{r}'_2, \hat{R}') = \sum_{\substack{m_{j_1}, m_{j_2} \\ m_{j_1}, M-m_j}} Y_{j_1}^{m_{j_1}}(\hat{r}'_1)\, Y_{j_2}^{m_{j_2}}(\hat{r}'_2)\, Y_j^{M-m_j}(\hat{R}') \cdot \tag{3.165}$$

$$\cdot\, (\ell_j m_j\, M - m_j | \ell_j JM) \cdot (j_1, j_2\, m_{j_1}\, m_{j_2} | j_1\, j_2\, j\, m_j)$$

where the terms in brackets are Clebsch-Gordan coefficients [42] . Moreover: $M-m_j=m_\ell$ and: $m_j = m_{j_1} + m_{j_2}$. The sum (3.165) is over all the $\underline{j}$ and $\underline{\ell}$ projections.

In order to carry out the separation of variables in the scattering equations, the interaction potential should also be expanded in multipolar coefficients of the correct symmetry. In the BF-frame of reference one can write, for instance, that:

$$V_{BF}(\underline{r}_1, \underline{r}_2, \underline{R}) = \sum_{\lambda_1, \lambda_2, \lambda} V^{\lambda}_{\lambda_1, \lambda_2}(r_1, r_2, R)\, P_{\lambda_1}(\cos\theta_1)\, P_{\lambda_2}(\cos\theta_2)\, P_{\lambda}(\cos\theta_{12}) \tag{3.166}$$

where:

$$\cos\theta_1 = \hat{r}_1 \cdot \hat{R} \quad ; \quad \cos\theta_2 = \hat{r}_2 \cdot \hat{R} \quad ; \quad \cos\theta_{12} = \hat{r}_1 \cdot \hat{r}_2 \tag{3.167}$$

provide the internal angular coordinates of two molecules. Eq.(3.166) can alternatively be written by expressing the angular part in terms of Racah coefficients [47] and referring the potential to the SF-frame:

$$V_{SF}(\underline{r}'_1, \underline{r}'_2, \underline{R}') = \sum_{\substack{\ell_1, \ell_2, \ell \\ m_1, m_2, m}} V^{\ell}_{\ell_1, \ell_2}(r'_1, r'_2, R')(\ell_1 m_1 \ell_2 m_2 \mid \ell_1 \ell_2 \ell\, m) \cdot \tag{3.168}$$

$$\cdot\, Y_{\ell_1}^{m_1}(\hat{r}'_1) \cdot Y_{\ell_2}^{m_2}(\hat{r}'_2)\, Y_{\ell}^{m}(\hat{R}')$$

where the Clebsch-Gordan coefficients [42] are used to express the coupling terms of the angular part of eq. (3.166) [51] :

$$P_{\lambda_i}(\cos\theta_i) = \frac{4\pi}{2\lambda_i+1}\sum_{m_i} Y_{\lambda_i}^{m_i}(\hat{r}'_i)\, Y_{\lambda_i}^{m_i}(\hat{R}') \ ; \quad i = 1, 2$$
$$P_{\lambda}(\cos\theta_{12}) = \frac{4\pi}{2\lambda+1}\sum_{m} Y_{\lambda}^{m}(\hat{r}'_1)\, Y_{\lambda}^{m}(\hat{r}'_2) \tag{3.169}$$

The corresponding radial coefficients are therefore related to each other via the following transformation:

$$V^{\ell}_{\ell_1\ell_2}(r'_1, r'_2, R') = \sum_{\lambda_1,\lambda_2,\lambda} V^{\lambda}_{\lambda_1\lambda_2}(r_1, r_2, R) \left\{ \frac{(4\pi)^{3/2}}{[(2\ell_1+1)(2\ell_2+1)]^{\frac{1}{2}}} \right\} \cdot$$
$$\cdot (-)^{\lambda_1+\ell_2+\ell} \begin{pmatrix} \lambda_1 & \lambda & \ell_1 \\ 0 & 0 & 0 \end{pmatrix} \cdot \begin{pmatrix} \lambda_2 & \lambda & \ell_2 \\ 0 & 0 & 0 \end{pmatrix} \cdot \begin{pmatrix} \lambda_1 & \lambda_2 & \lambda \\ 0 & 0 & 0 \end{pmatrix} \cdot \begin{Bmatrix} \lambda_1 & \lambda_2 & \ell \\ \lambda_2 & \lambda_1 & \lambda \end{Bmatrix} . \tag{3.170}$$

where the 3-j symbols have been defined as in eq. (3.82).

By applying the SF treatment of the previous Section, one can now use the wavefunction expansion of (3.163) in the usual eigenvalue equation with the Hamiltonian of eq. (3.159) and thus determine the coefficients C's that satisfy the boundary conditions. Hence:

$$\Psi_{tot} \underset{R\to\infty}{\sim} \exp(i\, \underline{k}_{\alpha} \cdot \underline{R}')\, \chi_1(\gamma_1; r_1)\, \chi_2(\gamma_2; r_2) \cdot Y_{j_1}^{m_1}(\hat{r}'_1) \cdot Y_{j_2}^{m_2}(\hat{r}'_2) +$$
$$+ \sum_{\substack{j'_1, j'_2 \\ m'_1, m'_2 \\ v'_1, v'_2}} \exp[i\, k_{\alpha'} R]/R\ f(j_1 m_1\, j_2 m_2\, v_1 v_2,\, j'_1 m'_1\, j'_2 m'_2\, v'_1 v'_2 \mid \hat{R}') \cdot$$
$$\cdot\, \chi_1(\gamma'_1; r_1)\, \chi_2(\gamma'_2; r_2)\, Y_{j'_1}^{m'_1}(\hat{r}'_1)\, Y_{j'_2}^{m'_2}(\hat{r}'_2) \tag{3.171}$$

where the wavevectors $\underline{k}_{\alpha}$ and $\underline{k}_{\alpha'}$ are those given in eq.(3.162). The corresponding differential and integral cross sections are defined through the above scattering amplitude in a way entirely similar to the one shown for atom-molecule encounters and reported by eq.s (3.74) and (3.77). The radial functions u's satisfy the following set of coupled differential equations:

$$\left\{ \frac{d^2}{dR'^2} - \frac{\ell'(\ell'+1)}{R'^2} + k^2_{\alpha'} \right\} u^{JMj_1j_2jv_1v_2\ell}_{j_1'j_2'j'v_1'v_2'\ell'}(R') = \tag{3.172}$$

$$= 2M \sum_{\substack{\text{all"} \\ \text{indeces}}} < j_1'j_2'j'v_1'v_2'\ell'J \,|V|\, j_1''j_2''j''v_1''v_2''\ell''J > \cdot u^{JMj_1j_2jv_1v_2\ell}_{j_1''j_2''j''v_1''v_2''\ell''}(R')$$

The whole physics of the collisional distribution of flux over the energy-allowed final channels is controlled by the anisotropic potential matrix elements which can be obtained in detail by using standard angular momentum coupling techniques, as recently presented in an excellent review [48] :

$$< j_1j_2j\; v_1v_2\ell\; J \,|V|\, j_1'j_2'j'v_1'v_2'\ell'\; J > =$$

$$= \sum_{\underline{\ell}_1,\underline{\ell}_2,\underline{\ell}} (4\pi)^{-3/2}(-)^{j_1+j_2+j_1'+J} \left\{ [\underline{\ell}]^2 + [\underline{\ell}_1,\underline{\ell}_2,j_1,j_2,j,\ell,j_1',j_2',j_1'\ell'] \right\}^{\frac{1}{2}} \cdot$$

$$\cdot \begin{pmatrix} \underline{\ell} & \ell' & \ell \\ 0 & 0 & 0 \end{pmatrix} \cdot \begin{pmatrix} \underline{\ell}_1 & j_1' & j_1 \\ 0 & 0 & 0 \end{pmatrix} \cdot \begin{pmatrix} \underline{\ell}_2 & j_2' & j_2 \\ 0 & 0 & 0 \end{pmatrix} \cdot \tag{3.173}$$

$$\cdot \begin{Bmatrix} \ell' & \ell & \underline{\ell} \\ j & j' & J \end{Bmatrix} \cdot \begin{Bmatrix} j' & j_2' & j_1' \\ j & j_2 & j_1 \\ \underline{\ell} & \underline{\ell}_2 & \underline{\ell}_1 \end{Bmatrix} \cdot \int dr_1\, dr_2\, \chi_1(v_1j_1;\, r_1')\, \chi_2(v_2j_2;\, r_2') \cdot$$

$$\cdot V^{\underline{\ell}}_{\underline{\ell}_1\underline{\ell}_2}(r_1',\, r_2',\, R')\, \chi_1(v_1'j_1';\, r_1')\, \chi_2(v_2'j_2';\, r_2)$$

The last integral on the r.h.s. contains the specifie coupling afforded by coefficient of the potential expansion of eq.(3.168). Its indeces, over which the above sumation extends have been underlined to avoid confusion with the coupling vector ℓ defined in (3.164).

The last angular term an the r.h.s. is a 9-j symbol; moreover, the compact notation implied by the square brackets has been used, i.e.:

$$[a_1,\, a_2,\, a_3,\, \ldots\ldots\, a_n] = (2a_1 + 1) \cdot (2a_2 + 1) \cdot \ldots\ldots\, (2a_n + 1) \tag{3.174}$$

The asymptotic behaviour of the functions allows the definition of the scattering S-matrix for each total angular momentum state $|JM>$ that in the present representation is conserved throughout the collision:

$$u(R') \underset{R'\to\infty}{\sim} \delta_{j_1 j'_1} \cdot \delta_{j_2 j'_2} \cdot \delta_{jj'} \cdot \delta_{\ell\ell'} \cdot \delta_{v_1 v'_1} \cdot \delta_{v_2 v'_2} \exp[-i(k_\alpha R' - \ell\pi/2)]$$

(3.175)

$$- \left(\frac{k_\alpha}{k_{\alpha'}}\right)^{\frac{1}{2}} \cdot S^J(j'_1 j'_2 j' \ell' v'_1 v'_2, j_1 j_2 j\ \ell\ v_1 v_2) \cdot \exp[i(k_{\alpha'} R' - \ell\pi/2)]$$

The usual matching, as defined in the previous section, of the incoming part of Ψ_{tot} with the required asymptotic form of (3.193) will determine the coefficients C once the radial functions of (3.171) have allowed us to obtain the S-matrix structure.

Such a matrix will then yield the detailed expression of the scattering amplitude that can be written aut as follows:

$$q(j_1\ m_{j_1}\ j_2\ m_{j_2}\ v_1\ v_2,\ j'_1\ m'_{j'_1}\ j'_2\ m_{j'_2}\ v'_1\ v'_2 \mid \hat{R}') =$$

$$= \pi^{\frac{1}{2}} \frac{i}{(k_\alpha\ k_{\alpha'})^{\frac{1}{2}}} \cdot \sum_{\substack{J\ell\ell' m\ell' \\ jj' m_j m_{j'}}} i^{(\ell - \ell')} [\ell]^{\frac{1}{2}} (j_1\ m_{j_1}\ j_2\ m_{j_2} \mid j_1\ j_2\ jm_j) \cdot$$

(3.176)

$$\cdot (\ell o\ j\ m_j | \ell_j JM) \cdot (j'_1 m_{j'_1}\ j'_2 m_{j'_2} | j'_1 j'_2 j' m_{j'}) \cdot (j' m_{j'}\ \ell' m_{\ell'} | j' \ell'\ JM) \cdot$$

$$\cdot Y_{\ell'}^{m_{\ell'}}(\hat{R}') \cdot \left\{ \delta_{(j_1 j'_1,\ j_2 j'_2,\ jj',\ \ell\ell',\ v_1 v'_1,\ v_2 v'_2)} - S^J(j'_1 j'_2 j' \ell' v'_1 v'_2,\ j_1 j_2 j \ell v_1 v_2) \right\}$$

This in turn yields the total inelastic cross section summed over final $m_{j'_i}$ values and averaged over all the initial m_{j_i} values:

$$\sigma(j_1 j_2 v_1 v_2 \to j'_1 j'_2 v'_1 v'_2) = \frac{\pi}{k_\alpha^2 [j_1, j_2]} \cdot \sum_{\substack{J,\ell,\ell' \\ j'_1, j_2}} [J] \left|\{\delta - S^J\}\right|^2 \quad (3.177)$$

The number of coupled equations is now clearly much larger that for the atom-molecule case since it is now given by: $N_C^J = N_1 \cdot N_2$, where:

$$N_i = (n_i^{max} + 1) \cdot (j_i^{max} + 1)^2 \qquad (3.178)$$

and the indeces refer to the highest rotational and vibrational levels included in the expansion (3.163) for each index i = 1, 2.

In the following Chapter we will therefore examine some the physical situations which can lead to simplified formulations of the extremely complex problem of dealing with several coupled equations. The ultimate decision, however, has to be related to the specific system under consideration and to the nature of the various coefficients appearing in the potential expansion (3.168).

Let us now turn to a problem which has not get been examined so far, i.e. the external requirements on w.f.'s and their effects on the sets of coupled equations that need to be solved. One can begin by considering the simplest situation of atom-atom collisions involving identical particles. The total wavefunctions is written down as a direct product of various contributions:

$$\Psi_{tot} = F_\Gamma(\underline{R})\, \chi_\Gamma(\underline{S}_A, \underline{S}_B)\, \Psi_\Gamma^{el}(\underline{r}_i\, \underline{s}_i;\, \underline{R}) \qquad (3.179)$$

where the F_Γ is the radial function, χ_Γ the nuclear spin function with coordinates $\underline{S}_A$, $\underline{S}_B$ and the electromic interaction Ψ_Γ^{el}, computed within the familiar B.O. scheme for a given electronic state Γ, depends on electronic coordinates $(\underline{r}_i, \underline{s}_i)$ for each $\underline{R}$. The exchange of position vectors $\underline{R}_A$, $\underline{R}_B$ defined in an SF frame of reference yields: $\underline{R} = -\underline{R}$. The partial wave expansion of $F_\Gamma(\underline{R})$, on the other hand, is given, for each wavevector value $\underset{\sim}{k}$, by the familiar expression:

$$\Gamma_\Gamma\, (\underline{R}) = \sum_\ell i^\ell\, (2\ell+1)\, f_\ell(R)\, P_\ell(\hat{R} \cdot \hat{k}) \qquad (3.180)$$

and since: $P_\ell(\hat{R} \cdot \hat{k}) = (-)^\ell\, P_\ell(-\hat{R} \cdot \hat{k})$ one obtains contributions to (3.180) both from the even and old values of ℓ:

$$\begin{aligned} F_\Gamma^g(R) &= F_\Gamma^g(-R) \\ F_\Gamma^u(R) &= -F_\Gamma^u(-R) \end{aligned} \qquad (3.181)$$

In other words, half of the partial waves of relative motion are gerade (ℓ even) and half are ungerade (ℓ odd) with respect to the interchange of the space coordinates of the nuclei.

The interaction potential generated by the electronic part, Ψ_Γ, follows the symmetry under inversion of the irreducible representation of $C_{\infty h}$ to which Γ belongs. Hence, for Σ_g^+, Σ_u^-, Π_g and Δ_g interaction curves, one finds, that:

$$\Psi_\Gamma^{el}(\underline{r}_i\ \underline{s}_i;\ R) = \Psi_\Gamma^{el}(\underline{r}_i\underline{s}_i;\ -R) \qquad (3.182a)$$

while for the Σ_g^-, Σ_u^+, Π_u and Δ_u curves:

$$\Psi_\Gamma^{el}(\underline{r}_i\ \underline{s}_i;\ R) = -\Psi_\Gamma^{el}(\underline{r}_i\underline{s}_i;\ R) \qquad (3.182b)$$

For the nuclear spin wavefunctions, one can construct symmetric or antisymmetric wavefunctions according to standard rules that apply to particles with arbitrary instrinsic spins. These rules will provide different relative weights according to the type of statics that is obeyed by the overall system and that depends on the number of nucleons present. For an even number of Fermions, the colliding systems obey the Bose - Einstein statistics under exchange of the nuclear position coordinates, while for an odd number of Fermions, they obey the Fermi-Dirac statistics [12]. Let us start with with the scattering between two ground state He^4 atoms which come together as a $^1\Sigma_g^+$ molecular state. There is zero nuclear spin and thus the Pauli principle requires the total wavefunction to be symmetric in the interchange of the nuclei. Eq.(3.182) tells us that Ψ_Γ^{el} is symmetric and hence $F_\Gamma(R)$ must be symmetric. It then follows that the incident wave must be given by:

$$F_\Gamma^g(R) = \sqrt{2} \sum_{\ell \text{even}} i^\ell (2\ell+1)\, j_\ell(kR)\, P_\ell(\hat{R}\cdot\hat{k}) \qquad (3.183)$$

which i the form of the expansion for a plane wave, except that now only even ℓ values are included. The factor $\sqrt{2}$ is introduced so that its incident flux density will be the same as that of $e^{i\underline{k}\cdot\underline{R}}$, hence it compensates for the removal of half the flux originating from the neglect of the odd ℓ values.

In the case of systems where more than one potential curve plays a role in the collision, and where one has no role of the Pauli principle because of zero total spin, both contributions have to appear in the total wavefunction. For example, if in the previous case one of the He^4 atoms is an ion, there is an additional molecular state for the He_2^+ molecule, the $^2\Sigma_u^+$, which is ungerade and which therefore brings back the odd ℓ contributions:

$$\Psi_{tot} = F^g_{^2\Sigma_g^+}(R)\,\Psi^{el}_{^2\Sigma_g^+}(\underline{r}_i\,\underline{s}_i;\,R) + F^u_{^2\Sigma_g^+}(R)\,\Psi^{el}_{^2\Sigma_g^+}(\underline{r}_i\,\underline{s}_i;\,R) \tag{3.184}$$

Finally, when the total nuclear spin is $\neq 0$, the Pauli principle weighs the g and u contributions according to the proper spin statistics. An example could be given by the elastic scattering of two ground state H^1 atoms. If one disregards the repulsive $^3\Sigma_u^-$ state, as the scattering is considered at low relative speeds and the internal degrees of freedom are included in the F's one can write then:

$$\begin{aligned}\Psi_{tot} = \frac{1}{2}\; & F^g_{^1\Sigma_g^+}(\underline{R})\,\Psi^{el}_{^1\Sigma_g^+}(\underline{r}_i\,\underline{s}_i;\,\underline{R})\,\chi_A(\underline{S}_A,\,\underline{S}_B) + \\ + \frac{1}{2}\; & F^u_{^1\Sigma_g^+}(\underline{R})\,\Psi^{el}_{^1\Sigma_g^+}(\underline{r}_i\,\underline{s}_i;\,\underline{R})\,\Sigma\,\chi_s(\underline{S}_A,\,\underline{S}_B)\end{aligned} \tag{3.185}$$

where the Σ symbol for the nuclear spin function χ_s indicates a sum over the three spin functions of ortho-H_2.

The Pauli principle then requires the application of Bosons statistics for the interchange of the nuclear coordinates of the H_2 molecule. Hence, since the vibrational wavefunctions are always symmetric, the rotational behaviour on the interchange will control the symmetry via the following relationship (for a spherical top):

$$\varphi^{rot}_{JM}(\hat{\mathbf{R}}) = (-)^J\,\varphi^{rot}_{JM}(-\hat{\mathbf{R}}) \tag{3.186}$$

From eq. (3.185) it therefore results that the para -H_2 molecule is populated by even rotational states, while the ortho-H_2 molecule is populated by odd rotational states.

When one is dealing with collisions between identical molecules, the relevant intermolecular potential has been calculated in principle from an antisymmetrical electronic wavefunction and is usually the lowest-lying potential energy surface. For closed-shell systems it is also very often of A_1 symmetry, which means that it does not change under the interchange of the molecular skeleton. Eq.(3.179) therefore concentrates on the nuclear spin wavefunctions and on the spatial wavefunction that is given by a product of translational, vibrational and rotational wavefunctions. For low energy collisions one is also usually disregarding the interchange of nuclei between molecules [50]. One also assumes that the resultant total nuclear spin quantum numbers remain constant throughout the collision.

If I_1 and I_2 are such numbers for the two colliding molecules and $I_1 = I_2$, we can treat the two molecules as dissimilar particles and there is no role played by the Pauli principle. If $I_1 = I_2 = I$, then there are $(2I + 1)^2$ possible nuclear spin states for the total system, of which $(I + 1) \cdot (2I + 1)$ are symmetric and $I \cdot (2I + 1)$ are antisymmetric with respect to interchange of the skeletons. In eq. (3.185) for instance, this rule gives us one H_2 state which is antisymmetric (the singlet state of the para-H_2) and three states that are symmetric in the ortho-H_2 molecule. The observed cross section is then the spin statistics weighted linear combination of the cross section σ_g and σ_u, where the g and u symbols refer to the symmetry of the spatial part of the wavefunction. The values of the weights will thus depend on whether the systems follow the Fermions or the Bosons statistics.

For Bosons, the possibilities (or weights) are given by the ratios of the previously discussed multiplicities:

$$W_g = \frac{(I + 1)}{(2I + 1)} \quad ; \quad W_u = \frac{I}{(2I + 1)} \tag{3.187a}$$

While for Fermions these weights become instead:

$$W_g = \frac{I}{(2I + 1)} \quad ; \quad W_u = \frac{(I + 1)}{(2I + 1)} \tag{3.187b}$$

Thus, the resulting differential cross section for elastic scattering of two H^1 atoms in their ${}^1\Sigma_g^+$ state is given by:

$$\frac{d\sigma}{d\Omega} = \frac{1}{4} \left| f^g_{{}^1\Sigma_g^+}(\hat{R}') \right|^2 + \frac{3}{4} \left| f^u_{{}^1\Sigma_g^+}(\hat{R}') \right|^2 \tag{3.188}$$

and there is no interference between any of the amplitudes because of the orthogonality of the electron and nuclear spin functions. In a general case, if the incident molecules are unpolarized, the observed cross section can thus be written as:

$$\sigma_{obs} = W_g \cdot \sigma_g + W_u \cdot \sigma_u \tag{3.189}$$

where the σ_g and σ_u contributions are obtained from the properly symmetrized wavefunctions $\Psi^{\pm}_{tot}$.

An alternative procedure is to treat the molecules as distringuishable in order to obtain an unsymmetrized scattering amplitude and to produce the example of eq.(3.188). It is however computationally easier to calculate q^g and q^u directly from Ψ^+_{tot} and Ψ^-_{tot} since their properly symmetrized angular parts form mutually orthogonal sets of functions which yield coupled equations of lower dimensionality when compared to the full, unsymmetrized equations.

The symmetrized wavefunction for two identical molecules can be obtained from expansion (3.163) by writing:

$$\Psi^{\pm}_{tot}(\underline{r}'_1, \underline{r}'_2, \underline{R}') = \sum{}' (k_\alpha R')^{-1} \; C^{\pm JM}_{\substack{j_1 j_2 j \\ v_1 v_2 v}} \cdot u^{\pm\, JM j_1 j_2 j\, v_1 v_2 \ell}_{j'_1 j'_2 j' v'_1 v'_2 \ell'}(R') \cdot \tag{3.190}$$
$$\cdot\; I^{\pm\, JM}_{\substack{j'_1 j'_2 j' \ell' \\ v'_1\, v'_2}} (\underline{r}'_1, \underline{r}'_2, \hat{R}')$$

where:

$$I^{\pm JM}_{\substack{j_1 j_2 j \\ v_1 v_2 \ell}} = \left\{ \frac{1}{\sqrt{2(1 + \delta_{v_1 v_2}\delta_{j_1 j_2})}} \right\} \cdot \Big\{ \chi_1(\gamma_1; r'_1)\, \chi_2(\gamma_2; r'_2) \cdot \tag{3.191}$$
$$\cdot\; \mathcal{J}^{JM}_{j_1 j_2 j \ell}(\hat{r}'_1, \hat{r}'_2, \hat{R}') \pm (-)^{j_1 + j_2 - j + \ell} \cdot \chi_2 \cdot \chi_1 \cdot \mathcal{J}^{JM}_{j_2 j_1 j \ell}(\hat{r}'_1, \hat{r}'_2, \hat{R}') \Big\}$$

and the summation in eq. (3.191) is indicated as $\sum{}'$ to be carried out over well ordered values of the parameters: $v_1 < v_2$ or $v_1 = v_2$ and $j_1 \leq j_2$. This means that [42]: $|a_1 a_2>$ corresponds to: $v_1 < v_2$ or $v_1 = v_2$ and $j_1 \leq j_2$, while $|a_2 a_1>$ corresponds to: $v_1 > v_2$ or $v_1 = v_2$ and $j_1 > j_2$.

The corresponding I coefficients vanish in the expansion (3.190) between states of different parity $(-)^{j_1+j_2+\ell}$ [49] and the coupled equations (3.172) now become:

$$\left\{ \frac{d^2}{dR'^2} - \frac{\ell'(\ell'+1)}{R'^2} + k^2_{\alpha\ell} \right\} u^{\pm JM}(j_1 j_2 j v_1 v_2 \ell | R') =$$

$$= 2\mathcal{M} \sum_{\substack{\text{primed}\\ \text{indeces}}}{}' < j_1 j_2 j v_1 v_2 \ell | V | j_1' j_2' j' v_1' v_2' \ell' >^{\pm}_{J} \cdot \tag{3.192}$$

$$\cdot\, u^{\pm JM}(j_1' j_2' j' v_1' v_2' v' \ell' | R')$$

where the symmetrized matrix elements vanish between channels of different parity. One can rewrite in a more campact form the interaction potential given by the multipolar expasion in the SF frame of eq. (3.168):

$$V(\underline{r}_1', \underline{r}_2', \underline{R}') = \sum_L V_L(r_1, r_2, R)\, \mathcal{P}_L(\hat{r}_1', \hat{r}_2', \hat{R}') \tag{3.193}$$

where $L(\mu)$ represents the triple index $\ell_1\ell_2\ell(m_1 m_2 m)$ and the $\mathcal{P}_L$ are given by:

$$\mathcal{P}_L(\hat{r}_1', \hat{r}_2', \hat{R}') = \sum_\mu (\ell_1 m_1\ \ell_2 m_2 | \ell_1 \ell_2 \ell\ m)\, Y_{\ell_1}^{m_1}(\hat{r}_1') \cdot \tag{3.194}$$

$$\cdot\, Y_{\ell_2}^{m_2}(\hat{r}_2') \cdot Y_\ell^m(\hat{R}')$$

Moreover, L must be an even integer for identical molecules and the radial coefficients must satisfy the relation [51]:

$$V^\ell_{\ell_1\ell_2} = (-)^{\ell_1+\ell_2}\, V^\ell_{\ell_2\ell_1}$$

If one now defines $\bar{L}$ as: $(\ell_2\ell_1\ell)$, eq. (3.193) can be rewritten as:

$$V(\underline{r}_1', \underline{r}_2', \underline{R}') = \sum_L V_L \cdot \left\{ \mathcal{P}_L + (-)^{\ell_1+\ell_2} \mathcal{P}_{\bar{L}} \right\} \tag{3.196}$$

and then the coupling matrix elements of eq. (3.173) can also be rewritten in a more compact form that allows for the further inclusion of symmetry effects:

$$< j_1 j_2 j\ v_1 v_2 \ell | V | j_1' j_2' j\ v_1' v_2' \ell >_J =$$

$$= \sum_L \left\{ f_L^J(j_1 j_2 j,\ j_1' j_2' j_1' \ell\ell') + (-)^{\ell_1 + \ell_2}\ \bar{f}_L^J\ (j_1 j_2 j,\ j_1' j_2' j_1' \ell\ell') \right\} \cdot$$

$$\cdot \iint dr_1\ dr_2\ \chi_1(v_1 j_1;\ r_1)\ \chi_2(v_2 j_2;\ r_2)\ V_L(r_1,\ r_2,\ R)\ \chi_1(v_1' j_1';\ r_1') \cdot \qquad (3.197)$$

$$\cdot\ \chi_2(r_2'\ j_2';\ r_2) = \sum_L\ A_L^J(j_1 j_2 j,\ v_1 v_2 \ell;\ j_1'\ j_2' j\ ,\ v_1'\ v_2'\ \ell')$$

where the f's are given by the 3-j, 6-j and 9-j symbols and coefficients of the r.h.s. of eq. (3.173).

The properly simmetrized coupling matrix elements of the radial equations (3.192) can now obtained via the previously defined expressions:

$$< j_1 j_2 j\ v_1 v_2 \ell | V | j_1' j_2' j' v_1' v_2' \ell' >_J^{\pm} =$$

$$= \{(1 + \delta_{v_1 v_2} \cdot \delta_{j_1 j_2}) \cdot (1 + \delta_{v_1' v_2'} \cdot \delta_{j_1' j_2'})\}^{-\frac{1}{2}} \cdot \qquad (3.196)$$

$$\cdot \sum_L\ A_L^J(j_1 j_2 j, v_1 v_2 \ell;\ j_1' j_2' j; v_1' v_2' \ell') + (-)^{j_1 + j_2 - j + \ell}\ \bar{A}_L^J\ (j_2 j_1 j,\ v_1 v_2 v_1 \ell; j_1' j_2' j', v_1' v_2' \ell')\}$$

where the rotational and vibrational levels obey the previously discussed criterion of a well ordered set.

The asymptotic behaviour of the radial solution $u^{\pm}(R)$ defines the corresponding scattering matrix $\underline{\underline{S}}^{\pm}$ in a manner entirely similar to eq.(3.175). Thus, after averaging over initial m_{j_i} values and summing over the final $m_{j'_i}$ values, one recovers an expression for the inelastic vibro-rotational cross section in the symmetrized scheme for indentical particles, in analogy with eq.(3.177):

$$\sigma_{g,u}(j_1 j_2\ v_1 v_2 \rightarrow j_1' j_2'\ v_1' v_2') = \frac{\pi}{k_\alpha^2\ [j_1,\ j_2]} \cdot (1 + \delta_{j_1 j_2} \cdot \delta_{v_1 v_2}) \cdot$$

$$\cdot\ (1 + \delta_{j_1' j_2'} \cdot \delta_{v_1' v_2'}) \cdot \sum_{Jjj'\ell\ell'} [J] \cdot\ |(\underset{\sim}{\delta} - \underset{\sim}{S}^J)|^2 \qquad (3.199)$$

where all the indeces over which the last sum on the r.h.s. of (3.199) runs have been omitted for simplicity in writing the standard S-matrix element $\underset{\sim}{S}^{J}$.

The final cross sections are in turn given by statistically weighing of the exchange parities as in eq.(3.189). For instance, we have already seen that hydrogen molecules are either p-H_2 if I = 0 or o-H_2 if the nuclear spin I = 1. The proper total symmetry requires p-H_2 in order to have even rotational quantum numbers and o-H_2 to have odd j-values. Furthermore these forms interconvert at a negligible rate in thermal energy collisions and, for all pratical purposes, can be considered as dynamically distinguishable species. Thus, collisions between p-H_2 and o-H_2 can be treated as collisions between different particles. Collisions between p-H_2 molecules, on the other hand, involve identical, non-distinguishable, particles and the corresponding statistical weights are:

$$w_g^{para} = 1 \qquad \text{and} \qquad w_u^{para} = 0 \tag{3.200}$$

That is, for p-H_2 the total symmetry requirements allows only even exchange symmetry, which implies that only this subset of coupled equations need be considered. In collisions between two o-H_2 molecules, the statistical weights are given by:

$$w_g^{ortho} = 2/3 \qquad \text{and} \qquad w_u^{ortho} = 1/3 \tag{3.201}$$

and both the even and odd symmetry close-coupled equations need to be considered and solved.

3.8 - Applications

The purely quantum-mechanical methods discussed in the previous Sections of this Chapt. have received an increased amount of attention in the last few years, where attempts were made at evaluating integral and differential cross sections (partial and total) for purely rotational inelasticity or for rotavibrational excitations.

The general drawback of such an approach, when rigorously ab initio treatments are used, is that it usually needs to be restricted to relatively low collision energies at which relatively few quantum internal states of the molecules are excited. Moreover, for each partial wave, the time τ_c for solving the close-coupled (CC) equations depends on the number N_c of channels, since $\tau_c \propto N_c^s$ with s between 2 and 3. When considering only open channels, N_c is also determined by the multiplicity and becomes: $N_c \propto N_j^2 \cdot N_v$, where N_j is the number of open rotational states that are assumed to be the same for each of the N_v open vibrational states. It then follows that the computing time for pure rotational excitation increases at least with j_{max}^4, where j_{max} is the highest included rotational state. One of the largest calculations carried out recently have included up to 118 channels [52]. With this upper limit in mind, it becomes apparent that to solve the complete set ot CC equations in order to reach acceptable convergence one has to limit the treatment to light systems, which exhibit widely spaced energy levels in their vibro-rotational structures and require relatively few partial waves for their expansion of the relevant continium states.

Some of the more important and increasingly more used approximations, within a quantum mechanical framework, are therefore discussed in the following Chapter in greater detail, while we outline here some of the cases where accurate, ab initio solutions of the full CC equations have been attempted.

(i) - The He + H_2 system

The He - H_2 system provides the simplest example of an anisotropic interaction between two neutral closed-shell partners and the evaluation of its potential energy surface (PES) has therefore received considerable attention from both experimentalists and theoreticians. SCF studies of the surface have been published by Roberts [53] and by Krauss and Miess [54] and they obviously provided a purely repulsive behaviour. A further calculation with the inclusion of a limited Cl yielded again a purely repulsive form [55]. The use of the IEPA model has instead provided a mixed - form potential [56] whose essential features were later reproduced by an orthogonalized multistructure VB method [57].

From scattering calculations one might begin to assess the range of geometries which need to be explored in order to attain acceptable results for lew temperature relaxation studies. The general implication has been that the $V(\underline{r}, \underline{R})$, with R being the H_2 bond distance, needs to be known at r values as large as 8-10 a_0 and, at any rate, the chosen values must include the classical turning points of the vibrational levels between which the interaction induces a transition.

The excitation of rotational states relies obviously on the knowledge of the anisotropic part of the PES. From a standard multipolar expansion as in eq. (3.80), if one truncates it after the second term, the anisotropy of the interaction can be expressed as:

$$\gamma(R, r) = V_2(R, r) \Big/ V_0(R, r) \qquad (3.202)$$

The ratio γ cannot be easily obtained experimentally, since, unlike other rare gas complexes with H_2, no rotational levels have yet been detected spectroscopically for He-H_2. It is, in fact, only a spectroscopic method that could give us the γ-dependence on R. The theoretical treatment via oscillator strength sum rules provides an asymptotic value of the anisotropy of ~ 0.0903 [58], while <u>ab-initio</u> calculations yield ~ 0.0907 [59]. The computed surfaces allow to deduce γ-values for finite values of r. The extrapolation to $r \to \infty$, however, does not lead to agreement with the accurately predicted asymptotic values.

The Van der Waals geometry favoured by the empirical estimates [60] predicts in fact a C_{2v} structure to be ~ 10% more stable than the linear form, while the IEPA calculation [56] yield a linear Van der Waals complex as ~ 50% more stable than the corresponding C_{2v} structure.

The structure of the interaction potential indicates, however, that the He-H_2 system exhibits rather weak anisotropy, with strong short-range coupling of the internal molecular states and a rapidly decreasing long-range coupling with increasing r values.

Some of the earliest close-coupling calculations were performed for the purely rotational excitation of H_2 via the interaction potential of Gordon and Secrest (GS) [55], which exhibits a rather large anisotropy factor at the H_2 equilibrium geometry, and via the harmonic oscillator approximation [61] . Their results for the $\sigma_{0\to 2}$ rotational excitation are larger than those produced later by the Krauss and Mies P.E.S. (KM) and for Morse oscillators used to describe the target vibrational states [62]. This latter surface shows a much more spherical behaviour but the corresponding scattering results indicate a significant interdependence of vibrational inelasticity on the amount of flux going into rotational channels. Thus, cross sections for vibrational excitations are found to decrease when accompanied by rotational inelasticity, while vibrational de-excitation increases with the presence of rotational excitation. Due to the small basis set used in the CC expansion of [62], their results are expected to be still far from convergence, while being also affected by the rather unrealistic value of the KM surface for the γ parameter.

Further calculation [63] of differential cross sections (DCS) indicate that the elastic ones exhibit almost all of their intensity within the forward 30° in the C.M. scattering angle, this being true even for the elastic cross sections of vibrationally excited targets. The rotationally inelastic angular distributions are, hovewer, strongly peaked in the backward direction at low collision energies ($E_{coll} \leq 0.9$ e V), while becoming nearly isotropic and then forward peaked with higher collision energies.

The vibrationally inelastic DCS also present a similar behaviour, since they start with a backward scattered form at low energies and provide an increasing buildup of the forward peaks when the energy increases. The shift to forward scattering was, in fact, found to be already complete at $E_{coll} \sim 2.0$ eV. With increasing rotational excitations ($\Delta j > 2$) the angular distributions are also found to be increasingly more backscattered. Since the KM P.E.S. is completely repulsive, to transfer increasing amount of $E_{coll.}$ into internal energy requires a larger number of "head-on" type of collisions. Thus, for the scattered region beyond $\theta_{CM} \sim 60°$ the inelastic cross sections are larger than the elastic ones in spite of the weakness of this coupling potential. The lack of higher order anisotropic contributions also explains the rapid decrease in magnitude of rotationally inelastic cross sections with increasing values of Δj.

A comparison with inelastic calculations performed within a classical framework shows that the classical DCS for the $j = 0 \rightarrow 2$ transitions is zero for $\theta_{CM} < 30°$ and then rises up sharply to the quantum mechanical values, where agreement stays good up to larger angles except at 180°. the $j = 0 \rightarrow 4$ DCS also exhibits a similar behaviour. This comparision suggests that whereas the classical torque in the weak potential well is apparently too small to produce a transition, the quantum mechanical uncertainy allows such a transition to happen. This is also consistent with the behaviour of stronger potentials exhibited by ligh systems with ionic interactions, like the Li^+ and H^+ case discussed later on.

Finally, it should be mentioned that the experimental behaviour of relaxation times of this system can only be explained by the inclusion of the attractive Van der Waals tail in the PES; this point however will be discussed in greater detail in a later Chapter.

To summarize what discussed thus far for the He-H_2 system, the results of a nearly exact quantal calculation for He atoms impinging on para-H_2 molecules ($j = 0$) at a relative collision energy of 1.09 e V are reported in Fig.3.3 [68]. The purely elastic cross sections for targets in their ground rotational and vibrational states ($00 \rightarrow 00$) and the rotationally inelastic cross sections ($00 \rightarrow 02$) are plotted in the scale of the left side of the figure. The potential used in the GS P.E.S. [55] with a correction to its long range part to bring it into agreement with elastic scattering experiments.

The inelastic cross sections that contain vibrational excitations or vibro-rotational excitations of the target molecule are plotted on the right-side scale. They are smaller than the previous ones by four orders of magnitude and exhibit the backward peaking behaviour that had been found in low-energy calculations performed within the correct CC treatment [63].

(ii) - The Li^+ - H_2 system

Another system which has been extensively examined in recent years, both from the experimental and theoretical viewpoints, has involved an ion-molecule pair of partners, where both the anisotropic character and the strength of the interaction are expected to be much larger.

The relevant PES used in the scattering studies was obtained from different treatments that were prompted by the early experiments of the Göttingen group [64] and which involved the $Li^+(^1S) + H_2(X^1\Sigma_g^+)$ system. Two of them involved calculations within a conventional SCF framework [65, 66], while the third one used a CI formulation [66]. In each case the standard Legendre polynomials potential expansion with only even order coefficients was performed, and a further fitting in the outer regions ($r \geq 12.0$ a.u.) to the correct,experimental values of the asymptotic terms given by perturbation theory in the non overlapping region [65] was also performed.

The CI-PES exhibits a spherical component which lies consistently slightly below the SCF-PES curve, as expected from the small energy lowering obtained in this system by the inclusion of correlation energy. The V_2 and V_4 components are

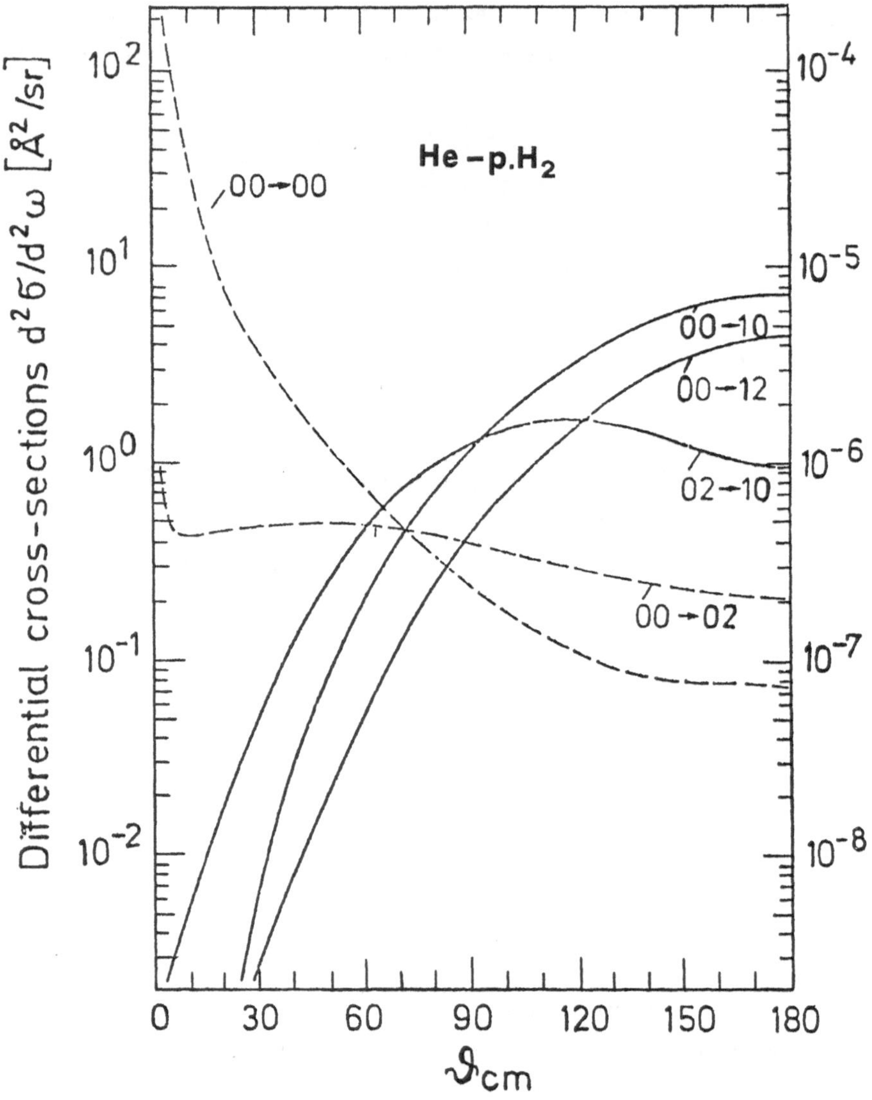

Fig. 3.3

Calculated differential elastic and inelastic cross sections ($|v_i j_i\rangle \rightarrow |v_f j_f\rangle$) for the scattering of H_2 atom from p-H_2 molecule at a relative collision energy E_{CM} = 1.09 eV [from J.P. Toennies, Chem.Soc.Rev., 3, 407 (1974)]. The dashed (cont.)

(Fig.3, cont.) curves are plotted via the left-side scale, while the continuous-line curves refer to the right-side scale.

instead very close to one another, thus confirming the small correlation effect on the surface character.

Various forms of the Schrödinger equation describing the internal molecular motions were also used, e.g.:

$$\left\{ -\left(\frac{h^2}{2\mu}\right) \nabla_R^2 - \varepsilon_i \right\} Y_j^{m_j}(\hat{R}) = 0 \qquad (3.203)$$

where:

$$\varepsilon_j = [j(j+1)]^2 h \cdot c \cdot B \qquad (3.204a)$$

or:

$$\varepsilon_j = j(j+1)\, h \cdot c \cdot B_0 - [j(j+1)]^2\, h \cdot c \cdot D_0 \qquad (3.204b)$$

where μ is the molecular reduced mass, B the rotational constant and B_0, D_0 respectively refer to the rotational constant and centrifugal dimostration correction constant for the v=0 molecular state. The eq. s (3.204a and b) then describe the rigid rotor (RR) equation and the energy corrected rigid rotor equation (ECRR).

Moreover, a more general vibrating rotor (VR) equation was also employed for the H_2 target, with ε_{vj} and χ_{vj} describing its vibrational-rotational energy levels and wavefunctions respectively:

$$\left\{ -\frac{h^2}{2\mu} \nabla_R^2 + V_0(R) - \varepsilon_{vj} \right\} \chi_{vj}(R)\, Y_j^{m_j}(\hat{R}) = 0 \qquad (3.205)$$

where $V_0(R)$ is the most accurate diatomic potential functions available; usually it corresponds to the Kolos-Wolniewicz potential that included adiabatic corrections [67].

For the scattering calculations at fixed-R values, one is obviously sampling the PES associated with purely rotational excitation processes. Due to the strongly interacting partners with much larger anisotropic components than the previous neutral system, the mean well-depth of the attractive part of the potential is about -150 m eV, but it changes according to the orientation angle from -50 meV up to about -250 meV. Close coupling calculations of the differential inelastic cross sections have been carried out on both the surfaces described before and by different groups [69-74]. Furthermore, classical Montecarlo trajectory calculations have also been done [75] using the same PES discussed before and at the same collision energies of the quantal calculations. Some of the above calculations also dealt with simultaneous vibrational rotational excitation process [73-75] whereby the complexity of a CC treatment is greatly increased.

One interesting result refers to the relative magnitudes of purely rotational excitation cross sections [70] with respect to the purely vibrational ones, the formers being considerably larger than the latters. This result is related to the relative values of the coupling integral appearing in eq.s (3.119) and (3.158); pairs of rotational eigenfunctions formed of functions with differenthj overlap significantly with each other and the anisotropic V_2 term has a large weighting effect over a range of r values up to $r \sim 4.0\ a_0$. The orthogonality of vibrational eigenfunctions, on the other hand, leads to nonvanishing contributions only for the domain of r in which V_0 is the main term that varies significantly, and for the present system this domain is $\leqslant 3.2\ a_0$.

The computational studies of numerical convergence also pointed out that for this strongly interacting system one can obtain cross sections converged to within a few percent only by including alla channels in the CC expansion which arise from the next two parity-allowed states beyond the final state of the transition [73]. The multiquantum transitions are always much larger than those found with less anisotropic systems and exhibit a considerably more undulatory structure.

In both quantum and classical calculations the elastic and rotationally inelastic ($\Delta j = 2$) cross sections show a definite rainbow. The quantum results, however, show a primary rainbow between 20° and 25° and a secondary rainbow at about 8°. The classical results exhibit instead only one rainbow at $\sim 40°$ [75]. The fast quantum mechanical oscillations are also given by the CC calculations and obviously are missing in a classical treatment. The results of Figure 3.4 are a clear example of these findings.

The relatively large integral cross sections for low-order rotational inelasticity ($0 \rightarrow 2,4$) are largely independent of the extent of the Legendre expansion of the potential while the higher V_λ contributions are still affecting the converged value of the ($0 \rightarrow 1$) vibrational excitation. The quality of the chosen PES does not affect drastically the computed cross sections: the near-HF limit results provide an excellent description of the interaction that governs low-order rotatio-

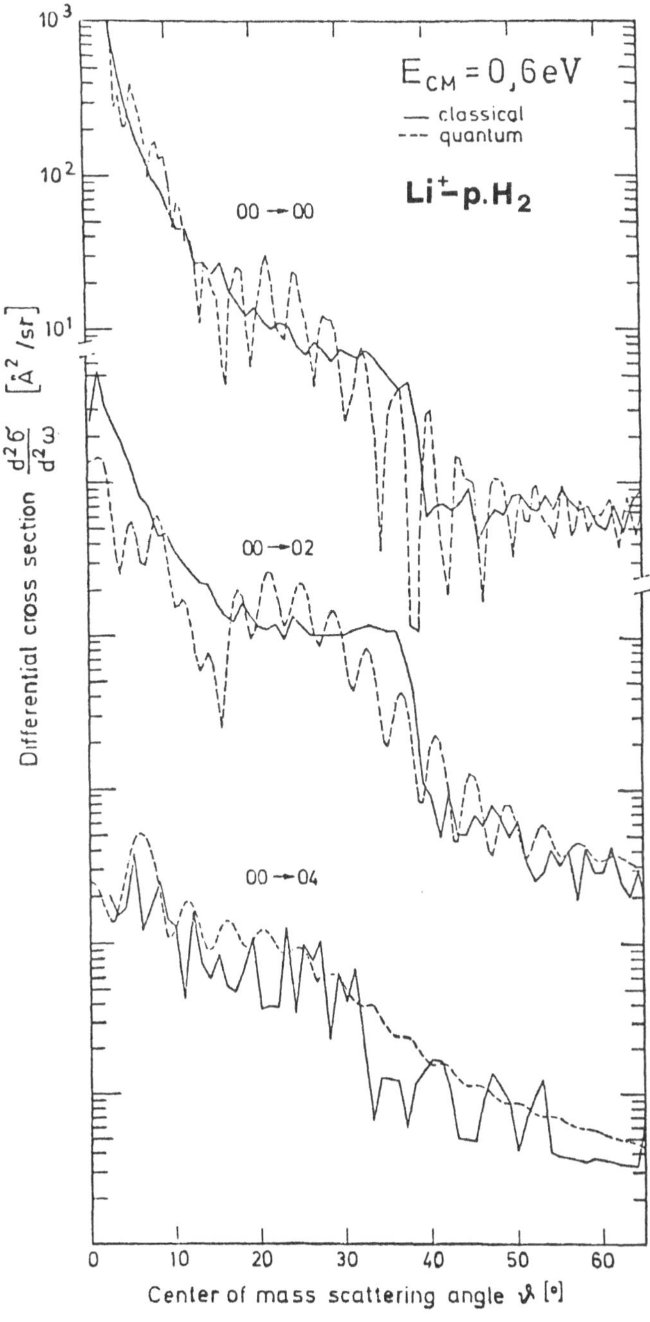

Fig. 3.4

Comparison of calculated differential cross sections for purely rotational excitation in Li^{+}-p-H_2 collisions. [from (cont.)

(Fig.4, cont.) J.P.Toennies, Chem.Soc.Rev., 3, 407 (1974)]. The dashed curves have been calculated via CC treatment with the PES of ref.[65], while the solid-line curves are from classical mechanics calculations [75].

nal transitions and fairly reliable results for multiquantum rotational jumps and vibrational (0 → 1) excitations.

The DCS exhibits almost invariably forward scattering behaviour and a rather steep descent to a minimum around $\theta_{CM} \sim 90°$. At low collision energies ($E_{CM} \sim 0.6$ e V) the inelastic DCS follow rather closely the elastic DCS and indicate strong reorientation effects of the H_2 molecules during collision [75]. The purely repulsive nature of the V_2 term provides strong inelastic coupling over a wide range of scattering angles and the rotational excitation show a rather broad distribution over final rotational states, with the maximum of this distribution shifted to larger Δj with increasing scattering angles.

For vibrational excitation processes one also finds that both impinging directions at $\theta = 0°$ and $\theta = 90°$ with respect to the H_2 bond lead to bond contraction with respect to the decreasing r values. Only for very close distances at $\theta = 90°$ ($r \leqslant 2.0\ a_0$) the atom-atom repulsion leads to an expansion of the H_2 bond, while the electron density increase is always a factor that tends to pull together the molecular protons.

Finally the DCS needed to form the ratios of inelastic angular distributions are presented in Figure 3.5 to be compared with the way measurements have been reported. The agreement of the various calculations is rather good, and it is also in good agreement with five of the six measured points, a fact attributed to a discrepancy wich is experimental in origin [68].

(iii) The $H + H_2$ calculations

The rotational excitation of H_2 molecules by the impact of H atoms has many physically interesting aspects because of its involvement in a variety of relaxation processes, including those of astrophysical interest like the cooling of interstellar clouds [76]. Furthermore, it has provided the prototype of the inelastic and reactive scattering treatment of molecules interacting with neutral particles.

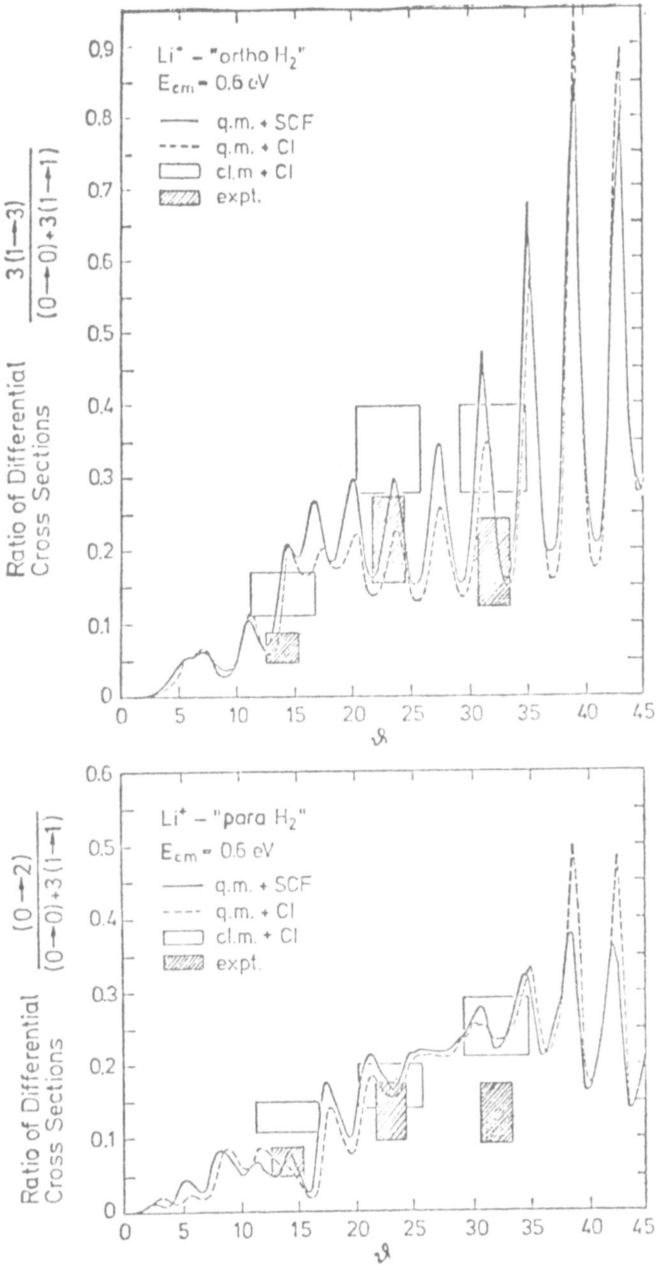

Fig. 3.5

Comparison of measured relative cross sections for purely rotationally inelastic processes. The top diagram refers to (cont.)

(Fig.5, cont.) $j = 1 \rightarrow 3$ transitions, while the bottom one is for the $j = 0 \rightarrow 2$ transitions [from J.Schaefer and W.A.Lester Jr, J.Chem.Phys., 62, 1913 (1975)]. The shaded boxes refer to experiments while the open boxes are from classical mechanics results [75].

Earlier attempts of obtaining a potential determined from first order perturbation calculations and corrected with long-range Van der Waals interactions were done to compute rotational excitations via approximate methods [77]. Semiempirical potential surfaces plus the long-range corrections provided a more consistent evaluation of the V_2 coefficient in the multipolar expansion [78], which was then used for purely rotational excitation calculations with the distorted wave Born approximation. An interesting feature of this V_2 term is that is drops from positive to negative at $r \sim 2.0\ a_0$, rises rapidly after the minimum and it becomes positive again at $r \sim 3.7\ a_0$. Only at further outer values of r (~ 7.0 a.u.) it become negative again. A different PES, i.e. the one obtained by Porter and Karplus [79] was used in CC calculations of purely rotational inelasticity studies of the same system at low collision energies [80]. Their inelastic, ($0 \rightarrow 2$), cross sections were slightly smaller than previous CC results [81] but about 10 times larger than the distorted wave results obtained with the Tang's potential (TP). The corresponding differential cross sections exhibited a strongly oscillatory, forward scattering behaviour.

A further comparison of the various PES forms already suggested in the literature has been recently carried out for the rotational excitation process via a different numerical treatment of the CC equations [82]. Drastic differences are found with the previous results [80] even when using the same PES: only small-angle oscillations are present in the DCS and they all smoothly raise to a strong backward-peaking behaviour [83]. A further comparison with previous inelastic DCS and total cross section calculation that had been performed via the TP potential [84] was also presented by Choi and Tang [83] and they found essentially exact agreement with that previous work. Even the comparison with the numerically not-converged results of ref. [81] shows good internal agreement, apart from the different level of numerical convergence.

The TP potential generally indicates that the largest contribution to the rotational transition comes from collisions with relatively large impact parameters, which will in turn cause forward peaks in the corresponding DCS values. In spite of the generally good agreement, however, the inelastic results show a strong dependence on the nature of the P.E.S. chosen to describe the anisotropic behaviour. Recent non-empirical calculations on the long-range behaviour of the H_3 system [86] seem to confirm the special nature of the V_2 term, with its changing of sign at smaller r's, which becomes the controlling factor in generating the special behaviour of inelastic DCS of this system.

Finally, an extensive three-dimensional study of non-reactive $H + H_2$ collisions that used the PK potential surface has been very recently performed by Schatz ans Kupperman [86] and the general backward-peaking behaviour of the previous calculations was confirmed for the rotationally inelastic DCS.

The corresponding elastic ones, on the order hand, exhibited a strongly forward-peaked form with a nearly monotonic decrease with increasing scattering angle.

REFERENCES

[1] R.D. Levine, J. Chem. Phys., 49, (1968) 51.

[2] R.D. Levine et. al. J. Chem. Phys., 50, (1969) 1694.

[3] W.R. Thorson, J. Chem. Phys., 34, (1961) 1744.

[4] W.R. Thorson, J. Chem. Phys., 39, (1963) 1431.

[5] I.M. Mills, Thearet. Chem., Vol.I, pg 110 (Chem. Soc. Specialist Per. Rep., London 1974.

[6] M.E. Rose, *Elementary theory of Angular Momenthen* (I Wiley and Sants duex, N.Y., 1957).

[7] J.H. Var Vleck, *The theory of electric and magnetic susceptibilities*, (Oxford U.P., 1932).

[8] M. Jacob and G.C. Wick, Ann. Phys.7, (1959) 404.

[9] C.F. Curtiss, J.O. Hirschefelder and F.T. Adler, J.Chem. Phys. 18, (1950) 1638.

[10] C.F. Curtiss and F.T. Adler, J. Chem. Phys. 20, (1952) 240.

[11] C.F. Curtiss, J.Chem. Phys., 21, (1953) 2045.

[12] G. Gioumousis and C.F. Curtiss, J. Chem. Phys. 29, (1958) 996.

[13] G. Gioumousis and QC.F. Curtiss, J. Math; Phys. 2, (1961) 96.

[14] G.Gioumousis, J. Math. Phys., 2, (1961) 723.

[15] A.M. Arthurs and A. Dalgarn, Proc. Roy Soc. A 256, (1960) 540.

[16] K. Takayanagi, Progr. Theoret. Phys. (Kyoto) Suppl. 25 (1963) 1.

[17] W.D. Davison, Discussions Faraday Soc. 33 (1962) 71.

[18] R.B. Bernstein et al., Proc. Roy. Soc. A274 (1963) 427.

[19] J.M. Blatt and L.C. Bierdenham, Rev. Mod. Phys. 24 (1952) 258.

[20] D.M. Brink and G.R. Setchler, *Angular Momentum* (Clarendon, Press, Oxford 1962).

[21] S. Green, J. Chem. Phys. 64 (1976) 3463

[22] L.S. Rodberg and R.M. Thaler, *Introduction to Quantum Theory of Scattering* (Academic Press, N.Y. 1967).

[23] K.P. Lawley and J. Chem. Phys. 43 (1965) 2943.

[24] H. Klar, Z. Phys. 228 (1969) 59.

[25] M. Tamir and M. Shapiro, Chem. Phys. Lett. 31 (1975) 166.

[26] A.R. Edmonds, *Angular Momentum in Quantum Mechanics* (Prunceton U.P., 1960).

[27] R.T. Pack and J.O. Hirschfelder, J. Chem. Phys., 49 (1968) 4009.

[28] R.T. Pack, J. Chem. Phys., 60, (1974) 633.
[29] W.A. Lester Jr., in *Modern Thoeretical Chemistry* Vol. 1, pag. 1 (Plenum Press, N.Y. 1976).
[30] G.G. Balint-Kurti in *MTB Int. Rev. Sci.* Phys. Chem. Ser. 2, 1, (1975) 285.
[31] D. Kivelson and E.B. Wilson, J. Chem. Phys. 20, (1952) 1575.
[32] G.W. King, R.M. Hainer and P.C. Cross, J. Chem. Phys. 11, (1943) 27.
[33] E.B. Wilson Jr. and J.B. Howard, J. Chem. Phys. 4, (1936) 260.
[34] B.T. Darling and D.M. Dennison, Phys. Rev. 57, (1940) 128.
[35] J.K.G. Watson, Mol. Phys., 15, (1968) 479.
[36] I.M. Mills and J.L. Duncan, J. Mol. Spectrosc. 9, (1962) 244.
[37] M.H. Alexander, J. Chem. Phys., 61, (1974) 5167.
[38] F.A. Gianturco et al., J. Chim. Phys. 74, (1977) 437.
[39] J. Schäefer and W.A. Lester Jr., Chem. Phys. Lett. 20, (1973) 575.
[40] C.B. Moore, Ann. Rev. Phys. Chem., 22, (1971) 387.
[41] J.D. Rapp and T. Kassal, Chem. Rev., 69, (1969) 61.
[42] K. Takayanagi, Adv. At. Mol. Phys. 1, (1965) 149.
[43] C.F. Hansen and W.E. Pearson, J. Chem. Phys. 53, (1970) 3557.
[44] R.D. Sharma and C.A. Brau, J. Chem. Phys. 50, (1969) 924.
[45] M.E. Riley and A. Kupperma, Chem. Phys. Lett., 1, (1968) 537.
[46] D.J. Diestler and V. McKoy, J. Chem. Phys., 48, (1968) 2951.
[47] W.D. Davison, Disc. Faraday Soc., 33, (1962) 71.
[48] H. Rabitz, in *Modern Theoretical Chemistry* Vol. 1, pag. 33 (W.H. Miller Ed., Plenum Press, N.Y. 1976).
[49] A.E. De Pristo and M.H. Alexander J. Chem. Phys. 66, (1977) 1334.
[50] G. Zarur and M. Rabitz, J. Chem. Phys. 60, (1974) 2057.
[51] S. Green, J. Chem. Phys., 62, (1975) 2271.
[52] J. Schäefer and W.A. Lester Jr., J. Chem. Phys. 62, (1975) 1913.
[53] C.S. Roberts, Phys. Rev. 131, (1963) 203.
[54] M. Krauss and E.M. Mies, J. Chem. Phys. 42, (1965) 2703.
[55] M.D. Gordon and D. Secret, J. Chem. Phys. 52, (1970) 120.
[56] B. Tsalpine and W. Kurzelnigg, Chem. Phys. Letters, 23, (1973) 173?.
[57] P.J.M. Geurts et al., Chem. Phys. Letters, 35, (1975) 44.
[58] G. Victor and A. Dalgarno, J. Chem. Phys., 53, (1970) 1316.
[59] W. Mayer, Chem. Phys. 17, (1976) 27.
[60] R. Shafer and R.G. Gordon, J. Chem. Phys., 58, (1973) 5422.

[61] W. Eastes and D. Secret, J. Chem. Phys., 56, (1972) 640.

[62] P.McGuire and D.A. Mirha, Int. J. Quantum Chem., 6, (1972) 111.

[63] P. McGuire, J. Chem. Phys. 62, (1975) 525.

[64] J.P. Toennies et al., Chem. Phys. Lett. 18, (1973) 87 and references quoted therein.

[65] W.A. Lester Jr. J. Chem. Phys. 57, (1972) 3028.

[66] W. Kurzelnigg et al., J. Chem. Phys., 1, (1973) 27.

[67] W. Kolos and L. Wolniewicz, J. Chem. Phys., 41 (1964) 3663.

[68] J.P. Toennies, Chem. Soc. Rev., 3, (1974) 407.

[69] J. Schaefer and W.A. Lester Jr. Chem. Phys. Lett., 20, (1973) 575.

[70] W.A. Lester Jr. and J. Schaefer, J. Chem. Phys., 59, (1973) 3676.

[71] W.A. Lester Jr. and J. Schaefer, J. Chem. Phys., 60, (1974) 1672.

[72] J. Schaefer, W.A. Lester Jr., D. Kouri and C.A. Wells, Chem. Phys. Lett. 24, (1974) 185.

[73] J. Schaefer and W.A. Lester Jr., J. Chem. Phys., 62, (1975) 1913.

[74] P. McGuire, Chem. Phys. 4, (1974) 249.

[75] G.D. Barg, G.M. Kendal and J.P. Toennies, Chem. Phys., 16, (1976) 243.

[76] B. Field, W.B. Someville and K. Dressler, Ann. Rev. Astron. Astrophys., 4, (1966) 207.

[77] K. Takyanagi and S. Nishimura, Publs. Astron. Soc. Japan, 12, (1960) 77.

[78] K.T. Tang, Phys. Rev. 187, (1969) 122.

[79] R.N. Porter and M. Karplus, J. Chem. Phys., 40, (1964) 1105.

[80] G. Walken Jr., W.H. Miller and M. Karplus, J. Chem. Phys. 56, (1972) 4930.

[81] A.C. Allison and A. Dalgarno, Proc. Phys. Soc. (London) 90, (1967) 609.

[82] B.H. Choi and K.T. Tang, J. Chem. Phys. 63, (1975) 1775.

[83] B.H. Choi and K.T. Tang, J. Chem. Phys., 63, (1975) 1783.

[84] E.F. Hoyes, C.A. Wells and D.J. Kouri, Phys. Rev. A4, (1971) 1017.

[85] J. Norbeck, P.R. Certain and K.T. Tang, J. Chem. Phys., 63, (1975) 590.

[86] G.C. Schatz and A. Kuppermann, J. Chem. Phys., 65, (1976) 4668.

4. DIMENSIONALITY REDUCTION METHODS FOR ROTO-VIBRATIONAL CROSS SECTION CALCULATIONS

4.1 Introduction

In the previous Chapter we established the fundamental theory describing physical and chemical aspects, on a microscopic level, of the elastic and inelastic collisions between molecules in their ground electronic states. The whole formulation of that theory was given by quantum mechanics based on the Schrödinger equation and it was made clear that the most systematic approximations used to provide possible numerical treatments involve the expansion of the total wavefunction for the system under study in terms of a unperturbed basis set of known, exact or nearly exact, functions.

In the specific cases of interest here, this basis set is generally taken to be the vibration-rotation states of the molecular target, vector-coupled with the rotational angular momentum states associated with the relative orbital motion of the projectile. From the analysis presented in the previous Chapter it followed that the number of possible basis functions for a given total angular momentum state labelled by J (and molecular vibro-rotational state labelled by $|v\ j\ \rangle$) is either $(2j + 1)$ or $(2J + 1)$, depending on whether j is smaller or larger than J. Furthermore, since most of the expansion procedures do not include basis functions for the relative scattering distance, the coefficients of the coupled set described above are functions of the scattering distance rather than costants and thus give rise to coupled equations, either differential or integral equations.

This is all well known but has been repeated to remind us of the basic difficulty that originates with this procedure, i.e. the rapid proliferation of the number of states involved in the expansion, even for the simplest problems that deal with H_2 as a molecular target. It is generally the rotational portion of the expansion that gives the prime difficulty, because many fewer vibrational states need to be considered in order to deal with the most common experimental conditions. For example, most molecules at room temperature have only the lowest one or two vibrational states populated for each normal vibrational mode, while each rotational manifold has an order of ten or more rotational states that are occupied. This number j_{max} of rotational states may actually not be too difficult to handle, but the number of projection states m_j is indeed a problem. There are in fact (2j + 1) of such projection states for each j level, which thus yield:

$$\sum_{j=o}^{j_{max}} (2j + 1) = (j_{max} + 1)^2 \qquad (4.1)$$

as a total number of $|j\ m_j>$ states for $o \leq j \leq j_{max}$. This number may easily exceed 100 and for a triatomic molecules target or for two colliding molecules (diatomics), one may possibly have to deal with an enormous number of states N:

$$N = \prod^{m} (n_{max,i} + 1) \cdot (j_{max,k} + 1)^2 \qquad (4.2)$$

where the label m describes the total number of normal modes of the polyatomic or of the two colliding molecules; $n_{max,i}$ is the highest occupied vibrational state within the ith mode and the number of occupied rotational states is assumed to be independent of the actual n_{max} value in the kth mode.

In therefore becomes apparent that the rotational states are the root of the problem and that any physical approximation that strives for a partial or drastic dimensionality reduction of the relevant coupled equations will essentially have to deal with this aspect.

Since the quality and reliability of such reductions will ultimately depend on their performances, as tested against either more accurate calculations (when they exist) or against accurate experiments, in the following Sections we will discuss some of the approximate schemes that have been most tested in the last few years and which have thus allowed a more firm understanding of their range of validity in terms of special features of the interaction potential or of specific couplings that are either disregarded or overemphasized within the relevant reductive scheme. Although it is illuminating to see how the underlying physics can be rephrased or simplified in various ways, the shortcuts are essentially worth using only when they enhance the informational return of our computational effort (per unit of computer time) without exceedingly blurring the physics one is trying to describe.

By looking at the final form of the coupled equations (CC) as given by the SF representation of eq.(3.58), one can write then in a more compact, matrix formulation:

$$\left\{ \mathbb{1} \frac{d^2}{dR'^2} + \mathbb{K}^2 - \frac{\mathbb{L}^2}{R'^2} - 2\mathcal{M}\,\mathbb{V}^J \right\} \mathbb{u}^J(R') = 0 \tag{4.3}$$

where $\mathbb{1}$ is a unit matrix, $\mathbb{u}^J(R')$ the solution matrix and the two diagonal matrix on the l.h.s. are given [1] by the following matrix elements (no vibrational states included):

$$k^2_{j\ell,j'\ell'} = k^2_j \cdot \delta_{jj'} \cdot \delta_{\ell\ell'} \tag{4.4a}$$

$$L^2_{j\ell,j'\ell'} = \ell(\ell + 1) \cdot \delta_{jj'} \cdot \delta_{\ell\ell'} \tag{4.4b}$$

and, for pure rotational excitations, the matrix element of the potential matrix, is given by:

$$V^J_{j\ell,j'\ell'} = \sum_\lambda V_\lambda(R')\, f_\lambda(j\ell, j'\ell'; J) \tag{4.5}$$

where the standard polynomial expansion of the interaction potential was used (eq. (3.80)) and the f_λ's are the Percival-Seaton coefficients of eq. (3.82). The usefulness of eq.(4.3) is to present the matrix elements responsible for the coupling clearly, hence to allow separate simplifications of them on the basis of their relative importance.

Thus, when the entire, or nearly entire, coupling between final channels comes from the potential surface, one has a situation of a Potential-Dominated regime where simplifications can be introduced in the structure of the $\mathbb{K}^2$ and/or $\mathbb{L}^2$ matrices. On the other hand, when centrifugal potential terms, or the importance ot the relative-motion kinematics, take over the relevant coupling with respect to the potential coupling terms, other simplifications can be introduced in the structure of these matrices, obviously affecting in a different way the consequent structure of the $\mathbb{V}^J$ matrix [2]. Thus, in the following Sections we will be discussing more in detail what simplifications can be introduced in the $\mathbb{L}^2$ and $\mathbb{V}^J$ matrices (or in the $\mathbb{K}^2$ matrix) and what are the physical justifications for doing it under different scattering conditions or for different interacting systems. Let us begin with one reductive scheme that has received a great deal of attention recently and that has been tested both against rigorous CC results and directly against experimental findings. This particular approach assumes a Potential-Dominated scattering process and hence focusses its attention on simplifying the $\mathbb{L}^2$ matrix.

4.2- The CS approach

By the CS approximation we mean here the Coupled-States or j_z-conserving approximation first proposed by McGuire [3] and independently formulated by Pack [4] and by McGuire and Kouri [5]. In its original form this decoupling approach consists in replacing the orbital angular momentum operator $\mathcal{L}^2$, in a BF reference frame, by a constant parameter $\bar{\ell}(\bar{\ell}+1)$ which, in most applications, was then chosen to be total angular momentum J or J - j.

Since the choice for the parameter $\bar{\ell}$, and its bearing on the simplified coupled equations, can be given in various ways and remains, to some extent, an arbitrary choice that is ultimately controlled by its success in agreeing with CC results or with available experiments, let us examine in some detail how the CS approach simplifies the strong coupling between rotational states implied by the correct CC equations and how the different choices for $\bar{\ell}$ affects the final expressions for opacity functions (partial integral cross sections), differential cross sections and magnetic transitions.

Following the original derivation of ref. [5], it is more convenient to write down the CC equations in the R-helicity representation, or within the BF frame as discussed in Section 5 of Chapter 3. For the purely rigid rotor target interacting with a structureless atom, the total hamiltonian in the BF frame takes the familiar form:

$$\mathcal{H}_{BF} = -\frac{1}{2\mathcal{M}}\frac{1}{R}\frac{\partial^2}{\partial R^2}R + \frac{\mathcal{L}^2}{2\mathcal{M}R^2} - \frac{1}{2\mu}\nabla_r^2 + V(\underline{r}, R) \qquad (4.6)$$

where $\mathcal{M}$ is the atom-diatomic reduced mass, μ the molecular reduced mass and the rotational operator describes the isolated molecular motion with the quantization axis taken along the BF z-axis, i.e. along $\hat{R}$.

If the above operator is applied to the trial wavefunction that has been expanded over a set of BF molecular states [eq.(3.101)], one sees that it would lead to equations with a purely radial dependence on R, if it were not for the $\mathcal{L}^2$ operator. Hence the lack of separation of the angular part of $\underline{R}'$ is purely kinematic in origin and does not depend on the nature of the potential. Whenever such a potential is primarily responsible for inelastic, $(j \rightarrow j')$ transitions one can then suggest a simpler form for the $\mathcal{L}^2$ eigenvalues and write:

$$\frac{\mathcal{L}^2}{2\mathcal{M}\, R^2} \sim \bar{\ell}(\bar{\ell} + 1)/2\mathcal{M}\, R^2 \tag{4.7}$$

so that the Schrödinger equation in the BF reference acquires the simpler form:

$$\left\{ -\frac{1}{2\mathcal{M}} \frac{1}{R} \frac{\partial^2}{\partial R^2} R + \frac{\bar{\ell}(\bar{\ell} + 1)}{2\mathcal{M} R^2} - \frac{1}{2\mu} \nabla_r^2 + V(\underline{r}, R) \right\} \psi_{\bar{\ell}}^{BF} = E\, \psi_{\bar{\ell}}^{BF} \tag{4.8}$$

where $\psi_{\bar{\ell}}^{BF}(\underline{r}, R)$ is the total wavefunction in the BF frame. Several approximations are involved in eq. (4.8) and they will become more clear if one starts from the rigorous eigenvalues structure of $\mathcal{L}^2$ and introduces the approximation (4.7) in the more explicit expressions already obtained in Section 3.5. We will thus see that one firstly neglects in eq. (4.8) any intermultiplet couplings that arise when $\mathcal{L}^2$ is rigorously given by $(\underset{\sim}{J} - \underset{\sim}{j})^2$, and then one also makes $\bar{\ell}$ independent of j so that the approximately diagonal $\mathcal{L}^2$ operator no longer depends on J and j.

Let us remind ourselves that, from the derivation of Section 3.5, the rigorous CC equations in the BF reference can be explicitely written as follows (for a rigid rotor target), with the meaning of the BF operators similar to the ones already defined in eq.(3.117 - 118):

$$h^{Jj'}_{\Omega',\, \Omega'\pm 1}(R)\, u^{Jj\Omega}_{j',\, \Omega'\pm 1}(R) + \left\{ \frac{d^2}{dR^2} - h^{Jj'}_{\Omega',\, \Omega'}(R) + k^2_{j'} \right\} u^{Jj\Omega}_{j',\Omega'}(R) = $$
$$= 2\mathcal{M} \sum_{j'',\Omega''} \langle j'\Omega' \,|V|\, j''\Omega'' \rangle_J \cdot u^{Jj\Omega}_{j'',\Omega''}(R) \tag{4.9}$$

where the potential coupling matrix elements can be rewritten via the well known multipolar expansion of $V(\underline{r}, \mathbf{R})$, and the familiar index Ω is the projection of the molecular rotational angular momentum j and of the total angular momentum J along the BF z-axis, i.e. along $\hat{R}$. Thus, as seen in eq. (3.119) and (3.120), one can write that:

$$
\begin{aligned}
< j\Omega\ |V|\ j'\Omega' >^J &= \sum_\lambda V_\lambda(R) \cdot g_\lambda(j\Omega,\ j'\Omega') \\
&= \sum_\lambda V_\lambda(R) \int d\hat{r}\ Y_j^{*\,\Omega}(\hat{r})\ Y_\lambda^0(\hat{r})\ Y_{j'}^{\Omega'}(\hat{r})
\end{aligned}
\qquad (4.10)
$$

Hence the last integral over $\hat{r}$ can be written as:

$$
\begin{aligned}
g_\lambda(j\Omega,\ j'\Omega') = (-)^\Omega \cdot \delta_{\Omega,\Omega'} \cdot \left\{ \frac{2\lambda + 1}{4\pi} \right\}^{\frac{1}{2}} \cdot \left\{ (2j + 1)(2j + 1) \right\}^{\frac{1}{2}} \cdot \\
\cdot \begin{pmatrix} j & \lambda & j' \\ 0 & 0 & 0 \end{pmatrix} \begin{pmatrix} j & \lambda & j' \\ -\Omega & 0 & \Omega' \end{pmatrix}
\end{aligned}
\qquad (4.11)
$$

The coupling is then rigorously independent of ℓ and is zero between states with different j_z components. As we mentioned before, this arises from the fact that the BF z-component of the projectile angular momentum is zero, which causes the projection of j to be equal to M, the projection of J. In this representation, therefore, the potential coupling is diagonal in Ω', as already implicitely shown in eq.(3.121), and conserves the j_z value, while the off-diagonal coupling in Ω comes from the first term on the l.h.s. of (4.9) which was explicitely given in eq.(3.118b). It is the dynamical part of the Hamiltonian that causes $\Delta\Omega$ transitions, as opposed to the SF representation, where the coupling between $|jm_j>$ states and the angular momentum state ℓ is completely contained in the potential coupling matrix elements and not partly separated as in eq.(4.11). Thus, only 3-j symbols are needed in eq.(4.11), as opposed to the 6-j symbols of eq.(3.82).

Now the explicit expressions of the dynamical terms just mentioned are given by (see Section 3.5 eq.(3.117)):

$$
h^{Jj}_{\Omega,\Omega\pm1}(R) = \left\{ [J(J + 1) - \Omega(\Omega \pm 1)]^{\frac{1}{2}} \cdot [j(j + 1) - \Omega(\Omega \pm 1)]^{\frac{1}{2}} \right\} \cdot R^{-2} \qquad (4.12a)
$$

and:

$$
h^{Jj}_{\Omega,\Omega}(R) = [J(J + 1) - 2\Omega^2 + j(j + 1)] \cdot R^{-2} \qquad (4.12b)
$$

The CS simplification of eq. (4.7) now allows us to state that, in the rigorous CC equations of the BF representation, one could neglect the intermultiplet couplings of eq.(4.12a) and further approximate eq.(4.12b) by writing the following assumptions:

$$h^{Jj}_{\Omega,\Omega\pm 1} = 0 \qquad ; \qquad h^{Jj}_{\Omega,\Omega}(R) = \frac{J(J+1)}{R^2} \tag{4.13}$$

Alternatively one can write down the diagonal centrifugal terms in its full form as given in (4.12b), although this choice slightly complicates the correct setting of the boundary conditions in the physical space [17].

The BF coupled equations within the formulation of the centrifugal sudden (CS) approximation of (4.13) are now given by:

$$\left\{ \frac{d^2}{dR^2} - \frac{J(J+1)}{R^2} + k^2_{j'} \right\} u^{Jj\Omega}_{j',\Omega}(R) = 2\mathcal{M} \sum_{j''} \langle j'\Omega|V|j''\Omega\rangle_J \cdot u^{Jj\Omega}_{j'',\Omega}(R) \tag{4.14}$$

where now Ω is conserved and the equations still depend on Ω. A separate set of (4.14) in fact exists for each Ω value consistent with $|\Omega| \leqslant \min(j, j')$, although the size of the set of equations is reduced by the fact the potential matrix of (4.10) is symmetric under a change of sign of Ω, hence only depends on $|\Omega|$.

Equations (4.14) presents powerful analogies with the partial wave equations encountered in Chapter 1 for the treatment of a central potential. The BF potential is in fact independent of R, thus the scattering in the SF will be independent of the polar angle and will depend only on the azymuthal angle. The CS approximation therefore surmises that the differences between SF and BF frames can be disregarded and the wanted SF solution is simply given by a separable BF solution of eq. (4.84) [5]:

$$\Psi^{SF}_{JM}(\underline{r}', \underline{R}') \simeq Y^{M}_{J}(\hat{R}')\, \Psi^{BF}_{J}(\underline{r}, R) \tag{4.15}$$

It is clear that the approximation given by eq.(4.15) should work best when the scattering is dominated by a backward direction, since then the initial and final directions of the rotating z axis (BF) will be the same. Hence inelastic forward scattaring, dominated by the long-range anisotropic terms of the potential, will present the largest discrepancy between the approximated and the exact results [7]. The approximate CS equations (4.14) represent now the entire collision process as a sum over Ω of individual transitions $|j\Omega\rangle \rightarrow |j'\Omega\rangle$. This is exactly correct in the limit $j \sim 0$. In the limit of large j values the off-diagonal, Coriolis coupling (4.12a) is the largest for $\Omega = 0$, which corresponds physically to the impinging projectile being in the plane of the molecular rotor, while it could be neglected as Ω approaches the value of $\pm$ j.

In approximating the diagonal potential as in eq. (4.13) one is in fact performing an averaging over all the Ω values of the exact centrifugal potential in the large-j limit [7] and the decoupled equations give equal weighting to each Ω value when calculating the cross sections as indicated above. The diagonal effective potential (ℓ-averaging) in fact produces oly one effective classical turning point for each $|j\ \Omega\rangle$ combination, while there would be a different one for each combination in the exact eq.s (4.9).

By looking at the physical implications of the CS treatment, one can then suggest that potential surface with strong short-range coupling and weaker long-range interactions should favour the backward scattering region, hence should perform the best under conditions (4.13). Moreover if the centrifugal interaction extends over a large region of partial waves (large J values for the lowest Δ j transitions), the Potential-Dominated situation might begin to be balanced by the kinematics [1], hence one should expect a worsening of the general agreement between CS results and CC results.

Taking now the incident plane wave in the SF + z' direction, the projections of J and j on this z' axis are initially equal, i.e. $\mathbf{M} \doteq m_j$, and also, since the initial $\hat{R}$ direction is in the -z direction (SF), it is: $\mathbf{M} = -\Omega$, as discussed in the previous Chapter. In the decoupled equations, however, the approximate centrifugal term also gives: $\mathbf{M} = \Omega'$, hence the T-matrix in the BF representation [eq.(3.159)] becomes simpler and is block-diagonalized in $|\Omega|$ [1]:

$$T^{J}_{j\mathbf{M},j'\Omega'} = \delta_{\mathbf{M}\Omega'} \cdot T^{J\Omega'}_{j,j'} \tag{4.16}$$

This allows one to write the SF expression for the scattering amplitude in a form simpler than the one given by (3.136) or by (3.82):

$$f^{CS}(jm_j \rightarrow j'm_{j'}|\hat{R}') = \frac{\pi^{\frac{1}{2}}}{k_j} \sum_{J,\ell} (i)^{j-j'+1}(-)^{m_j+m_{j'}} \delta_{m_j,\Omega} \frac{(2J+1)}{(2\ell+1)^{\frac{1}{2}}} \cdot$$

$$\cdot \; C(Jj'\ell;\, m_j \; -m_{j'} \; 0) \cdot C(Jj'\ell;\, m_j \; -m_{j'},\, m_j \; -m_{j'}) \; \cdot \tag{4.17}$$

$$\cdot \; T^{J\Omega}_{j\,j'} \cdot Y^{m_j - m_{j'}}_{\ell}(\hat{R}')$$

where is also: $m_{\ell'} = m_j - m_{j'}$.

If one further neglects to rotate back into the SF frame the final rotor state $Y^{\Omega'}_{j'}(\hat{r})$, on the ground that one is disregarding the differences between SF frame and BF frame according to eq.(4.15), then it is also true that: $m_j = m_{j'}$ [6] and the result of (4.17) further simplifies as f llows:

$$f^{CS}(jm_j \rightarrow j'm_{j'}|\theta_{R'}) = \frac{\delta_{m_j m_{j'}}}{2i\,k_j} \sum_J (2J+1)\, T^{Jm_j}_{j,j'} \cdot P_J(\cos\theta_{R'}) \tag{4.18}$$

from which the differential cross sections are given as:

$$\frac{d\sigma}{d\hat{R}'}(jm_j \rightarrow j'm_j) \;=\; |f^{CS}(jm_j \rightarrow jm_j|\,\theta_{R'})|^2 \tag{4.19}$$

The corresponding integral cross sections are then obtained by summing over $m_{j'}$ and averaging with equal weighting over m_j:

$$\sigma(j \rightarrow j'\;;\; k_j^2) = \frac{\pi}{k_j^2} \sum_J \sigma^J_{j\rightarrow j'}(k_j^2) \tag{4.20}$$

where the partial inelastic opacities are given by:

$$\sigma^J_{j\rightarrow j'}(k_j^2) = \frac{(2J+1)}{(2j+1)} \cdot \delta_{m_j m_{j'}} \cdot \sum_{m_j=-\mathrm{Min}(j,j')}^{\mathrm{Min}(j,j')} |T^{Jmj}_{j,j'}|^2 \tag{4.21}$$

where Min (j, j') indicates the smaller value between j and j'.

In the previous presentation, the arbitrary parameter $\bar{\ell}$ of (4.7) was tied to the J value of each set of coupled equations. The treatment has been however generalized in the recent literature to include a variety of possibilities that have been presented in a very interesting series of recent papers [8-12]. Before discussing them more in detail, however, it is worth mentioning an equivalent decoupling model that has been suggested starting from a different form of helicity representation [13].

In the previous BF frame, in fact, the z axis of the reference frame is taken to be along the rotating vector $\hat{R}$ and the orbital angular momentum can be written as:

$$\underline{\ell} = \underline{\mathbf{R}} \times \mathbf{P}_{\mathbf{R}} = R \cdot \hat{z} \times \mathbf{P}_{R} \tag{4.22}$$

hence:

$$\ell_z = 0 \quad \text{and:} \quad (\underline{P}_R)_z = P_{\mathbf{R}} \tag{4.23}$$

which means that $\underline{\ell}$ has a vanishing projection on $\hat{z}$ and the projection of the momentum $\underline{P}_R$ on the z axis is the customary radial momentum P_R. This is called the R-helicity representation and is different from the P-helicity frame of representation, in which the wave vector $\underline{k}$ is taken instead along the BF z-axis [14] and hence the orbital angular momenta are quantized along the direction of the linear momentum P_K. The projection of J onto the R-helicity z-axis was given by Ω before, while the P-helicity defines its projection on $\underline{k}$ as μ.

Thus, the matrix elements of the $\mathbb{L}^2$ of eq.(4.3) are now given by expressions entirely analogous to eq.s (4.12):

$$\begin{aligned}
\langle JM\, j'\mu' | \mathcal{L}^2 | JM\, j'\mu'' \rangle &= J(J+1) + j'(j'+1) - 2\mu''^2 \;; \qquad \mu' = \mu'' \,, \\
&= -\{[J(J+1) - \mu''(\mu'' \pm 1)] \cdot \\
&\quad \cdot [j'(j'+1) - \mu''(\mu'' \pm 1)]\}^{\frac{1}{2}} \;; \qquad \mu' = \mu'' \pm 1 \\
&= 0 \qquad\qquad \text{otherwise.}
\end{aligned} \tag{4.24}$$

By using the standard multipolar expansion of the potential, as in (4.10), one then recovers the exact equivalent of eq.(4.11), where the Ω-index has been substituted by the new μ-index. This result therefore shows that the P-helicity basis provides coupling potentials diagonal in μ and is completely equivalent to the R-helicity of above [1,15].

The suggested decoupling, called the P-helicity decoupling (PHD) [13], disregarded the off-diagonal elements in eq.(4.24), but fully retained the diagonal terms that appear in it. The reduced coupled equations that follows are entirely equivalent to eq.s (4.14), both from the points of view of the necessary asymptotic matching conditions [1] and of the recovery of the PHD equations from the former by matrix rotations [16].

The computational saving introduced by the CS decoupling can readily be understood if one remembers that, as stated in the Introduction to this Chapter, the number of fully coupled equations is given by the square of the number of states included in the expansion. The numerical algorithms on the other hand vary in general, for small sets of equations, as the square of their number. Hence, if τ is the time required to solve one equation, the time to solve (rotational) CC equations is: $t_{CC} = (j_{max} + 1)^4 \cdot \tau$. In the case of CS equations, the number of equations in each system is equal to the number of rotor levels included in the expansion, j_{max}. Hence, the corresponding computational time, t_{CS}, will be approximately given [5] by: $t_{CS}\ (j_{max} + 1)^3 \cdot \frac{\tau}{3}$, which becomes a substantial computational saving even for relatively small values of j_{max}. The CS treatment is obviously totally unchanged when vibrational excitations are also included, since the corresponding CS equations can be rewritten in a way analogous to the one shown for the CC equations in Chapter 3. Finally, it is interesting to note that the CS approximation can be obtained directly in the SF representation if one approximates the centrifugal matrix elements over the SF angular coefficients of (3.55) via the CS prescription of (4.7):

$$\begin{aligned} < \mathcal{Y}^{JM}_{j\ell} | \mathbf{\ell}^2 | \mathcal{Y}^{JM}_{j'\ell'} > & \ell(\ell + 1) \cdot \delta_{jj'} \cdot \delta_{\ell\ell'} \\ & \sim J(J+1) \cdot \delta_{jj'} \cdot \delta_{\ell\ell'} \end{aligned} \tag{4.25}$$

where the identification: $J = \bar{\ell}$, as discussed before, has been used. On the other hand, it has been shown that the SF angular eigenfunctions of (3.55) and the corresponding BF eigenfunctions of (3.144) are related by the unitary transformation [17]:

$$\mathcal{Y}^{JM\eta}_{j\ell}(\hat{r}', \hat{R}') = \sum_{|\Omega|} \mathcal{P}^{JM\eta}_{j\ell|\Omega|} \cdot \mathcal{Y}^{JM\eta}_{j|\Omega|}(\hat{r}', \hat{R}') \tag{4.26}$$

where tha parity index $\eta = (-)^{j+\ell}$ and:

$$\mathcal{P}^{JM\eta}_{j\ell|\Omega|} = (-)^{J+|\Omega|} \{2(2\ell + 1)\}^{\frac{1}{2}} \begin{pmatrix} j & J & \ell \\ |\Omega| & -|\Omega| & 0 \end{pmatrix} (1 + \delta_{|\Omega|,0})^{-\frac{1}{2}} \tag{4.27}$$

It then follows that eq. (4.25) can be rewritten by substituting the SF angular eigenfunction with the unitary transformation of the BF angular eigenfunctions [17]. Due to the orthogonality relations between 3-j coefficients, the new BF centrifugal matrix elements become:

$$< \mathcal{Y}^{JM\eta}_{j|\Omega|} \, |\mathbf{L}^2| \, \mathcal{Y}^{JM\eta}_{j'|\Omega'|} > = J\,(J+1) \cdot \delta_{jj} \cdot \delta_{|\Omega|,|\Omega'|} \quad (4.28)$$

Thus, the making of the average-ℓ approximation of (4.25) in the SF frame leads to the CS equations without any further approximation. One clearly sees from (4.28) that the kinematic matrix of the centrifugal terms is block-diagonalized into smaller subsystems; for each of which $|\Omega|$ is a good quantum number. The transformation (4.27) obviously applies to the total wavefunctions, and to the T-matrix elements indicated in (4.16). One can therefore explicitely obtain the direct expressions that transform the CS elements of the S-matrix from the BF to the SF representations via the operators of eq. (4.27):

$$\left\{ S^{J\eta}_{j\ell,j'\ell'} \right\}^{CS} = \sum_{|\Omega|} [(2\ell+1)(2\ell'+1)]^{\frac{1}{2}} \cdot \begin{pmatrix} \ell & j & J \\ o & |\Omega| & -|\Omega| \end{pmatrix} \cdot \begin{pmatrix} \ell' & j' & J \\ o & |\Omega| & -|\Omega| \end{pmatrix} \cdot \left\{ S^{J\eta}_{j|\Omega|,j'|\Omega|} \right\}^{CS} \quad (4.29)$$

Different expressions can be obtained if the CS conditions of (4.25) are replaced by other choices for $\bar{\ell}$ [18] as we will briefly discuss below.

In the derivations presented thus far, the j_z- conserving or Centrifugal sudden approximation has been obtained entirely in the BF reference frame by simplifying the $\mathbf{L}^2$ matrix elements via either the first or both conditions (4.13), which make an explicit choice for the arbitrary parameter $\bar{\ell}$ of (4.7).

It was further shown that simple unitary transformations allow one to relate BF S-matrix and SF S-matrix, within the CS approximation as defined above.

However, attracting great interest has been the search for a way of obtaining the simple formulae (4.18)-(4.21) for inelastic amplitudes, differential cross sections and opacities without the further approximation (4.15). Moreover, it would be also interesting to start directly in the SF representation and retrieve from there the simple equation (4.18) while giving to $\bar{\ell}$ the more reasonable meaning of an <u>orbital</u> angular momentum and not of a <u>total</u> angular momentum.

Recently Parker and Pack [8], and independently Shimoni and Kouri [18] have analyzed the expression for the scattering amplitude (4.17) as given in the BF frame with the additional CS approximation (4.7). They have shown that, if $\bar{\ell}$ is chosen to be independent of J and taken to be equal to the final orbital angular momentum ℓ', the Clebsh-Gordan coefficients can be analitically summed over J and the scattering amplitude directly reduces to the expression (4.18) without the further assumption (4.15). Thus, one can write:

$$f^{CS}_{\ell_f}(jm_j \to j'm_{j'}|\hat{R}) = \frac{i(-)^{j+j'}}{2(k_j\, k_{j'})^{\frac{1}{2}}}\, \delta_{m_j m_{j'}} \sum_{\ell'} (2\ell'+1)\, T^{\ell' m_j}_{j,j'} \cdot P_{\ell'}(\cos\theta_{R'}) \qquad (4.30)$$

where the index ℓ_f reminds us of the ℓ'-association chosen in equation (4.30) and where an additional phase factor appears with respect to the original expression (4.18), in order to take properly into account the correct asymptotic matching with the SF physical amplitudes [8,23]. The new T-matrix is then labelled differently and is given by:

$$T^{\ell' m_j}_{j,j'} = \delta_{jj'} - S^{\ell' m_j}_{j\to j'} \qquad (4.31)$$

and since $|\Omega| = m_j$, the general CS S-matrix is related to the correct BF S^J-matrix by the relation:

$$S^J(j'\Omega'|j-\Omega) \sim \delta_{\Omega,\Omega'}\,(i)^{j+j'}(-)^{J-\bar{\ell}}\, S^{\bar{\ell}\Omega}_{j\to j'} \qquad (4.32)$$

The corresponding integral cross sections can now be written as:

$$\sigma_{\ell_f}(j\to j'\,;\,k_j^2) = \frac{\pi}{k_j^2}\sum_{\ell'} \sigma^{\ell'}_{j\to j'}(k_j^2) \qquad (4.33)$$

where the averaged transition probabilities or inelastic opacities are given by:

$$\sigma^{\ell'}_{j\to j'} = \frac{(2\ell'+1)}{(2j+1)} \sum_{m_j=-j_m}^{j_m} |\,T^{\ell' m_j}_{j\to j'}\,|^2 \qquad (4.34)$$

where the m_j-sum runs over the same indeces as in (4.21) i.e. is limited between-j_m and j_m with: $j_m = \min(j,j')$.

The above opacities, however, are those of a well defined orbital angular momentum, while the CC opacities refer to a given J value. Thus, one can alternatively use eq. (4.17), let the J-label of the T matrix be the more general $\bar{\ell}$

and then substitutes it with ℓ' without performing the sum of the Clebsch-Gordan coefficients [18].

The obtained scattering amplitude can then be used in the same way as above to yield differential cross sections and integral cross sections like those of in (4.33):

$$\sigma_{\ell_f}(j \to j'; k_j^2) = \frac{\pi}{k_j^2} \sum_J \sigma^J_{j\to j'}(\bar{\ell} = \ell_f) \tag{4.35}$$

where the new, total angular momentum partial opacities are now given by [8]:

$$\sigma^J_{j\to j'}(\bar{\ell} = \ell_f) = \frac{1}{2j+1} \sum_{\ell' m_j} (2\ell'+1)\, C^2(\ell' j' J;\, o\, m_{j'}, -m_j)\, |T^{\ell'|\Omega|}_{j\to j'}|^2 \tag{4.36}$$

The results outlined in the above equations, and generalized with a further parity requirement on the triangular relations between angular momenta [18-20], are rather important in that they clearly show the physical origin of the CS approximation without explicit reference to BF-oriented approximations. The arbitrary parameter appearing in (4.7) is an angular momentum quantum number and its identification with the final ℓ value, ℓ', directly produces simple expressions for integral cross sections. The corresponding T-matrix does not any longer need to be diagonal in Ω, but becomes simply labelled through its initial value because the other Ω' index of Section 5.3 has been summed over by the averaging implied in the CS scheme. Several other types of averaging factors have also discussed within this context [18, 20] and tested numerical results. Regarding this last point, in fact, it has also been shown [21] that the BF scattering amplitude within the CS approximation is indeed not-diagonal in Ω when ℓ is identified by ℓ' and allows then to obtain $\Delta\Omega$ magnetic transitions.

If one now turns to an SF formulation of the decoupling of eq.(4.7), one comes across several interesting results which have been obtained with various choices for ℓ and which also show the relationship of the CS scheme with the more restrictive Sudden Approximation discussed in the following Section [21].

First of all, by following the work of Secrest [10], one sees that the substitution of the correct $\mathbf{L}^2$ matrix in the Arthurs-Dalgarno SF representation with the constant element of prescription (4.7) yields equations that separate the dependence between the **R**'orientation $(\theta_{R'}, \phi_{R'})$ and the $\underline{r}'$ orientation $(\theta_{r'}, \phi_{r'})$ [11]:

$$\left\{ \frac{d^2}{dR^2} - \frac{\bar{\ell}(\bar{\ell}+1)}{R^2} + k_{j'}^2 \right\} u_{jj'}^{m_j}(R) = 2\,\mathbf{M} \sum_{j''} V_{j'j''}^{m_j}(R) \cdot u_{jj''}^{m_j}(R) \qquad (4.37)$$

where the coupling potential depends, as discussed before, only on the relative $\hat{R}'\cdot\hat{r}'$ angle γ:

$$V_{jj'}^{m_j}(R) = 2\pi \int \sin\gamma \, d\gamma \, Y_{j'}^{m_j *}(\gamma, o)\, V(R, \gamma)\, Y_{j''}^{m_j}(\gamma, o) \qquad (4.38)$$

Thus, it produces exactly the results of the **BF** formulation of (4.14). Both the radial solutio s of (4.37) and the corresponding **S**-matrix elements are related by a continuous angular transformation to the approximated, physical expression of the SF from of the Scattering matrix at a given orientation:

$$S^{\bar{\ell}}(jm_j \to j'm_{j'};\, \theta''_{R'}, \phi''_{R'}) = \sum_{m_j} \mathcal{D}^{j'}_{m_{j'},m_j}(\theta''_{R'}, \phi''_{R'}, o) \cdot S^{\bar{\ell} m_\ell}_{j\to j'}\, \mathcal{D}^{j*}_{m_{j'},m_j}(\theta''_{R'}, \phi''_{R'}, o) \qquad (4.39)$$

The SF scattering amplitude at the final scattering angle is then given by integrating (4.39) over all the possible orientations in the decoupled $|\ell j\rangle$ scheme [11]:

$$f^{CS}(jm_j \to j'm_{j'}|\hat{R}') = \sum_{\ell,\ell',m_{\ell'}} (-)^{\ell-\bar{\ell}} \cdot i \cdot \left\{ \frac{(2\ell+1)\pi}{k_j k_{j'}} \right\}^{\frac{1}{2}} \cdot \int d\hat{R}'' \left\{ Y_\ell^0(\hat{R}'')[\delta_{jj'}\delta_{m_j m_{j'}} - S^{\bar{\ell}}(jm_j \to jm_{j'};\, \hat{R}'')] \cdot Y_{\ell'}^{m_{\ell'}}(\hat{R}'') \right\} \cdot Y_{\ell'}^{m_{\ell'}}(\hat{R}') \,. \qquad (4.40)$$

If $\bar{\ell}$ is now chosen to be ℓ', the ℓ-summation is done analytically in (4.40) and the integration over $\theta''_{R'}$ and $\phi''_{R'}$ can be done trivially [11] , which then yields, after simple passages, to the much simpler expression of eq.(4.30) for the scattering amplitude. Thus, both SF and BF formulations of the CS decoupling scheme yield the original expression of ref.[5] by directly choosing $\bar{\ell} = \ell'$ and without any other approximation. The same result has also been obtained for the SF derivation of the Sudden Approximation and with the choice of $\bar{\ell} = \ell'$.

One important difference between SF and BF derivations, however, is that in the SF formulation the scattering amplitude is diagonal in m_j, i.e. the magnetic transitions are not allowed, while the averaged cross sections (integral and differential) are given by the same expressions. The BF expression on the other hand, is non-diagonal in $\Delta\Omega$ with the above choise of $\bar{\ell}$[18, 21] and so the R-helicity magnetic transitions may then be obtained.

Another choise for $\bar{\ell}$, in order to simplify eq.(4.40), is given by setting it equal to ℓ, the initial orbital angular momentum. In this case the summations over ℓ' and $m_{\ell'}$ can be carried out trivially, as may also the following **R**" integration. The scattering amplitude then becomes:

$$f^{CS}_{\ell_i}(jm_j \rightarrow j'm_{j'}|\hat{R}') = \frac{i}{2(k_j k_{j'})^{\frac{1}{2}}} \sum_{\ell} (2\ell + 1) \cdot \left[\delta_{jj'} \cdot \delta_{m_j m_{j'}} - S^{\ell}(jm_j \rightarrow j'm_{j'};\, \theta_{\mathbf{R}'}, \phi_{\mathbf{R}'})\right] P_{\ell}(\cos\theta_{R'}) \tag{4.41}$$

where ℓ_i is the analogous of the ℓ_f index in eq.(4.30).

By using the transformation (4.39) one then obtains a very interesting relationship between the simpler formula (4.30) and the result of (4.41) [11]:

$$f^{CS}_{\ell_i}(jm_j \rightarrow j'm_{j'}|\hat{R}') = \sum_{\lambda,\lambda'} \mathcal{D}^{j'*}_{\lambda' m_{j'}}(\hat{R}')\, f^{CS}_{\ell_f}(j\lambda \rightarrow j'\lambda'|\hat{R}')\, \mathcal{D}^{j}_{\lambda m_j}(\hat{R}') \tag{4.42}$$

Since the corresponding T-matrix of (4.30) is diagonal in m_ℓ eq.(4.42) may be rewritten as follows:

$$f^{CS}_{\ell_i}(jm_j \rightarrow j'm_{j'}|\hat{R}') = \frac{i}{2(k_j\, k_{j'})^{\frac{1}{2}}} \sum_{\ell} (2\ell + 1) \left\{ \sum_{\lambda} \mathcal{D}^{j'*}_{\lambda m_{j'}}(\hat{R}') \cdot T^{\ell\lambda}_{jj'} \cdot \mathcal{D}^{j}_{\lambda m_j}(\hat{R}') \right\} P_{\ell}(\cos\theta_{R'}) \tag{4.43}$$

An important consequence of this result is that, because of the non-diagonal nature of the $\mathcal{D}$- rotation matrices, the SF amplitude obtained via the CS approximation with the $\bar{\ell} = \ell$ prescription is in general non-zero for $m_j \neq m_j$, hence one can write down polarization differential cross sections which, as opposed to (4.19), do not vanish for $\Delta m_j \neq 0$.

It therefore appears that, in the SF representation, the CS scattering amplitudes obtained with the two angular momentum choices of ℓ_i and ℓ_f for the parameter $\bar{\ell}$ are related only through a rotation. In the simple $f_{\ell_f}^{CS}$ expression of eq. (4.30), in fact, both projections of j and j' are quantized along the direction of the incident momentum $\underline{k}_j$, while equation (4.43) represents a situation in which the projection of both rotational quantum numbers are taken along the direction of the final momentum $\underline{k}_{j'}$,[22].

Finally it is interesting to note that the approximate expression (4.43) for the scattering amplitude, once used eq.(4.19), and then degeneracy averaged, yields a differential cross section equation that is exactly the same as the one given by (4.30), i.e. the original CS expression of ref.[5]. This follows trivially from the unitarity of the rotation $\mathcal{D}$-matrices:

$$\frac{d\sigma}{d\Omega}(j \to j'|\hat{\mathbf{R}}') = \frac{k_j'}{k_j(2j+1)} \sum_{m_j m_{j'}} | f_{\ell_i}^{CS}(jm_j \to j'm_{j'}|\hat{\mathbf{R}}')|^2$$
$$= \frac{1}{4k_j^2(2j+1)} \sum_{m_j} | \sum_{\ell} (2\ell+1) T_{jj'}^{\ell m_j} P_\ell(\cos\theta_{\mathbf{R}'})|^2 \qquad (4.44)$$

Hence the corresponding degeneracy-averaged integral cross section of eq. (4.33) is given by:

$$\sigma_{\ell_i}(j \to j'; k_j^2) = \frac{\pi}{k_j^2} \sum_{\ell} \sigma_{j\to j}^{\ell}(k_j^2) \qquad (4.45)$$

Thus the degeneracy-averaged cross sections are identical for the choices of $\bar{\ell}$. This however is not exactly the case for the corresponding opacities. In fact following the same precedure used to yield eq.(4.36), one now obtains a slightly different expression for the total angular momentum partial opacities [21]:

$$\sigma^{J}_{j\to j'}\,(\bar{\ell} = \ell_i) = \frac{1}{2j+1} \sum_{\ell m_j} (2\ell + 1)\, C^2(\ell\; j\; J;\; 0\; m_j - m_j)\;\cdot \tag{4.46}$$
$$\cdot\; |\, T^{\ell|\Omega|}_{j\to j'}|^2$$

which is now a weighted sum over T-matrix elements indexed with the initial channel orbital angular momentum. The result of (4.36), on the other hand, shows a weighted average over ℓ', which will change for each $(j \to j')$ transition, hence giving more weight in the averaging to $\Delta j > o$ excitation precesses. Since these opacities are not physically observable quantities, however, the differences could be shown to be important only through numerical investigations as has recently been studied for the case of $H^+ - H_2$ collisions [21].

Another interesting result shown for CS amplitudes, this by Schinke and Mc Guire [21], is that the $\bar{\ell} = \ell_i$ choice in the SF representation can be transformed into the $\mathbf{R}$-helicity representation (BF frame), where one also finds that $\Delta\Omega$ transitions are allowed as were the Δm_j transitions in the SF frame of reference.

In all our previous discussions, the various dynamical observables have been given, in either frames, via the T-matrix as obtained in the BF formulation shown in eq.(4.17) and phase corrected via eq.(4.32) [8]. It is also of some interest, however, to obtain as well the corresponding approximate expressio for the $T^J(j\ell \to j'\ell')$ of the SF Arthurs-Dalgarno formulation, via the Centrifugal Sudden reduction that we have discussed thus far. From the usual form of the relationship between $\underset{\approx}{S}$ and $\underset{\approx}{T}$ matrices:

$$T^J_{CS}(j\ell \to j'\ell') = \delta_{jj'} \cdot \delta_{\ell\ell'} - S^J_{CS}(j\ell \to j'\ell') \tag{4.47}$$

one can obtain an SF T-matrix after reducing and simplifying the CC equation (3.58) with the CS condition (4.7) and with the correction of the required phases from (4.32) [21]:

$$T^J_{CS}(j\ell \to j'\ell') = (-)^{j+j'_i-\ell-\ell'-2\bar{\ell}} \sum_{\Omega} \left\{ \frac{(2\ell'+1)(2\ell+1)}{(2J+1)^2} \right\}^{\frac{1}{2}} \cdot \tag{4.48}$$
$$\cdot\; C(j\;\ell\;J;\;\Omega\;o\;\Omega)\; C(j'\ell'J;\;\Omega\;o\;\Omega)\; T^{J\Omega}_{j\to j'}$$

where the index Ω = min (j, j', J). The state-to-state inelastic integral cross sections, the degeneracy-averaged differential and integral cross sections can now be given by expressions analogous to eq.s(4.36) and (4.46), but this time with the T^J_{CS} $(j\ell \rightarrow j'\ell')$ matrix element of (4.48). For instance, in the case of $\bar{\ell} = \ell_i$ by including the phases in the expression (4.47) and by recalling the correct CC, SF scattering amplitude of Chapter 3:

$$f^{CC}(jm_j \rightarrow j'm_{j'}|\hat{R}') = \left\{\frac{\pi}{k_j\, k_{j'}}\right\}^{\frac{1}{2}} \sum_{J\ell\ell' M} i^{\ell-\ell'+1}(2\ell+1)^{\frac{1}{2}} \cdot \tag{4.49}$$

$$\cdot\, C(j\ell J;\, m_j\, o\, M) \cdot C(j'\ell' J;\, m_{j'} m_{\ell'} J)\, T^J(j\ell \rightarrow j'\ell')\, Y_{\ell'}^{m_{\ell'}}(\hat{R}')$$

the use of (4.48) into (4.49) and the integration of (4.19) over the scattering angles, gives [24] now the following total cross section:

$$\sigma^{CS}_{jm_j \rightarrow j'm_{j'}}(\ell = \ell_i) = \sum_{\ell'} \Big| \sum_{\ell=|J-j|}^{J+j} \sum_{J=|\ell'-j'|}^{\ell'+j'} i^{\ell} \left\{\frac{\pi(2\ell+1)}{k_j k_{j'}}\right\}^{\frac{1}{2}} \cdot \tag{4.50}$$

$$\cdot\, C(j\,\ell\, J;\, m_j\, o\, m_j)\, C(j'\ell' J;\, m_{j'}\, m_j - m_{j'}\, m_j) \cdot T^J_{CS}(j\ell \rightarrow j'\ell')\Big|^2$$

One sees that the transitions with $\Delta m_j \neq 0$ take place through a Clebsh-Gordan-weighted sum of the $\Delta\Omega$ = 0 transitions given by the T-matrix elements on the r.h.s. of equation (4.48).

With the substitution of $\bar{\ell} = \ell'$ (or ℓ_f) in eq.(4.48), the phase becomes independent of ℓ and the sum in the T-matrix can be done analytically both on ℓ and J. The use of (4.48) into (4.49) thus leads to the approximate expression (4.30) and gives zero for the state-to-state integral cross sections of (4.50), since the J-summation in (4.49) gives $m_{\ell'} = 0 = m_j - m_{j'}$ thus forcing the well known selection rule indicated by (4.30) for the CS-SF inelastic magnetic transitions.

These amplitudes however, as discussed before, are non-diagonal in the R-helicity projection quantum number, in that they conserve the sum over Ω' values rather than the Ω value itself [18], hence they allow $\Delta\Omega$ transitions, with Ω being now the j-projection onto the relative distance $\underline{R}'$ ($\equiv \hat{z}$ in the BF frame).

Finally for completeness the third possible choice, $\bar{\ell}$ = J, should also be considered.

This choice allows us to rewrite (4.48) as:

$$T^{J}_{CS}(j\ell \to j'\ell' | \bar{\ell} = J) = (-)^{j+j'} \, i^{\ell'-\ell} \sum_{\Omega} \left\{ \frac{(2\ell'+1)(2\ell+1)}{(2J+1)^2} \right\}^{\frac{1}{2}} \cdot \tag{4.51}$$

$$\cdot \, C(j\ \ell\ J;\ \Omega\ o\ \Omega)\ C(j'\ell' J;\ \Omega\ o\ \Omega) \cdot \tilde{T}^{J\ell'\Omega}_{j\to j'}$$

where a phase-corrected T-matrix has been defined as being given by:

$$T^{J\ell'\Omega}_{j\to j'} = \delta_{jj'} \cdot \delta_{\ell\ell'} - (-)^{\ell'+J} \, S^{J}_{CS}(j\ell \to j'\ell') \tag{4.52}$$

The substitution of (4.51) into the CC expression (4.49) provides an approximate expression of the scattering amplitude, in the SF representation for $\bar{\ell} = J$:

$$f^{CS}_{J}(jm_j \to j'm_{j'} | \hat{R}') = \left[\frac{\pi}{k_j \, k_{j'}} \right]^{\frac{1}{2}} \sum_{J\ell\ell'M} i(-)^{j+j'} (2\ell+1)^{\frac{1}{2}} \cdot$$

$$\cdot \, C(j\ell J;\ m_j\ o\ M)\ C(j'\ell' J;\ m_{j'}\ m_{\ell'}\ M) \cdot \tag{4.53}$$

$$\cdot \, T^{J}_{CS}\,(j\ell \to j'\ell' \mid \bar{\ell} = J) \cdot Y^{m_{\ell'}}_{\ell'}(\hat{R}')$$

since $\Omega = M = m_j$ and $m_j - m_{j'} = m_{\ell'}$, by substituting (4.51) into (4.53) one can perform analytically the sum over ℓ and hence obtain:

$$f^{CS}_{J}(jm_j \to j'm_{j'} | \hat{R}') = \frac{\pi^{\frac{1}{2}}}{k_j} \sum_{J\ell'} i(-)^{j+j'} (2\ell'+1)^{\frac{1}{2}} \cdot \tag{4.54}$$

$$\cdot \, C(j'\ell' J;\ m_{j'}\ m_{\ell'}\ m_j) \cdot C(j'\ell' J;\ m_j\ o\ m_j)\ T^{J\ell' m_j}_{j\to j'} \cdot Y^{m_{\ell'}}_{\ell'}(\hat{R}')$$

Now this expression is entirely equivalent to the original CS scattering amplitude of (4.17) where the implicit association of J to $\bar{\ell}$ had been performed in eq.(4.13). The different phases in (4.54) are in fact due to the T-matrix definition of (4.52).

This is an interesting fact, since one finds that the present ℓ-choice allows magnetic transition $\Delta m_j \neq 0$; and only the further disregard of the BF-to-SF transformation of the total wavefunction leads to the selection rule of (4.18). The last expression for the scattering amplitude can obviously be used to yield differential cross sections via eq.(4.19). The corresponding integral cross sections are then given, after the usual averaging over m_j degeneracy, by sums like (4.35) over the new expression for corresponding opacities [21]:

$$\sigma_J(j \to j'; k_j^2) = \frac{\pi}{k_j^2} \sum_J \sigma^J_{j\to j'} (\bar{\ell} = J) \tag{4.55}$$

and:

$$\sigma^J_{j\to j'}(\bar{\ell} = J) = \frac{1}{2j+1} \sum_{\ell'=|J-j'|}^{J+j'} \sum_{m_j=-m_n}^{m_n} (2\ell' + 1) \cdot$$
$$\cdot\, C^2(\ell' j' J; o\, m_j - m_j) \mid T^{J\ell'|\Omega|}_{j\to j'} \mid^2 \tag{4.56}$$

The last expression is different from eq.s (4.36) and (4.46) of the corresponding opacities with the two other choices of ℓ. The summation over m_j in (4.56) is in fact running over the index $\pm\, m_n$ = min (j', j', J) while the previous summations are over an index that runs between ± min (j, j') [25]. An appreciable difference with the previous results for inelastic integral cross sections should then show up when both j and j' are large, hence the sum (4.56) will be restricted by J, leading to smaller cross sections with the $\bar{\ell} = J$ choice.

It has also been shown that, where inelastic transitions (i.e. $j \neq j'$) are concerned, eq.(4.56), the phase factor in (4.52) goes out in the absolute square of the T-matrix elements [21], hence the integral cross section expression can be simplified by performing the ℓ' summation:

$$\sigma^J_{j\to j' \neq j} (\bar{\ell} = J) = \frac{(2J+1)}{(2j+1)} \sum_{|\Omega|=-m_n}^{m_n} \mid T^{J|\Omega|}_{j\to j'} \mid^2 \tag{4.57}$$

This last expression thus reveals even more clearly the differences of this choice of ℓ with those yielding eq.s (4.34) and (4.45).

Since the original formulation of the CS decoupling scheme appeared in the literature, several calculations have been performed on a variety and rather extensive comparisons have been made with rigorous CC results, with available experimental data or with different other approximations like IOSA (see Section 3 below) or the EP method (see Section 4 below). Table 1 collects the various systems examined and gives indication of the transitions computed. Where only Δj is quoted, only purely rotational excitations were discussed.

Moreover, because of the essentially arbitrary nature of the $\bar{\ell}$ parameter of (4.7) it has also become a point of further interest to test a variety of choices, both for degeneracy-averaged DCS and ICS, for elastic and inelastic opacities and for state-to-state magnetic transitions, in order to correlate them with the specific nature of the interaction at play.

From a qualitative, physical standpoint it was argued at the beginning of this Chapter that the CS approach should generally be good for collision systems with predominantly short-range anisotropy, where rotational excitation becomes a direct process in the BF frame. In follows that it would be reasonable to assume that the asymptotically equal distribution of each rotor state j in each substate m_j is not appreciably altered by $\Delta\Omega$ transitions caused by the long-range Coriolis coupling of eq.(4.12a).

This has been reasonably well confirmed by the low-j transitions with intermediate and weak long-range anisotropy [7] but the CS results markedly underestimate inelastic opacity functions in the large-J region for large-j transitions, as shown in He-CO [17, 34] and confirmed by the Li^+- H_2 calculations [42]. Moreover, the threshold-region behaviour, ofter characterized by numerous resonace effects controlled by the full m_j-coupling in eq.s(4.12), is also only qualitatively described by CS calculations of H+CO [32] and He + H_2 CO [35], even if the long-range anisotropy is not particularly strong in these systems.

The many studies performed on the He + pH_2 vibrational relaxation (e.g. see: ref. [43]) have often employed the CS method to yield the vibrational cross sections at low energies. Although a clearcut comparison with experiments is ofter blurred by the wide spreading of the results as a function of the potential surfaces employed, this weakly anisotropic system, with dominant short-range interactions, has proven to be well suited for a CS treatment and its (10 $\rightarrow$ 00) vibrational deexcitation cross sections are suggested to lie within 15% - 20% of the "converged" CC values [44]. It has also indicated that very little effect is caused on its cross section values by an Ω-dependent choise of the ℓ parameter in (4.7) [29], as also confirmed by the He + CO system [17].

Moreover, recent studies on the definite parity j_z-conserving coupled sta-

TABLE 1

Rotovibrational cross sections (Integral and/or Differential) computed within the CS approximation.

System	Processes considered (collision energy in eV)		Reference (year)
H + p H_2,D_2	Δj = 0,2,4,6 (0.5 → 1.5)	(T → R)	[26] (1975)
He + p H_2	Δj = ±0,2,4,6,8 Δv = -1 (0.1 → 1.2,4.2)	(T → R) (R,V → T)	[5] (1974) [6] (1975)
He + p H_2	Δv = -1 Δj = 0,2,...,10 (0.5 → 1.0) (0.0015 → 0.3)	(R,V → T)	[27] (1976) [28] (1975) [43] (1977); [29] (1976)
He + o H_2	Δv = -1 Δj = 2,4,6 (0.006 → 0.3)	(R,V → T)	[30] (1976)
H + CO	Δj = ±1,2,...,7 (0.01 → 0.05)	(R → T)	[31] (1975)
He + N_2	Δj = ±0,2 (0.005 → 0.05)	(R → T)	[32] (1975)
He + CO	Δj = 1,2,3,4 (0 → 0.019)	(T → R)	[33] (1977)
He + CO	Δj = 1,2,3,4 (0 → 0.015)	(T → R)	[34] (1976) [37] (1975) [40] (1977)
He + CO	Δj = ±1,2,...,6 (0.014)	(R → T)	[17] (1976)

Table 1 (cont.d)

He + HCN	$\Delta j = \pm 0,1,\ldots,5$ (0.008)	(R → T)	[7] (1976) [40] (1977) [50] (1975)
He + H_2CO	28 rotational transitions (0.017 → 0.086)	(T → R)	[35] (1975)
He + HCℓ	$\Delta j = 0,1,\ldots,4$ (0.006 → 0.037)	(T → R)	[36] (1975) [40] (1977)
Ar + N_2	$\Delta j = \pm 0,\ldots,14$ (0.066)	(R → T)	[7] (1976)
Ar + HCℓ	$\Delta j = 0,\ldots,4$ (0.077)	(T → R)	[38] (1977)
Ar + CsF	$\Delta j = 0,1,2$ (0.087)	(T → R)	[72] (1976)
He + NH_3	$\Delta j, \Delta k = 0,1,2,3,4$ (0.016 → 0.021)	(T → R)	|39| (1976)
Li^+- H_2	$\Delta j = 0,1,\ldots,4$ (06 → 1.2)	(T → R)	[42] (1974)
Li^+- CO	$\Delta j = 0,1,\ldots,4$ (1.0)	(T → R)	[41] (1978)

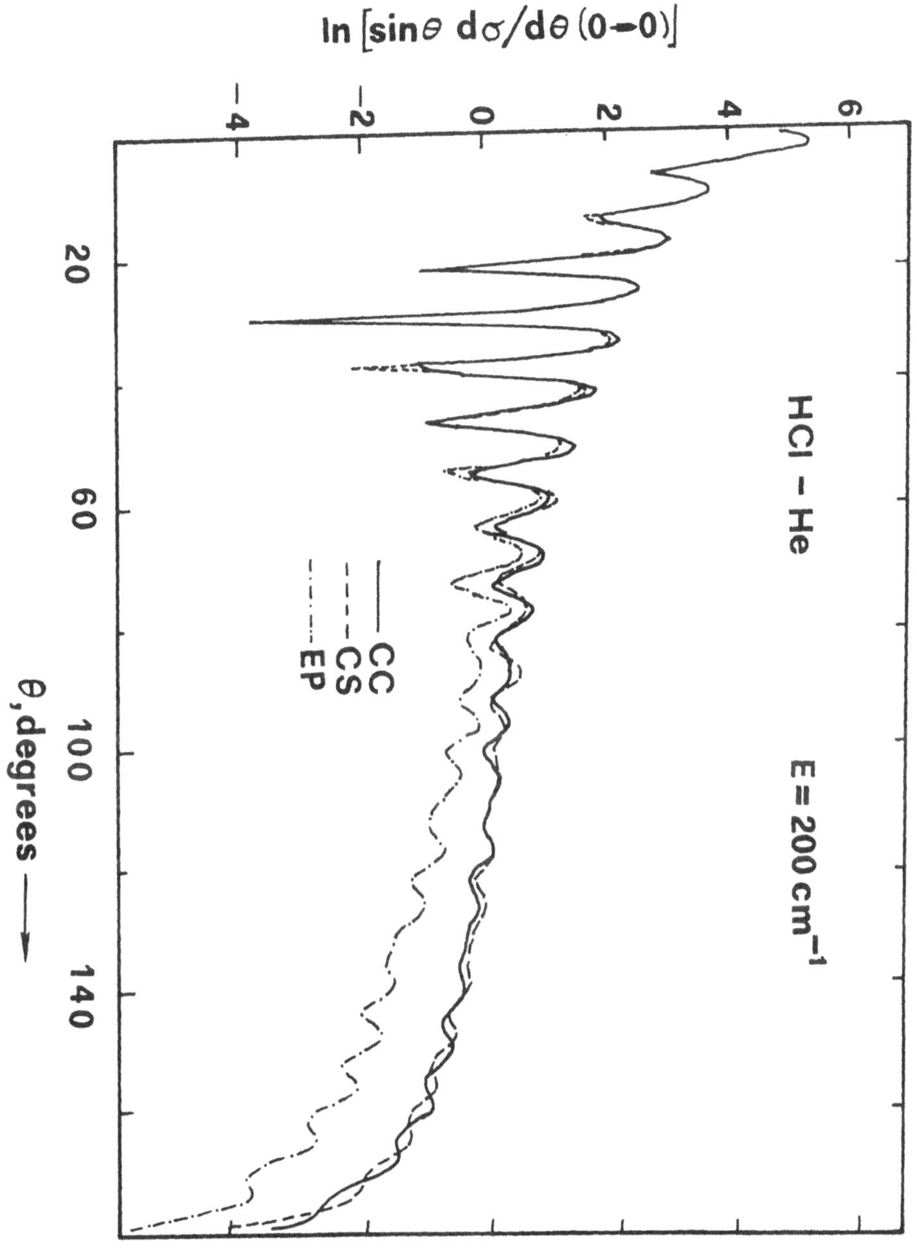

Fig. 4.1

Differential elastic scattering cross sections for He + HCℓ collisions at a total energy of 200 cm^{-1} [from S.Green and L.Monchick, J.Chem.Phys., 66, 3085 (1977)]. (cont.)

(Fig.1, cont.)

CC = close coupling calculations

CS = coupled states approximation

EP = effective potential method (see Section 4.4)

tes (DPj_z) [18 - 20] have performed some further numerical tests on low-energy, low-j rotational transitions in He + pH_2. They have shown differences that while most marked at the opacity level, became less drastic when summed over J to yield integral cross sections. Thus, parity can be ignored to compute the latter while the ℓ-choice seems to provide improved results when $\ell = \ell_f$, which is also the procedure, as seen before, that most directly leads to the original [5] formulation of the CS scattering amplitude.

An important result of these studies, however, is that CS T-matrix does not need to be diagonal in Ω and that one is here approximating a sum over final Ω' in contrast to the original formulation [4,5] of the CS equations in the BF reference system.

Furthermore, the CS approach has been tested in a series of studies [36, 37, 40, 45] over a wide variety of gas phase phenomena that depend on intermolecular forces solely via binary collision dynamics, like viscosity and diffusion [46], relaxation phenomena such as sound absorption and dispersion [47], NMR spin-lattice relaxation [48] and collision-induced spectral pressure broadening [49]. Thus, the He + HCl interaction, as obtained via the EGM approach (see Chapter 2), was found to yield gas kinetic cross sections that depend almost exclusively on the spherical average of the interaction itself [36], as opposed to the behaviour of the He + CO calculations [37]. Since the former is strongly repulsive at short range, the CS results are in good agreement with CC calculations for low-j transitions at thermal energies, a result also confirmed for the He + HCN system [50] which exhibits a similar dominance of its short-range interaction [7].

This is true however only for integral cross sections, both elastic and inelastic, while the elastic differential cross sections appear to be good at small angles (forward scattering) because of the optical theorem relationship with ICS but become increasingly poor at larger angles, especially for inelastic differential cross sections. Some examples are shown in Fugures 4.1 and 4.2 for the He projectiles with HCl and CO respectively [40]. One sees there that the CS agreement with CC results is surprisingly good over a wide range of angles for the elastic

DCS (Fig. 4.1), while the inelastic processes exhibit marked off-phase oscillatory behaviour at small angles (Fig. 4.2).

For gas transport cross sections, however, the CS results are rather poor in predicting diffusion cross sections, just because of their failure to predict large-angle scattering which are the dominant contributions, as shown in eq.(2.69). If, on the other hand, the further disregard of BF-SF transformation eq.(4.15) is employed [40], it gives rather good predictions of gas kinetics cross sections. The collision induced pressure broadening is also rather badly predicted by the CS calculations which are in general inaccurate in providing individual S-matrix elements, while recovering their agreement with CC results when summing over J to yield total integral cross sections [40]. Purely rotational excitation cross sections, as yielded at low energies by the CS approach, appeared however in nearly quantitative agreement with CC results performed with the only atom-symmetric top studied so far, the He + NH_3 rotationally inelastic collisions |39|.

It should be pointed out that the generally better agreement found for transport cross sections computed via the simple formula (4.18), as opposed to the correct transformation to SF frame with $\bar{\ell} = J$, has been suggested as an indication of the physically more realistic use of $\bar{\ell} = \ell_f$ as a parameter choice in eq.(4.7) [25].

Finally, the comparison of CC magnetically inelastic integral cross sections (Δm_j transitions) with the corresponding CS results for the He + HCl case [45] indicates that the $\bar{\ell} = \ell_f$ choice that produces integral cross sections diagonal in m_j agrees rather well with the close coupling calculations, while overestimating the $m_{j'}$ dependence of the magnetically averaged cross section [45]. It correctly predicts, however, thet the rotational inelasticity peaks around $m_j = 0$, i.e. that rotational excitation of systems with long-range interaction tends to preserve molecular orientation.

This result is also confirmed by the $\bar{\ell} = \ell_i$ choice for the parameter [51]. In this case, however, the variation of the cross sections with m_j is in the wrong direction, i.e. the CS ($\bar{\ell} = \ell_i$) cross sections increase with m_j increasing, while the CC and CS ($\bar{\ell} = \ell_f$) calculations both decrease with increasing m_j. The depolarization cross sections, $\sum_{m_{j'}} \sigma(jm_j \to j'm_{j'})$, on the other hand show a better agreement with CC results when going from $\bar{\ell} = \ell_f$ to $\bar{\ell} = \ell_i$, especially for transitions elastic in j [51], since in this case the CS approach treats forward and backward

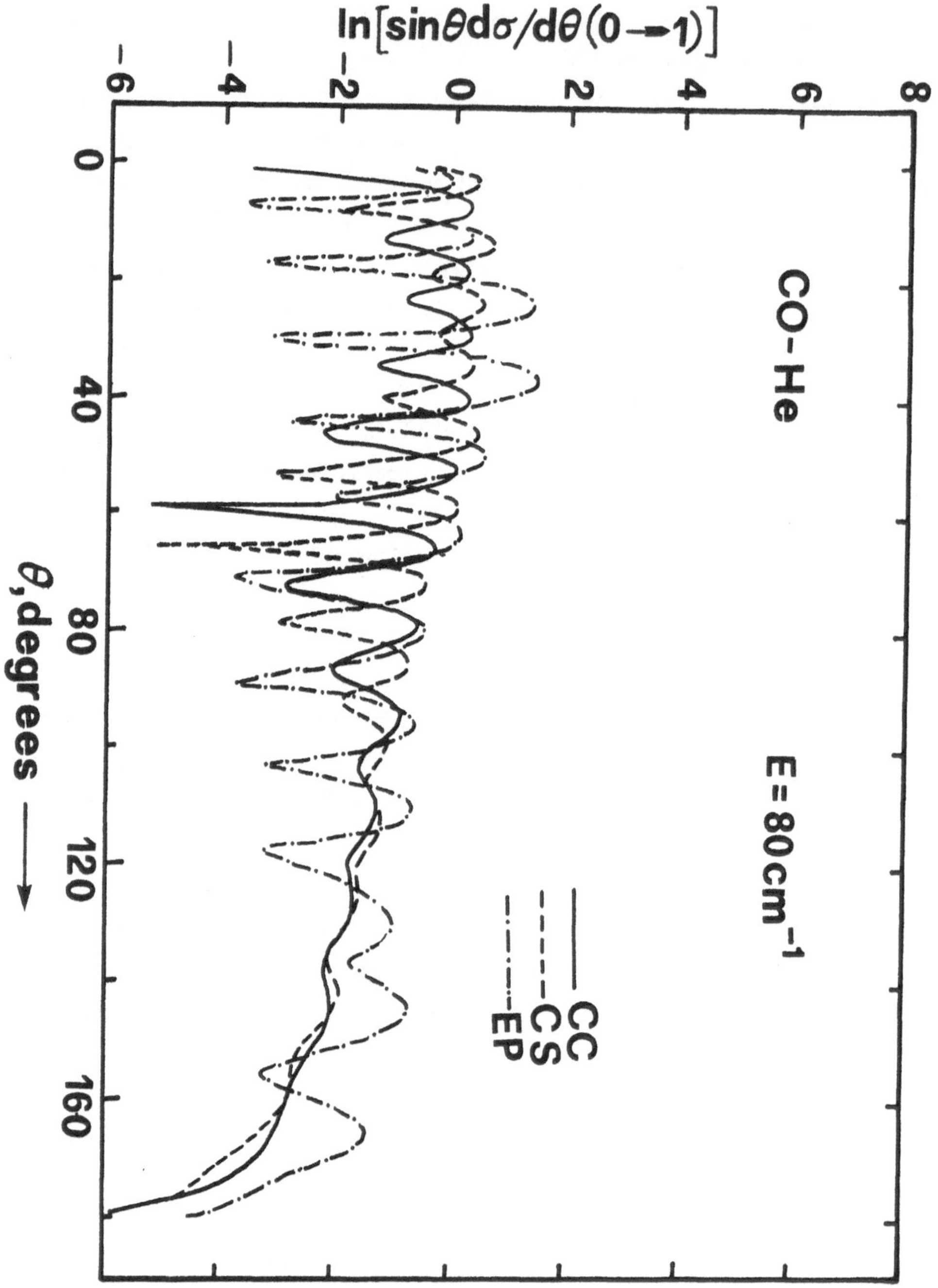

Fig. 4.2

Differential scattering cross sections for excitation of CO from $j = 0$ to $j' = 1$ by collision with He at a total energy (cont.)

(Fig.2, cont.) of 80 cm^{-1} [from S.Green and L.Monchick, J.Chem. Phys., 66, 3085 (1977)].

CC = close coupling calculations

CS = coupled states approximation

EP = effective potential method (see Section 4.4)

scattering equally well. The latter choice also eliminates the unphysical selection rule $\Delta m_j = 0$ found for the $\bar{\ell} = \ell_f$ case in the SF representations, as discussed before.

4.3-The Sudden Approximation methods

Another relevant, general suggestion which strives at reducing the number of coupled equations one has to solve, involves a wide class of atom-molecule scattering processes, although mainly been applied thus far to atom-diatomic systems. Its essential point is that, when looking at rotational excitation processes, or at vibrational excitation via anisotropic potentials, the whole physical evolution could be such that it is feasibly treated by considering mainly sudden or adiabatic collisions with respect to the rotationally inelastic channels. The idea is that the rotational periods of the molecular states of interest are large when compared to the specific collision time. Implicitely this means that the amount of energy transferred in a given ($j \rightarrow j'$) transition is small compared to the translational energy.

The consequence of this approach is then to treat the internal molecular states, and especially the rotational ones, as energy-degenerate, a fact which therefore reduces the inelastic collision process to solving primarily a single Schrödinger equation for the scattering of, say, an atom off a molecule with fixed orientation.

The earliest quantum treatments of this type originate in the parallel work of nuclear scattering [52] and in the suggestions of Takayanagi [53], Kramer and Bernstein [54] and Levine [55]. All these treatments are based on the assumption discussed above and define an infinite energy degeneracy of the individual rotational channel wavevectors:

$$k_j^2 \sim k_{\bar{j}}^2 \quad \text{(all j)} \tag{4.58}$$

The basic formulae derived for adiabatic scattering amplitudes in the context of nuclear physics [52] have been discussed extensively for electron-molecule collisions as well, where detailed expressions have been obtained for the rotational and vibrational excitations of homonuclear diatomics [56], polar diatomic [57] and polyatomic molecules [58].

The simple physical argument given in the above cases shows that, for non-resonant scattering of electrons from neutral molecules, the collision time is short compared to the rotational period or the vibrational period, provided the incident energy is greater than 0.1 e V. This fact suggests that one can neglect the kinetic energy of the nuclei and define a physically reasonable zeroth order Hamiltonian for the (N + 1) electron system:

$$\mathcal{H}_{el}^{(N+1)} \Psi(\underline{r}_N, \underline{r}', \underline{\mathbf{R}}') = \mathcal{E}(\underline{\mathbf{R}}') \Psi(\underline{r}_N, \underline{r}'; \underline{R}') \tag{4.59}$$

where it has been explicitely indicated that both Ψ and $\mathcal{E}$ depend parametrically on the space-fixed molecular orientation $\underline{R}'$.

The discrete solutions of eq.(4.59) represent the bound electronic states of the projectile plus the electrons of the target system, while the continuous solutions describe the scattering of an electron from a molecule in which the nuclei are fixed at a relative configuration $\underline{R}'$, and hence are usually called the fixed nuclei wavefunctions [56]. The exact form of the expansion for the correct scattering wavefunction of the total system (electron + molecule) can then be seen as a sort of generalization of the Born-Oppenheimer approximation discussed in Chapter 2. The target nuclear functions and bound electronic wavefunctions will in fact be added to the set of continuum eigensolutions of (4.53) in order to provide a complete set in the space of electronic coordinates for fixed $\underline{R}'$ geometries [59]. The standard procedure for substituting such an expansion into the correctly complete Schrödinger equation leads, after eliminating the electronic coordinates, to an infinite set of coupled equations for the nuclear vibro-rotational wavefunctions $\chi_{jv}(\underline{R}')$. The coupling is provided by a matrix operator that represents the dynamic interactions of the continum electronic part and the vibration-rotation of the molecule. The adiabatic nuclei approximation consists then on the assumption that such coupling is negligible since the continuum eigensolutions of (4.59) are only slowly varying functions of $\underline{R}'$.

In the resulting BF coordinate system the molecular axis for diatomics, or the principal axis in polyatomics, in chosen as the z-axis and decouples all equations with different ℓ_z projections for each angular momentum ℓ. Thus, coupling only occurs through the ℓ values and the S-matrix is obtained for each ℓ_z component. Such a matrix has been shown to be related to the corresponding SF S-matrix via a unitary transformation [60]. The corresponding solutions of the coupled equations for atom-molecule collisions were recently given in the literature and were briefly discussed in the previous Section [31].

In heavy-particle collision problems, the physical approximation discussed thus far provides the basis of what are called the Sudden Approximations and which generally fall into two classes. In one the approch usually adopted is the time-independent treatment whereby the decoupling is provided by the energy degeneracy condition of (4.48) combined with the centrifugal degeneracy, i.e. with the assumption:

$$\frac{\ell(\ell + 1)}{R^2} \simeq \frac{\bar{\ell}(\bar{\ell} + 1)}{R^2} \quad (\text{all } \ell) \tag{4.60}$$

already discussed in Section 2 of this Chapter [eq.(4.7)]. The corresponding phase shift can then be shown to be of infinite order in the anisotropy of the potential [10]. In the other class of treatments, the relative coordinates of the colliding partners follow a trajectory determined by a spherically averaged potential while the sum over all inelastic transitions appears as a time dependent perturbation [54, 61]. Such a treatment gives phase shifts which are of first order in the anisotropy of the potential, while the angle that the passing atom forms with the space-fixed molecule is variable. One can further simplify the treatment by adopting straight-line trajectories for the incoming projectile [62].

Applications of the time-independent formalism began with the early 70' [63, 64] and it was then shown that the first-order time-dependent treatments [54] are equivalent to taking the first two terms in an expansion of the phase shifts with regard to various contributions for increasing order of the anisotropy in the interaction potential. This leads to defining the Infinite Order Sudden Approximation (IOSA) by combining both conditions (4.58) and (4.60) and by formulating the results in either an SF set of coordinates [65, 66] or a BF set of coordinates [4,67]. In either case, the problem solved is one of a decoupled radial equation with the partner atoms held fixed in space at a given geometry, i.e. at a fixed value of the internal coordinate between the incoming direction and the only molecular bond that exists in the atom-diatomic instance.

To see the reductions involved more in detail, let us now start from the correct CC equations for a rotating-vibrating molecule, as discussed in Chapter 3, and treat rotational states as simply those given by the rigid rotor approximation. The SF representation of eq.(3.58) can easily be rewritten to include the vibrational degrees of freedom (in atomic units):

$$\left\{ \frac{d^2}{dR^2} + k^2_{j'v'} - \frac{\ell'(\ell'+1)}{R^2} \right\} u^{Jj\ell v}_{j'\ell'v'}(R) = \sum_{j''\ell''v''} 2\mathcal{M} < j'\ell'v'J|v|j''\ell''v''J > \cdot u^{Jj\ell v}_{j''\ell''v''}(R) \tag{4.61}$$

where:

$$k^2_{j'v'} = E_{coll} - \mathcal{E}_{j'v'} \tag{4.62}$$

are the channel wavenumber and energy and the total angular momentum representation has been used via the angular coefficients of eq. (3.55):

$$|j\ell JMv> = \mathcal{Y}^{JM}_{j\ell}(\hat{R}', \hat{r}') \cdot \chi^v_j(r) = \sum_{m_j m_\ell} C(j\ell J; m_j m_\ell M) \; Y^{m_j}_j(\hat{r}') \, Y^{m_\ell}_\ell(\hat{R}') \cdot \chi^v_j(r) \tag{4.63}$$

The corresponding reaction (scattering) amplitude is given by the simple generalization of eq.(3.71):

$$q(j\, m_j\, v \rightarrow j' m_{j'} v' | \hat{R}') = \sum_{\substack{J\, M \\ \ell \ell' m_{\ell'}}} i^{\ell-\ell'} [\pi(2\ell+1)]^{\frac{1}{2}} \cdot C(j\ell J; m_j\, 0\, M) \cdot C(j'\ell' J; m_{j'}\, m_{\ell'}\, M)\, T^J_{j'\ell'v',\, j\ell v} \cdot Y^{m_{\ell'}}_{\ell'}(\hat{R}') \tag{4.64}$$

$$T^J_{j'\ell'v',j\ell v} = \delta_{jj'} \cdot \delta_{\ell\ell'} \cdot \delta_{vv'} - S^J(j\ell v \to j'\ell'v') \tag{4.65}$$

The Sudden Approximation simplifies eq.(4.61) via the conditions (4.58) and (4.60), allowing one to obtain complete uncoupling by using a continuous, angle-dependent transformation, as shown for general potentials in refs.[10,66]:

$$\left\{ \frac{d^2}{dR^2} - \frac{\bar{\ell}(\bar{\ell}+1)}{R^2} + k^2_{v'\bar{j}} \right\} f^{\ell\overline{j}v}_{v'}(R', \hat{R}', \hat{r}') =$$

$$= 2\mathbf{M} \sum_{v''} < v'|V|v'' >_{\bar{j}} \cdot f^{\overline{\ell j v}}_{v''}(\mathbf{R}', \hat{R}', \hat{r}') \tag{4.66}$$

where the new radial wavefunctions now depend parametrically on the SF orientations of $\underline{R}'$ and $\underline{r}'$, i.e. they depend only on the internal angular coordinate $\gamma = \hat{R}' \cdot \hat{r}'$, shown in the potential multipolar expansion of eq. (3.80). The indeces $\hat{R}'$ and $\hat{r}'$ are then continuous indeces for the coupling matrix elements on the r.h.s. of (4.66) which are in turn diagonal in the new rotational index $\bar{j}$. The above equations are, of course, still rigorously close-coupling in the vibrational expansion. Since all the terms of the expansion (3.80) have been extended to all orders, the approximate solutions f constitute the complete or infinite order approximation (IOSA) already discussed [64].

These solutions must satisfy the usual boundary conditions:

$$f^{\overline{\ell j v}}_{v'}(R'; \gamma) \underset{R'\to o}{\sim} 0 \tag{4.67a}$$

and:

$$f^{\overline{\ell j v}}_{v'}(R'; \gamma) \underset{R'\to\infty}{\sim} \overline{k}^{-\frac{1}{2}}_{v'} \left\{ \delta_{vv'} \exp[-i(\overline{k}_v R' - \bar{\ell}\,\pi/2)] - \right.$$

$$\left. - S^{\overline{\ell j}}_{vv'}(\gamma) \exp[i(\overline{k}_{v'} R' - \bar{\ell}\,\pi/2)] \right\} \tag{4.67b}$$

It has been shown [31] that, by solving eq. (4.66) for the f functions one obtains an approximate solution of the CC equations (4.61) that includes an infinite rotational basis set that can be carried through for a potential V with any angle dependence [58].

The corresponding CC S-matrix elements can then be obtained in an approximate way via the IOSA matrix elements that appear in (4.67b):

$$S^{J}(j\ell v \rightarrow j'\ell'v') \sim i^{\ell+\ell'-2\bar{\ell}} < j'\ell'JM \mid S^{\overline{\ell j}}_{vv'}(\gamma) \mid j\ell\, JM > \qquad (4.68)$$

where the phase factor is needed to bring the approximate solution with the same phase as the radial part of the exact solution (3.61) [21], exactly as previously discussed in the SF treatment of the CS approach [11]. This approximate S-matrix conserves probability but is not quite symmetric for most j and ℓ [66]. It is moreover expected to be valid whenever the kinetic energy is large as compared with the rotational energy spacings, and the differences in centrifugal potentials of the coupled channels are small [68]. Such an approximation, as previously shown for the CS approximation, contains the arbitrary choices of ℓ and j, and it is clear that the accuracy of the results will also depend on the choice made for these parameters, a factor which controls the type of subtransitions ($\Delta m_j \neq 0$) that can be calculated within IOSA [21].

The corresponding IOSA-form of the T-matrix becomes, from eq.(4.65):

$$T^{\overline{\ell j}}_{v'v}(\gamma) = \delta_{v'v} - S^{\overline{\ell j}}_{v'v}(\gamma) \qquad (4.69)$$

and the corresponding SF scattering amplitude of (4.64) is approximated by the following expressions that makes use of eq.(4.68):

$$q(j'm_{j'}v' \rightarrow jm_j v \mid \hat{R}') \simeq \sum_{\substack{JM_J \\ \ell\ell' \\ m_{\ell'}}} (-)^{\ell-\bar{\ell}} [\pi(2\ell+1)]^{\frac{1}{2}} \cdot C(j\ell J; m_j o M) \cdot \qquad (4.70)$$

$$\cdot\, C(j'J; m_{j'} m_{\ell'} M) \cdot < j'\ell' M \mid T^{\overline{j\ell}}_{v'v}(\gamma) \mid j\ell\, JM > Y^{m_{\ell'}}_{\ell'}(\hat{R}')$$

Since the coupling elements on the l.h.s. of (4.70) contain a T-matrix which does not depend on orientation (i.e. M_J) and is therefore diagonal in J and $M_{J'}$, one can run the sums over J and M_J for the second Clebsch-Gordan coefficient on the r.h.s. of (4.70) and for the element on the left side of the T-matrix integral (and also over J' and $M_{J'}$ for which one can add sums without changing the results) and use the properties of the C.G. coefficients to rewrite eq.(4.70) in the following way:

$$\mathfrak{q}(jm_j v \to j'm_{j'}v' | \hat{R}') = \sum_{\ell\ell' m_{\ell'}} (-)^{\ell-\bar{\ell}} [\pi(2\ell + 1)]^{\frac{1}{2}} \cdot \cdot T^{IOSA}(\ell' m_{\ell'}, j'm_{j'} \mid \ell\, o, jm_j) \cdot Y_{\ell'}^{m_{\ell'}}(\hat{R}') \tag{4.71}$$

where:

$$T^{IOSA}(\ell' m_{\ell'}, j'm_{j'} \mid \ell\, o, jm_j) = \int d\hat{R}'\, d\hat{r}'\, Y_{\ell'}^{m_{\ell'}^{*}}(\hat{R}')\, Y_{j'}^{m_{j'}}(\hat{r}')\, T_{vv'}^{\overline{j\ell}}(\gamma)\, Y_j^{m_j}(\hat{r}')\, Y_\ell^{o}(\hat{R}') \tag{4.72}$$

Equation (4.71) now becomes the expression for the scattering amplitude in the IOSA approach and within an uncoupled angular momenta SF representation [10].

If one next chooses for the arbitrary parameter $\bar{\ell}$ the initial value of the partial wave angular momentum, ℓ_i, the elastic channel wavefunction in each set of coupled equations is given its right centrifugal potential. Moreover, eq.(4.71) can be further simplified in the following way by summing over ℓ' and $m_{\ell'}$ and by bringing back the average energy factor from eq.(3.70), corrected by the condition (4.58):

$$\mathfrak{q}_{IOSA}(jm_j v \to j'm_{j'}v' \mid \hat{R}') = \left(\frac{i}{2}\right) \cdot \left\{ \frac{1}{k_{\bar{j}v}\, k_{\bar{j}v'}} \right\}^{\frac{1}{2}} \cdot \cdot \sum_\ell (2\ell + 1) \int Y_{j'}^{m_{j'}}(\hat{r}')\, T_{vv'}^{\overline{j\ell}}(\gamma)\, P_\ell(\cos\theta_{R'})\, Y_j^{m_j}(\hat{r}')\, d\hat{r}' \tag{4.73}$$

Here $\theta_{R'}$ is the c.m. scattering angle and the summation on ℓ corresponds to the scattering amplitude formula for multichannel scattering via a spherical potential, as discussed in Chapter 1. The approximate prescription for computing inelastic scattering amplitudes within the IOSA scheme, $\mathfrak{q}_{IOSA}$'s, therefore requires one to hold the angle γ fixed, to carry out the vibrationally inelastic scattering calculations and then to perform a further integration over the γ values in order to get in the end a well-defined quantity.

The integral in question is of course a two dimensional integral over

$(\theta_{r'}, \phi_{r'})$ the orientation angles of the rigid molecule in the SF frame of reference. This can be further reduced by transforming the rotational wavefunctions, over which variables one is integrating, to a new set of integration coordinates which are measured with respect to the BF z-axis chosen to point along the direction of the incoming projectile $\hat{R}$ (see Fig. 3.2).

Eq. (4.73) may then be rewritten after a transformation of the Y's in the following way:

$$q_{IOSA}(jm_j v \to j'm_{j'} v'|\hat{R}') = \sum_{\mu\mu'} \mathcal{D}^{j'*}_{\mu' m_{j'}}(\hat{R}')\,\mathcal{D}^{j}_{\mu m_j}(\hat{R}') \cdot \left(\frac{i}{2}\right)\left(\frac{1}{k_{jv}\,k_{jv'}}\right)^{\frac{1}{2}}$$

$$\cdot \sum_{\ell} (2\ell + 1) \int Y^{\mu'*}_{j'}(\hat{r})\, T^{\bar{j}\ell}_{vv'}(\gamma)\, P_{\ell}(\cos\theta_{R'})\, Y^{\mu}_{j}(\hat{r})\, d\hat{r} \tag{4.74}$$

Now the two-dimensional integration is over (θ_r, ϕ_r), the molecular orientation with respect to the BF reference frame of Fig. 3.2. The T-matrix elements are however independent of ϕ_r since they only depend on $\hat{R}.\hat{r}$, and thus the integration over this angle yields $\delta_{\mu'\mu}$. Hence the first summation on the r.h.s. of (4.74) becomes a single sum over the index μ and the second sum of integrals becomes:

$$\sum_{\ell} (2\ell + 1)\, 2\pi \int Y^{\mu}_{j'}(\gamma, o)\, T^{\bar{j}\ell}_{vv'}(\gamma)\, P_{\ell}(\cos\theta_{R'})\, Y^{\mu}_{j}(\gamma, o) \sin\gamma \; d\gamma \tag{4.75}$$

This is now a one-dimensional integral over $\gamma = \theta_r$ and the $|j\mu>$ are just normalized associated Legendre polynomials. It is worh mentioning at this point that the $\bar{\ell} = \ell$ choice has led to making BF-IOSA amplitudes (i.e. the last integral examined) diagonal in the μ quantum number, a result similar to the CS approximation as discussed in the previous section of this Chapter.

The SF-amplitude however is not diagonal in m_j and can be used to compute $(\Delta m_j \neq o)$ transition within a given j-manifold [21].

An alternative approach is to choose for the parameter $\bar{\ell}$ the final angular momentum value, ℓ' which ensures that one radial function in (4.61) has its correct centrifugal potential, although it imposes the same ℓ' on all other connected radial functions. Within the CS approximation this choice has been shown to be the most direct consequence of condition (4.7) without the further reduction (4.15) (e.g. ref. [9]).

By thus putting $\bar{\ell} = \ell'$ into (4.71) and following various manipulations extensively discussed in the literature [8,9] one finally gets an alternative expression for (4.73):

$$q'_{IOSA}(jm_j v \rightarrow j'm_{j'}v' | \hat{R}') = \frac{i}{2} (-)^{j+j'} \delta_{m_j m_{j'}} \cdot (k_{\bar{j}v} k_{\bar{j}v'})^{-\frac{1}{2}} \cdot$$

$$\cdot \sum_{\ell'} (2\ell' + 1) 2\pi \int Y_j^{m_j}(\gamma, o)\, T_{vv'}^{\bar{j}\ell'}(\gamma)\, P_{\ell'}(\cos\theta_{R'})\, Y_{j'}^{m_j}(\gamma, o)\, d(\cos\gamma) \tag{4.76}$$

where again the integration to be carried in the BF frame is a one dimensional one over the internal coordinate γ. The m_j labels refer to the SF frame and this new amplitude is exactly equal to the BF frame formulation of the CS approach. Thus q'_{IOSA} is diagonal in m_j and cannot give Δm_j transitions, as opposed to the q_{IOSA} of eq.(4.74). The expression for the inelastic transition amplitude found in this last case is however simpler than the one of eq.(4.74) and, disregarding the energy factor differences, reproduces that originally suggested for the udden entrifugal approximation (CS) discussed in the previous section [5].

The degeneracy-averaged differential cross sections can also be obtained within the Sudden Approximation discussed here and it has been shown [58] that because of the unitarity of $\mathfrak{D}$ -functions both scattering amplitudes q_{IOSA} and q'_{IOSA} give identical results [21]. The correct CC expression has been in eq.(3.74). By using the result of eq. (4.74) and of eq.(4.75), or the final result of eq.(4.76) one obtains in the present case, from the properties of the $\mathfrak{D}$ -functions, that:

$$\sigma^{IOSA}(jv \rightarrow j'v' | \theta_{R'}) = \frac{1}{4(2j+1)k_{\bar{j}v}^2} \sum_{m_j} \Big| \sum_{\ell} 2\pi(2\ell + 1) \int_0^{2\pi} Y_{j'}^{m_j}(\gamma, o) \cdot$$

$$\cdot\; T_{vv'}^{\bar{j}\ell}(\gamma)\, P_{\ell}(\cos\theta_{R'}) \cdot Y_j^{m_j}(\gamma, o) \sin\gamma\, d\gamma \Big|^2 \tag{4.77}$$

Thus, although q_{IOSA} and q'_{IOSA} are different expressions, they provide the same formulae for degeneracy-averaged inelastic differential cross sections.

Instead of calculating the $\sigma(\theta_{R'})$ directly, one can expand that part of eq. (4.77) given by the following expression:

$$T^{\bar{j}}_{vv'}(\gamma|\theta_{R'}) = \sum_{\ell} (2\ell + 1)\, T^{\bar{j}\ell}_{vv'}(\gamma)\, P_{\ell}(\cos\theta_{R'}) \tag{4.78}$$

in Legendre polynomials of γ at each value of $\theta_{R'}$ [69]:

$$T^{\bar{j}}_{vv'}(\gamma|\theta_{R'}) = \sum_{\lambda} t^{\bar{j}\lambda}_{vv'}(\theta_{R'})\, P_{\lambda}(\cos\gamma) \tag{4.79}$$

where the coefficients are given by:

$$t^{\bar{j}\lambda}_{vv'}(\theta_{R'}) = (\lambda + \frac{1}{2}) \int_{-1}^{1} P_{\lambda}(\cos\gamma)\, T^{\bar{j}}_{vv'}(\gamma|\theta_{R'})\, d(\cos\gamma) \tag{4.80}$$

and can be obtained by Gaussian quadrature. The degeneracy-averaged differential cross secion of (4.77) then becomes:

$$\frac{d\sigma^{IOSA}}{d\hat{R}'}(j'v' \leftarrow jv) = \frac{1}{4k^2_{jv}} \sum_{\lambda} (2\lambda + 1)^{-1} \cdot C^2(j\lambda j';\, ooo) \cdot \cdot |t^{\bar{j}\lambda}_{vv'}(\theta_{R'})|^2 \tag{4.81}$$

where one manages to factor out the dynamic dependence of the cross sections in the τ's and the state dependence in the Clebsch Gordan coefficients [70].

The total integral cross sections can now be obtained by direct integration of eq.(4.77) over the $\theta_{R'}$ angles:

$$\sigma^{IOSA}(jv \rightarrow j'v') = \frac{\pi}{k^2_{jv}} \sum_{\ell} (2\ell + 1)\, \mathcal{P}_{\ell}(jv \rightarrow j'v') \tag{4.82}$$

where the IOSA opacity functions are given by:

$$\mathcal{P}_{\ell}(jv \rightarrow j'v') = \frac{1}{(2j + 1)} \sum_{m_j} \left| 2\pi \int_{o}^{2\pi} Y^{j'}_{m_j}(\gamma, o)\, T^{\bar{j}\ell}_{vv'}(\gamma)\, Y^{j}_{m_j}(\gamma, o) \sin\gamma\, d\gamma \right|^2 \tag{4.83}$$

If one wants to express the opacity functions in terms of total contributions from the correct CC expression of the scattering amplitude (4.64), the integrations over $\theta_{R'}$ have to be performed via the q_{IOSA} and q'_{IOSA} expressions. One then finds

[21] that the two choices of ℓ lead to different forms of J-opacity contributions, which contain different weighted sums over transition probabilities indexed by either ℓ or ℓ' as a channel index. The physical observables however, i.e. the total integral cross sections, are given by identical expressions summed over J.

A further choice for $\bar{\ell}$ is of course given by $\bar{\ell} = J$, but appreciable differences in the contributions to total integral cross sections appear in this case as already shown for CS approximations in eq. (4.56). Thus, although one can evaluate magnetic transitions, the average cross sections are likely to come out less accurate. There is of course no a priori method available to determine the best choice for $\bar{\ell}$. An empirically obvious criterion is that the computed cross sections agree within a reasonable degree of accuracy with the correctly evalued CC results and that reproduce the physics of the collision process under study at the collision energies for which the impulse approximation holds. In the end, therefore, the proper choice is likely to be determined by the particular nature of the interaction potential, the relative strengths of the magnetic transitions and the relative collision energy used in the experiments as compared to the interaction time. The $\bar{\ell} = \ell'$ choice provides the simplest expressions for a "spherical" potential scattering equation, while the $\bar{\ell} = \ell$ choice allows also to obtain Δm_j - transitions that are excluded in the former case. These alternatives perform a different 'averaging' of the correct SF T-matrix elements over the rotational multiplicities. For the former choice, in fact, $(2j' + 1)^{-1}$ values of $T^{j\bar{\ell}'}_{vv'}(\gamma)$ contribute to the integral (4.75), while for the latter there are $(2j + 1)^{-1}$ values of the integral that contribute to a given SF total angular momentum J. The total sum is, as seen before, the same in both cases. Moreover, since the physical conditions for the energy sudden degeneracy of levels (4.58) are valid at higher collision energies, the J values that are more important are the larger ones, hence it makes little difference, for low-rotational quantum number transitions, which "averaging" of the T-matrix elements is performed.

The applications to heavy systems such as Ar, TeF, $Ar\text{-}N_2$, $Ar\text{-}F_2$, $Ar\text{-}Cl_2$, $Ar\text{-}Br_2$ and $Ar\text{-}J_2$ [71] have yielded cross sections of useful accuracy in comparison with CC-results, although the agreement is worse for large J contributions. A comparison of IOSA results and CS results has also been performed for the $(j = 1 \rightarrow j' = 2)$ inelastic cross sections of Ar-CsF [72] and rather good agreement was obtained there with the $\bar{\ell} = \ell$ choice of parameter.

Another comparison of IOSA calculations and CS cross sections has recently been performed for H^+- pH_2 collisions at 3.7 e V and 10 e V and for various rotationally inelastic transitions [21]. Their results indicate the generally satisfactory agreement and validity of the present approach except in the case of long-range $\Delta j = 2$ transitions, which are dominated by charge-quadrupole interaction terms.

Returning now to the IOSA opacity functions form of eq.(4.83) it is interesting to also rewrite the T-matrix elements in a Legendre expansion form:

$$T^{\bar{j}\ell}_{vv'}(\gamma) = \sum_L \tau^{\bar{j}\ell L}_{vv'} \; P_L(\cos\gamma) \tag{4.84}$$

where, one obviously has that:

$$\tau^{\bar{j}\ell L}_{vv'} = (L + \frac{1}{2}) \int_{-1}^{1} T^{\bar{j}\ell}_{vv'}(\gamma) \; P_L \; (\cos\gamma) \; d\,(\cos\gamma) \tag{4.85}$$

then the $\mathcal{P}_\ell$'s can be written as:

$$\mathcal{P}_\ell(jv \rightarrow j'v') = \sum_L (2L+1)^{-1} \cdot C^2(jLj';\, o\; o\; o) \; |\tau^{\bar{j}\ell L}_{vv'}|^2 \tag{4.86}$$

The corresponding cross sections thus become:

$$\sigma^{IOSA}(jv \rightarrow j'v') = \sum_L (2L+1)^{-1} \cdot C^2(jLj';\, o\; o\; o) \; (\pi/k^2_{jv}) \cdot \\ \cdot \sum_\ell (2\ell+1) \; |\tau^{\bar{j}\ell L}_{vv'}|^2 \tag{4.87}$$

which, for transitions from the ground rotational state $(j = o)$ reduces to:

$$\sigma^{IOSA}(ov \rightarrow j'v') = \frac{\pi}{(2j'+1)k^2_{ov}} \sum_\ell (2\ell+1) \; |\tau^{\bar{j}\ell j'}_{vv'}|^2 \tag{4.88}$$

so that one can write in general [69, 73] that:

$$\sigma^{IOSA}(jv \rightarrow j'v') = \frac{k_{ov}^2}{k_{jv}^2} \sum_{j''} C^2(j\ j''\ j';\ o\ o\ o) \cdot \sigma(ov \rightarrow j''v'') \qquad (4.89)$$

where again one can factorize the whole matrix of cross sections and constructing it from one of its columns. In the analysis of many experiments, however, one is interested in obtaining at several collision energies the total scattered intensity, i.e. it becomes necessary to sum over final rotational states whenever the target molecule has been "prepared" in a specific initial vibro-rotational state:

$$I(jv \rightarrow v' \mid \theta_{R'}) = \sum_{j'} \sigma(jv \rightarrow j'v' \mid \theta_{R'}) \qquad (4.90)$$

If one now chooses $\bar{j}$ to be equal to the initial rotational state j, this assures one that the large elastic component in each set of coupled equations (4.66) will have the correct channel wavenumber Moreover, the correct expression of (3.74) for each individual cross section can be used in (4.90) to obtain a simplified from via eq.(4.77):

$$I(jv \rightarrow v' \mid \theta_{R'}) = \sum_{j'} \sigma^{IOSA}(jv \rightarrow j'v' \mid \theta_{R'}) =$$

$$= \sum_{j} \frac{1}{4(2j+1)k_{jv}^2} \sum_{m_j} \Big| \sum_{\ell} 2\pi(2\ell+1) \int Y_j^{m_j}(\gamma, o)\, T_{vv'}^{j\ell}(\gamma) \cdot \qquad (4.91)$$

$$\cdot\ P_\ell(\cos\theta_{R'})\ Y_j^{m_j}(\gamma, o)\, d(\cos\gamma)\ \Big|^2$$

by taking the m_j-sum inside the integral and using the spherical harmonics addition theorem, one rewrite the above equation as follows:

$$I_j(v \rightarrow v' \mid \theta_{R'}) = \frac{1}{4k_{jv}^2} \Big| \sum_{\ell} (2\ell+1) \int_{-1}^{1} T_{vv'}^{j\ell}(\gamma)\, P_\ell(\cos\theta_{R'}) \cdot$$

$$\cdot\ d(\cos\gamma) \Big|^2 = \frac{1}{2} \int_{-1}^{1} |f_j(v \rightarrow v';\ \theta_{R'})|^2\, d(\cos\gamma) \qquad (4.92)$$

where:

$$f_j(v \to v';\, \gamma |\, \theta_{R'}) \;=\; - \left(\frac{1}{2ik_{\bar{j}v}}\right) \sum_{\ell} \,(2\ell + 1)\; T^{\bar{j}\ell}_{vv'}(\gamma)\; P_\ell(\cos\theta_{R'}) \qquad (4.93)$$

This result would also have been found if one used the $\bar{\ell} = \ell'$ choice for the parameter appearing in the sudden centrifugal approximation (4.60). The eq. (4.93) now gives just the formula for the vibrationally inelastic scattering of sperical vibrators interacting via an isotropic potential. If represents the simple angle average of a spherical particle differential cross section calculated by holding the potential angle fixed.

The corresponding integral cross section in obtained from (4.93) by first integrating over $\theta_{R'}$ the γ-dependent spherical amplitude computed at each $\theta_{R'}$ value:

$$\begin{aligned} f_j(v \to v';\, \gamma) \;&=\; \int_{-1}^{1} f_j(v \to v';\, \gamma|\theta_{R'})\; d(\cos\theta_{R'}) \\ &=\; \left(\frac{\pi^{\frac{1}{2}}}{ik_{jv}}\right) \sum_{\ell}\; (2\ell + 1)^{\frac{1}{2}} \cdot T^{j\ell}_{vv'}(\gamma) \end{aligned} \qquad (4.94)$$

This is turn yields the spherical particle integral cross section for fixed γ[4]:

$$\sigma_j^{IOSA}(v \to v';\, \gamma) \;=\; \frac{\pi}{k_{jv}^2} \;\sum_{\ell}\; (2\ell + 1)\; |\; T^{j\ell}_{vv'}(\gamma)|^2 \qquad (4.95)$$

from which one easily obtains the integral cross section by integration over γ:

$$\sigma_j^{IOSA}(v \to v') \;=\; \frac{1}{2} \int_{-1}^{1} \sigma_j^{IOSA}(v \to v';\, \gamma)\; d(\cos\gamma) \qquad (4.96)$$

This is a remarkably simple result which essentially gives the vibrationally inelastic integral cross section for spherically interacting partners. The subscript j of the initial rotational state also appears as a superscript in the T-matrix of eq.(4.95), which depends on it through the weak and often negligible dependence of the vibrational wavefunctions χ_{jv} on their rotational quantum numbers and through the k_{jv} value which, often in experiments, is practically independent of j [74].

If one examines the even simpler case of purely rotational excitation in atom-molecule collisions, then the coupled equations (4.96) become a single uncoupled equation for each contributing partial wave and the corresponding elements of the S-matrix in (4.97b) acquire the "potential" scattering simplified form see in Chapter 1:

$$S^{\bar{\ell} k_{\bar{j}}}(\gamma) = \exp\left(2i\, \eta_{\bar{\ell}}^{k_{\bar{j}}}(\gamma) \right) \tag{4.97}$$

where the η's are the phase shifts for a fixed value of the potential angle γ and are labelled via the energy parameter of condition (4.58). The two choices for $\bar{\ell}$ discussed before allow one now to obtain even simpler expressions for the scattering amplitudes q_{IOSA} and q'_{IOSA} of eq.s(4.74) and (4.76) respectively.

The total differential cross sections from a given rotational state is now given, in analogy with eq.(4.92):

$$I_j(\theta_{R'}) = \frac{1}{2} \int_{-1}^{1} | f_{k_{\bar{j}}} (\gamma|\theta_{R'}) |^2 \, d(\cos\gamma) \tag{4.98}$$

where:

$$f_{k_{\bar{j}}}(\gamma\,|\theta_{R'}) = \left(-\frac{1}{2ik_{\bar{j}}} \right) \sum_{\ell} (2\ell + 1) \left\{ 1 - \exp\left[2i\eta_{\ell}^{k_j}(\gamma) \right] \right\} P_{\ell}(\cos\theta_{R'}) \tag{4.99}$$

and the $k_{\bar{j}}$ value now defines the collision kinetic energy, independent of j in most experimental cases.

The total integral cross section follows above and from eq.s(4.95) (4.96), i.e. one can write:

$$\sigma_{k_j}^{IOSA}(\gamma) = \frac{4\pi}{k_j^2} \sum_{\ell} (2\ell + 1) \sin^2 \eta_{\ell}^{k_j}(\gamma) \tag{4.100a}$$

and:

$$\sigma_{k_j}^{IOSA} = \frac{1}{2} \int_{-1}^{1} \sigma_{k_j}^{IOSA}(\gamma) \, d(\cos\gamma) \tag{4.100b}$$

The interesting aspect of this approximations relies on the drastic decoupling operated by conditions (4.58) and (4.60) which essentially yield cross section calculation formulae for spherical potentials and can thus be readily applied to several systems with a limited computational cost. If one further relates these quantum results with earlier classical equivalents [75] it then becomes possible to simplify greatly the calculation of trasport properties of dilute gas mixtures that one usually expresses as functions of collision integrals [76]. In fact, integral, differential, viscosity and diffusion cross sections have recently been calculated [74] via the IOSA approach and by using the computed potential from the EGM treatment discussed in Chapter 2. Only rotational inelasticity was considered there and the integral cross sections showed reasonable agreement, especially in the high energy region, with the corresponding experiments. The transport property cross sections appeared to be in even better agreement with measured data and showed great differences between the computations including the whole potential expansion in eq. (4.66) and those that only considered its spherical component. This is confirmed by their comparisons with experimental differential cross sections (DCS) presented in Fig. 4.3. The velocity-averaging is done around the most probable experimental relative energy and arbitrary units where used for the plots.

Considering the experimental uncertainty, as shown in the Figure for a few angles, the agreement is very good and the differences between the two computed cuves point once more to the important anisotropic affects contributing to experimentally averaged cross sections and which markedly quench the spherical potential oscillations generated by V_0 only. This result has already been pointed out by Cross [61] who showed, by using the semiclassical sudden approximation, that the anisotropic component of this intermolecular potentials causes both the slow rainbow oscillations and the rapid diffractions in the observed DCS to be quenched and shifted. The use of the IOSA approach to the rapid evaluation of DCS for experimentally observed systems via realistic anisotropic potentials can then provide a rather direct check of which features of atom - molecule interactions control the damping and shifting effects discussed here Such an approach also lends itself to a detailed evaluation of several cross-section matrices in order to determine in turn potential parameters through a "self consistent" interaction with experimental data [77].

Turning now to vibro-rotational excitation calculations, the IOSA approximation of the relevant coupled equations obviously provides an attractive instrument for analyzing the interplay of rotational and vibrational channels regarding

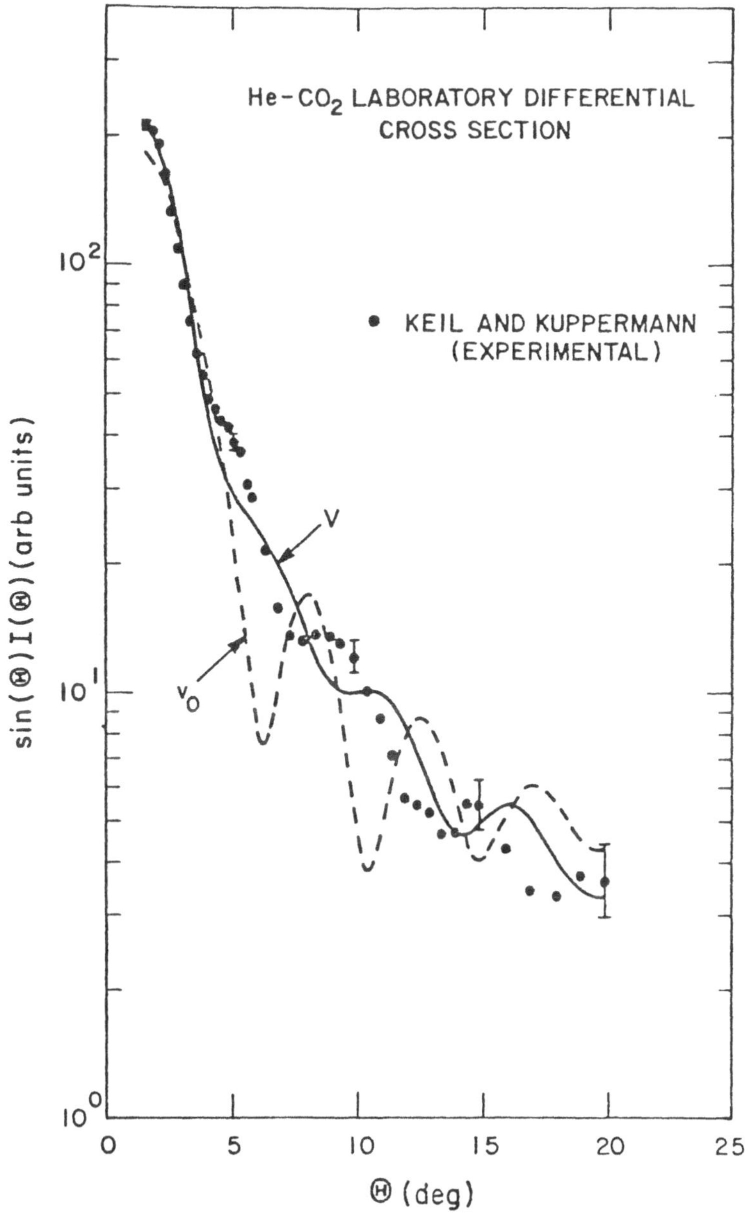

Fig. 4.3

Velocity averaged laboratory DCS for He + CO_2 at a collision energy of ~746 °K. The solid circles are experimental results, the solid line represents the DCS (IOSA) calculated via the full aniso- (cont.)

(Fig.3, cont.) tropic potential V, while the dashed line describes the same calculations but performed only via V_0, the spherical component of the total V. [from G.A.Parker and R.T.Pack, J.Chem.Phys., 68, 1585 (1978)].

Energy redistribution within the molecular degrees of freedom.

In eq.(4.66), in fact, the coupling between differential equations is achieved according to the number of vibrational target wavefunctions used for the total wavefunction expansion of eq.(4.63) and does not depend on the number of rotational states. Since rigorous CC calculations have been performed so far with a rather limited number of $\chi_v^j(r)$ functions, because of the rapid proliferation of the coupled channels (see above), the IOSA approach can go further into testing their effects on numerical accuracy of vibro-rotational cross sections.

Some results from such numerical experiments have recently been obtained for the He - H_2 system, where the H_2 target was treated as a vibrating rotor and the potential used was the KM potential discussed in the last section of Chapter 3 [78]. It was found there that the number of vibrational functions which need to be included at intermediate collision energies is rather large even for mildly anisotropic systems, as is the present one, for which up to six vibrational levels need to be taken in for the energy range from 1.2 to 4.2 e V. It was usually found that at least two levels above the final state of the transition considered has to be included [79], an interesting analogy with the rotationally inelastic results discussed in the previous Chapter.

In the case of the nearly-spherical interaction acting in the He - H_2 case, it has also been found that the use of purely spherical potential components V_0, the so-called Breathing Sphere Approximation (BSA), yields integral cross sections for a given $v \rightarrow v'$ transition that are only slightly smaller than the corresponding ones obtained by summing over the first few rotational states as in eq. (4.90). Even in this case one finds that the effect of anisotropic potential components on vibrational inelasticity is one of increasing the values for the corresponding cross sections [80], although the correct behaviour of flux rearrangement among the various final states is a strong function of the system under examination and little can yet be said from a general point of view.

Another interesting "observable" for vibrationally inelastic processes appears in eq.(4.95) where inelastic (V-T) integral cross section for spherical partners is presented for a given value of the relative angular coordinate γ. Calculations of $\sigma(v \to v'; \gamma)$ at various collision energies for molecule-molecule encounters via piecewise potentials showed in fact [81] that the efficiency in transferring energy into the molecular vibrational modes for neutral systems was maximal within a rather narrow cone along the bond direction, the cone becoming larger with increasing moments of inertia of the targets. This again indicates qualitatively that the energy degeneracy introduced by the sudden treatment of the rotational levels is more accurately reproduced, at fixed collision energy , by the heavier molecules with more closely spaced rotational levels and where little torque is applied by the lighter projectiles.

As a consequence then one expects that, with increasing near-degeneracy of neighbouring levels the angle dependence would become less pronounced for the corresponding vibrational excitations.

As an example, the results for ($0 \to 1$) vibrational excitation of Cl_2 molecules in H_2 - Cl_2 collisions is presented in Figure 4.4 for a relative collision energy of 2.0 e V. Five vibrational channels were used in the coupling expansion (4.66). One sees there that the γ values for which the contributions to inelasticity are still sezeable go till 30° - 40° from a maximum value for $\gamma = 0$ but are practically negligible at $\gamma = \pi/2$. The same happens with other neutral systems like O_2 and H_2. Here the potential used was essentially a 3-body potential from Lennard-Jones parameters. When the more anisotropic interaction of ion-diatomic case is examined [21] for purely rotational excitations, one finds that orientation effects yield larger phase shifts with larger angular differences from the collinear geometry.

Vibrational excitation, however, is generally expected to be more efficient, classically speaking, for along-the-bond encounters due to the "bond-dilution" effects from the projectile interaction and it is therefore interesting to see that this point is borne out within the IOSA approach [81] for H^+ - H_2 collisions. The preliminary results of [81] indicate that, with neutral partners, the lighter exhibits a narrower angular "cone" where vibrational inelastic stills contributes, while heavier systems present a broader "active" angular range. In the Ar-CO_2, for instance, the increased phase shifts found [71] for collinear encounters is also along the findings of [81]. It, is clear, however, that the use of more realistic

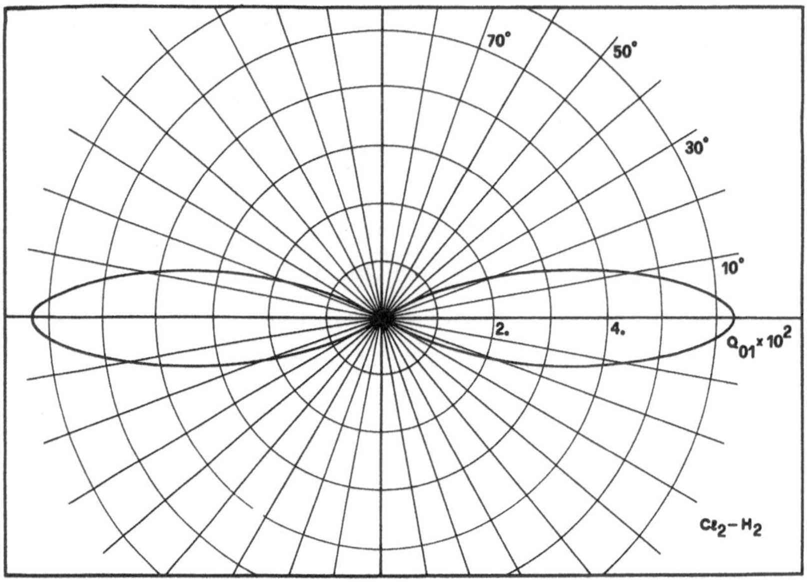

Fig. 4.4

Influence of the γ changes on the sudden approximation contributions to $(0 \rightarrow 1)$ vibrational inelasticity for $C\ell_2 + H_2$ collisions. The hydrogen molecule was treated as a structureless projectile, while the $C\ell_2$ vibrational states were expanded over Morse oscillators. E_{coll} = = 2.00 eV [from F.A.Gianturco and U.T.Lamanna, Chem.Phys.Lett., 6, 326 (1970)].

potentials is an essential need for a clear assessment of general rules. This is confirmed by the IOSA results that have been obtained for the vibrational inelastic (integral) cross sections of the integral in eq.(4.96) and computed via an ab initio evalued potential surface [83] for the Li + H_2 system [84]. Figure 4.5 presents in fact the $\sigma^{IOSA}_{j=o}(o \rightarrow 1; \gamma)$ as a function of the contributing partial waves and at various values of γ. One sees here that the largest contributions come from "on-the-bond" collisions (and they markedly increase this effect with increasing collision energies [84]) and also that the high-ℓ regions (inset) ex-

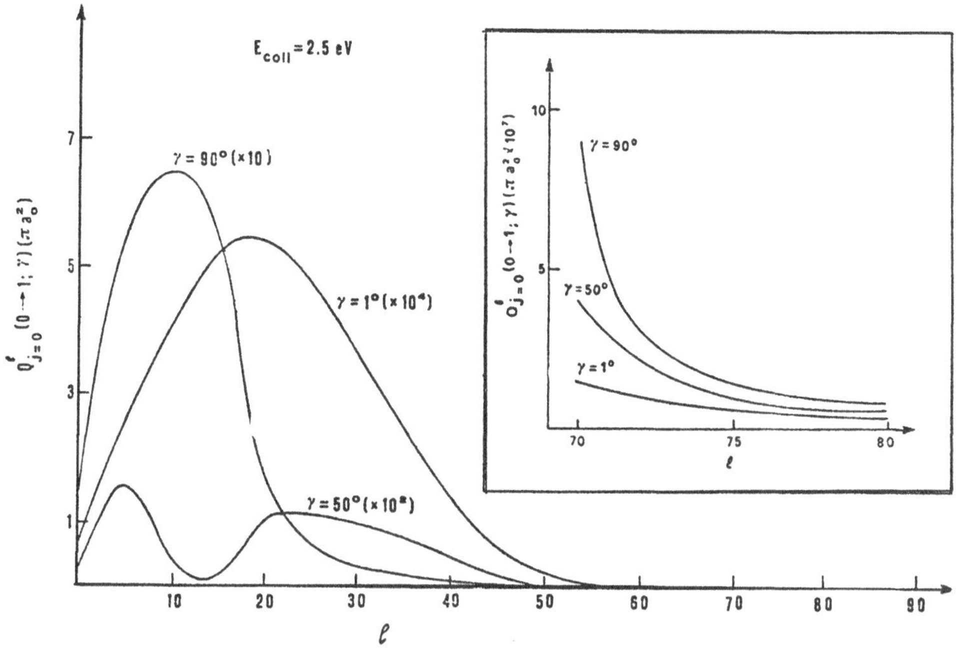

Fig. 4.5

Rotationally summed integral cross sections for the (0→1) vibrational excitation in $Li + H_2$ collisions. The various contributions at various γ values (= arccos $\hat{r}\cdot\hat{R}$) are plotted as a function of the partial wave parameter. The inset shows an enlarged plot of the high-ℓ contributions [F.A.Gianturco and U.T.Lamanna (1978)].

hibit a more isotropic behaviour with γ, beyond ℓ = 75, in fact, all angles tend to the same contribution.

This is therefore an interesting example of the different situations that take place it the low-ℓ, potential-dominated regions and in the large-ℓ region of centrifugal potential dominance. In the latter instance, in fact, the scattering process becomes essentially isotropic because of the large centrifugal barrier that cancels out the anisotropic features of the potential.

4.4 The Effective Potential Treatment

As we have discussed in the previous Sections, the crucial problem in reducing the dimensions of the rigorous CC equations originates from the multiplicity of the rotational states to be included in the CC expansion and from the large number of them needed even at moderate collision energies. As a consequence, the CS and IOSA approximations have assumed a physical situation where cross section values are dominated by the potential terms (of its multipolar expansion) and have strived for a convincingly approximate form of the centrifugal coupling in a total angular momentum representation.

An even more drastic reduction of the correct centrifugal coupling has been given by Rabitz and coworkers [85-89] and has been extensively compared in recent years with CC results, CS calculations and semiclassical approximations (see Table 2).

Their original derivation starts from the f_λ coefficients of eq.(4.5) that appear in the CC rigorous vibro-rotational equations for atom-molecule collisions as shown in eq.(4.66). As discussed in Chapter 3, they can be written (SF reference frame) as follows:

$$f_\lambda(j'\ell', j''\ell''; J) = (-)^{j''+j'-J} \left\{[j''] \cdot [j'] \cdot [\ell''] \cdot [\ell']\right\}^{\frac{1}{2}} \cdot$$
$$\cdot \begin{pmatrix} j'' & \lambda & j' \\ 0 & 0 & 0 \end{pmatrix} \cdot \begin{pmatrix} \ell'' & \lambda & \ell' \\ 0 & 0 & 0 \end{pmatrix} \cdot \begin{Bmatrix} J & \ell' & j' \\ \lambda & j'' & \ell'' \end{Bmatrix} \tag{4.101}$$

where the $[\alpha]$ symbols stand for $(2\alpha + 1)$ and the $\{...\}$ are 6-j coefficients (see Section 3).

The effective potential (EP) theory seeks a simplified procedure that connects a $|jv>$ state to a $|j'v'>$ state regardless of spatial orientation effects, i.e. regardless of the given magnetic states $m_j, m_{j'}$ involved in the rotational transitions. The labelling of each transition by such states appears, on the other hand, exclusively via the 6-j coefficient of (4.101), as can be easily seen if one uses a decoupled representation of the involved angular momenta [87].

Their effect is one of putting in some probability amplitude for making an m_j-dependent coupling between the (ℓ', ℓ'') orbitals that are already coupled via the potential anisotropy of order λ.

TABLE 2

Rotovibrational cross section calculations on atom(ion)-molecule and molecule-molecule systems performed via the EP model.

System	Processes considered (collision energy in eV)		Reference (year)
$He + H_2$ (o. and p.)	$\Delta j = \pm 0,2,4,6$ $(0.37 \rightarrow 0.87)$	$(T \rightarrow R)$	[88] (1974)
	$\Delta v = 1$; $\Delta j = 0,2,\ldots,6$ $(0.9 \rightarrow 1.5)$	$(R,V \rightarrow T)$	[89] (1975)
	$\Delta v = -1$; $\Delta j = 0,\ldots,8$ $(1.2 \rightarrow 2.0)$		[90] (1974)
	$\Delta v = -1$; $\Delta j = 0,\ldots,10$ $(1.2 \rightarrow 2.0)$		[91] (1975)
$H + H_2$	$\Delta j = \pm 0,2,4,6$ $(0.5 \rightarrow 1.5)$	$(T \rightarrow R)$	[92] (1975)
$H + CO$	$\Delta j = 1,2,\ldots,7$ $(0.01 \rightarrow 0.05)$	$(T \rightarrow R)$	[31] (1975)
$He + CO$	$\Delta j = 1,2,3,4$ $(0 \rightarrow 0.019)$	$(T \rightarrow R)$	[34] (1976)
$He + HC\ell$	$\Delta j = \pm 1,2,3$ $(0.004 \rightarrow 0.04)$	$(T \rightarrow R)$	[33] (1977)
$He + N_2$	$\Delta j = 0,2$ $(0.0015 \rightarrow 0.005)$	$(T \rightarrow R)$	[32] (1975)
$He + HCN$	$\Delta j = \pm 0,1,2$ $(0 \rightarrow 0.014)$	$(R \rightarrow T)$	[94] (1975) [33] (1977)
$H_2 + H_2$	$\Delta j = \pm 0,2,4$ $(0.05 \rightarrow 0.1)$	$(T \rightarrow R)$	[95] (1974)
	$\Delta j = \pm 0,\ldots,8$; $\Delta v = \pm 1$ $(0.5 \rightarrow 1.35)$	$(V,R \rightarrow T)$	[101] (1977)
HF - HF	$\lvert 00\rangle \rightarrow \lvert 11,02,22\rangle$; $\lvert 11\rangle \rightarrow \lvert 02\rangle$ $(0.06 \rightarrow 1.0)$	$(R \rightarrow T)$	[99] (1977)

Table 2 (cont.d)

$HD + H_2$	$\Delta j = \pm 0,1,2,3$ $(0.01 \rightarrow 0.11)$	$(T \rightarrow R)$	[96] (1975)
$N_2 - N_2$	$\Delta j = \pm 0,2,4$ $(0.025 \rightarrow 0.2)$	$(T \rightarrow R)$	[98] (1975)

Their uncoupled formulation, in fact, yields unity after summation over all the m_j and $m_{j'}$ [87].

What one needs, however, is an effective Hamiltonian $\mathcal{H}_{eff}$ that also includes the above wanted reduction in the corresponding definition of the spectral eigenstates of the unperturbed target, i.e. of the eigenvalues generated by the $\mathcal{H}_{int}$ Hamiltonian contribution of eq. (3.5). The unperturbed molecular hamiltonian needs, in fact to remain diagonal in the effective basis of vibrational eigenstates that are now independent of m_j: $|j m_j v\rangle \Rightarrow |jv\rangle$:

$$\langle v'j' \mid \mathcal{H}_{int}^{eff} \mid v_j \rangle = \delta_{jj'} \cdot \delta_{vv'} \, \varepsilon_{vj} \qquad (4.102)$$

where the energy level spacing should be the same as from the original $\mathcal{H}_{int}$ of the system. For the more general case of molecule collisions, the suggestion made by Rabitz and coworkers [87] was thus to write down the following matrix elements for the new effective Hamiltonian:

$$\langle j_1' \, j_2' \, v_1' \, v_2' \mid \mathcal{H}_{int}^{eff} \mid j_1 \, j_2 \, v_1 \, v_2 \rangle \sum_{\substack{\text{all } m_i \\ m_i'}} \mathbf{G}(j_1, j_2, j_1' \, j_2'; m_1, m_2, m_1' \, m_2' \mid \hat{r}') \cdot \qquad (4.103)$$

$$\cdot \langle j_1' \, m_1' \, j_2' \, m_2' \, v_1' \, v_2' \mid \mathcal{H} \mid j_1 \, m_1 \, j_2 \, m_2 \, v_1 \, v_2 \rangle$$

where the function G depends only on the rotational variables and their orientations, since the $\mathcal{H}_{eff}$ acts only on the roto-vibrational variables <u>without</u> orientation effects. For two molecules the index i = 1,2 in the summation over the functions $\mathbf{G}$ that acts as a weighting factor over orientations.

Since the new G-function should reflect the coupling structure of the correct Hamiltonian, or of the 6-j and 9-j symbols in (3.75), its simplifying effect should be related to such a structure. Thus, the constraint:

$$\sum_{m_j,\, m_2} G = 1 \tag{4.104}$$

leads to rewriting (4.103) as:

$$< j_1' \; j_2' \; v_1' \; v_2' \mid \mathcal{H}_{eff} \mid j_1 \; j_2 \; v_1 \; v_2 > \; =$$

$$= \left\{ - \frac{1}{2M} \nabla^2_{R'} + \varepsilon^{j_1}_{v_1} + \varepsilon^{j_2}_{v_2} \right\} \delta_{j_1 j_1'} \cdot \delta_{j_2 j_2'} \cdot \delta_{v_1 v_1'} \cdot \delta_{v_2 v_2'} \cdot \tag{4.105}$$

$$\cdot < j_1' \; j_2' \; v_1' \; v_2' \mid V_{eff}(R') \mid j_1 j_2 \; v_1 v_2 >$$

where now we have introduced the effective potential coupling between vibro-r tational states without explicit orientation effects, and one has therefore obtained only a dependence on $|\underline{R}'|$, the SF intermolecular distance or atom-molecule distance. Returning now to the atom-molecule case, the simplifying effect of the G-functions can be further defined on an ad hoc basis [87] by rewriting the correct CC coupling expression in the following way:

$$< j'\ell'v'J \mid V \mid j\ell vJ > \; = V^{J}_{\ell\ell'}(jv \rightarrow j'v' \mid \underline{R}') =$$

$$= \sum_{\lambda} V^{\lambda}_{vv'}(R') \cdot \tilde{f}_{\lambda} \, (j\ell,\, j'\ell';\, J) \tag{4.106}$$

$$= \sum_{\lambda} V^{\lambda,eff}_{vv'}(R') \cdot f^{eff}_{\lambda}(j,\, j')$$

where the vibrational coupling term is correctly given by:

$$V^{\lambda,eff}_{vv'}(R') \; = \; \int dr' \; \chi^{eff*}_{v'}(r') \; V_{\lambda}(r',\, R') \; \chi^{eff}_{v}(r') \tag{4.107}$$

with $\chi_v^{eff}(r')$ begin the $\mathcal{H}_{int}^{eff}$ vibrational eigenfunctions.

The angular coupling term f_λ^{eff} now is drastically different from its correct form (4.101) and is instead given by:

$$f_\lambda^{eff}(j, j') = (-)^{(|j-j'|+j+j')/2} \cdot \left\{(2j + 1)(2j' + 1)\right\}^{1/4} \cdot (2\lambda + 1)^{-\frac{1}{2}} \cdot \begin{pmatrix} j' & \lambda & j \\ 0 & 0 & 0 \end{pmatrix} \tag{4.108}$$

We see that the $\underline{j} + \underline{\ell}$ coupling has been eliminated and that the angular effective coefficients now depend only on the $|j\rangle \to |j'\rangle$ transitions without any orientational effect from the 6-j coefficients. The anisotropy of the potential, P_λ, only couples j and j' and is not acting on the two phase angles of the corresponding initial and final spherical harmonics of the rigid rotor target that provide the magnetic orientations.

In conclusion, the physical assumption of this method is essentially to consider the m_j-dependence of the coupling matrix elements of the correct CC treatment as weak, hence allowing one to average the full potential with a suitable weighting for each m_j index and to assume that the wavefunctions for different m_j values are m_j-independent to a first approximation.

It has been shown [1], in fact, that the assumed SF form of the coupling in (4.108) can also be obtained in a BF representation, where by the corresponding coefficients come out to be weighted averages over all possible Ω-orientations. The corresponding correct centrifugal coupling terms of eq.s (4.12) are simplified by the EP treatment in the sense that the off-diagonal contributions of (4.12a) are disregarded as in the CS approximation, while the diagonal full terms (4.12b) in the BF representation are written as follows:

$$h_{\Omega,\Omega}^{Jj}(R) \sim h_{eff}^{\ell,j}(R) = R^{-2} \cdot \delta_{jj'} \cdot \left\{ [\ell(\ell + 1) + j(j + 1)] - 2 \sum_{\alpha,\lambda} \alpha^2 \cdot [2\lambda + 1]^{\frac{1}{2}} \right\} \cdot \begin{pmatrix} j & j & 0 \\ \alpha & -\alpha & 0 \end{pmatrix} \cdot \begin{pmatrix} j & j & \lambda \\ \alpha & -\alpha & 0 \end{pmatrix} \tag{4.109}$$

which is clearly a weighted average of the exact contribution (4.12b). Thus, the

EP approach is another step beyond the CS decoupling technique. They both treat the centrifugal barrier within the same approximation, but only the EP approximation alters the coupling potential by also averaging its orientation dependence. The consequence of this averaging is to recover ℓ as a good quantum number, since one puts $\ell = J$ in (4.109), and to obtain complete decoupling in ℓ and only partial coupling over the vibrational states expansion with a weighting coefficient for each $|jv\rangle \to |j'v'\rangle$ matrix element.

Because of the simpler form of f_λ^{eff} one can now expand the wavefunction ψ as follows:

$$\psi(\underline{R}', r') = \sum_{v,\ell} \chi_v^{eff}(r') \cdot (R')^{-1} \cdot \tilde{u}_{vj}^{\ell}(R')\, P_\ell(\cos \theta_{R'}) \qquad (4.110)$$

The resulting Schrödinger equation: $\mathcal{H}_{eff}\, \psi = E_{tot}\, \psi$ now becomes much simpler than its CC expression (4.61) and can easily be obtained in the following form:

$$\frac{1}{2\mathcal{M}} \left\{ \frac{d^2}{dR'^2} + k_{jv}^2 - \frac{\ell(\ell+1)}{R'^2} \right\} \tilde{u}_{vj}^{\ell}(R') = \sum_{v'j'} \langle j'v'|V_{eff}|jv\rangle\, \tilde{u}_{v'j'}^{\ell}(R') \qquad (4.111)$$

This is again a strongly decoupled equation for which one seeks asymptotic solutions of the form:

$$\psi_{vj}(\underline{R}') \underset{R'\to\infty}{\sim} (2\pi)^{3/2} \{\delta_{vv'} \cdot \delta_{jj'} \exp(i\, k_{vj}\, z') + \tilde{f}_{vj\to v'j'}(\theta_{r'}) \cdot$$

$$\cdot\, (R')^{-1} \exp(i\, k_{v'j'}\, R')\} \qquad (4.112)$$

Now, since one can expand both the scattering amplitude and the plane wave in standard Legendre polynomial (see Chapter 1), the above asymptotic solution can be finally obtained in terms of S-matrix elements:

$$\psi_{vj}(\underline{R}') \underset{R'\to\infty}{\sim} \{(2\pi)^{3/2} \cdot 2i\, k_{vj}\, R'\}^{-1} \cdot \sum_{\ell} (2\ell + 1) \cdot P_{\ell}(\cos \theta_{R'}) \cdot$$

$$\cdot \Big\{ \delta_{jj'} \cdot \delta_{vv'} \exp[-i(k_{vj}R' - \ell\pi/2)] - \qquad (4.113)$$

$$- (\delta_{jj'} \cdot \delta_{vv'} + 2i\, k_{vj}\, \tilde{f}^{\ell}_{vj\to v'j'}) \exp[i(k_{vj}R' - \ell\pi/2)] \Big\}$$

where one has defined an S-matrix element as [95]:

$$S^{\ell}_{vj\to v'j'} = \left\{ \frac{k_{v'j'}}{k_{vj}} \right\}^{\frac{1}{2}} \cdot \{\delta_{jj'} \cdot \delta_{vv'} + 2i\, k_{vj}\, \tilde{f}^{\ell}_{vj\to v'j'}\} \qquad (4.114)$$

or:

$$\tilde{f}^{\ell}_{vj\to v'j'}(k_{vj}) = -\frac{1}{2i(k_{vj}\, k_{v'j'})^{\frac{1}{2}}} \cdot \{\delta_{jj'} \cdot \delta_{vv'} - S^{\ell}_{vj\to v'j'}(k_{vj})\} \qquad (4.115)$$

The last term on the r.h.s. of (4.115) defines the effective T-matrix which yields in the usual way the total, integral inelastic cross sections [87] for atom-molecule collisions:

$$\sigma^{EP}_{vj\to v'j'}(k^2_{vj}) = \frac{\pi}{k_{vj}^2} \sum_{\ell} (2\ell + 1) \mid T^{\ell}_{vj\to v'j'} \mid^2 \qquad (4.116)$$

Because of the forced averaging, however, the detailed balacing provided by this approach yields:

$$k^2_{vj}\, \sigma^{EP}_{vj\to v'j'} = k^2_{v'j'} \cdot \delta^{EP}_{v'j'\to vj} \qquad (4.117a)$$

as opposed to the correct result:

$$(2j + 1)\, k^2_{vj}\, \sigma^{CC}_{vj\to v'j'} = (2j' + 1)\, k^2_{v'j'}\, \sigma^{CC}_{v'j'\to vj} \qquad (4.117b)$$

Eq.s(4.117a) and (4.117b) differ by a factor (2j' + 1)/(2j + 1) which goes to one for $\Delta j \ll j$ and $j \gg 1$. For low-j transitions, however, a corrected expression that performs the proper "counting of states" has to be introduced [89]:

$$\sigma^{EP}_{vj\to v'j'} = \frac{\pi}{k_{vj}^2} \; [(2j' + 1)/(2j + 1)]^{\frac{1}{2}} \sum_{\ell} (2\ell + 1) \; |T^{\ell}_{vj\to v'j'}|^2 \qquad (4.118)$$

an expression which does not suffer from the above shortcoming.

One sees then that the coupled eq.s (4.111) are much smaller in number than the original CC equations, because of the simpler structure of the $f^{eff}_{\lambda}(j, j')$ coefficients. All the m_j-information has been washed out and only indirictly kept as a weighting factor of the vibrational coupling for each $j \to j'$ transition coupling. If v_{max} and j_{max} correspond to the maximum vibrational and rotational states included in the expansion (4.110), then the EP coupled equations will be given in number by $N^{\ell}_{EP} = (v_{max} + 1) \cdot (j_{max} + 1)$ for each partial wave used in the expansion. Thus, the corresponding relationship with the correct N^J_{CC} will be given, as discussed in Section 1 for pure rotational inelasticity by: $N^{\ell}_{EP} = (N^J_{CC})^{\frac{1}{2}}$ which is a very significant reduction, especially for large collision energies and closely spaced rotational levels that rapidly require very large j values. Moreover, the ratio of computational expenses will be given (as seen before) for a small number of eq.s :(CC expense/EP expense) $\sim (j_{max} + 1)^2$ for pure rotatioanl inelasticity which is even smaller than the CS ratio discussed in Section 2. Table 2 presents some or the EP calculations that have been performed thus far on various atom-molecule and molecule-molecule collisional processes. For the weakly anisotropic cases like He - H_2 and H-CO the method yields integral cross sections in semiquantitative agreement with CC results. For more anisotropic systems like He-HCN, the comparison with CC elastic cross sections provided agreement only within ~ 20%. For the EP approach a much slower basis convergence as opposed to CC results was also found, a result attributed to the greater EP rigidity in describing the detailed partitioning of flux among the physical $|jm_j\rangle$ channels, which are instead properly accounted for in the rigorous CC expansion.

The EP modelling of the interaction forces the coupling information to be distributed among fewer channels, thus enhancing the off-diagonal coupling contributions. Since the rate of convergence is roughly inversely proportional to the strength of this coupling, the EP approach yields overall slower convergence with anisotropic systems [94].

The threshold behaviour provided by the EP model is also generally unsatisfactory even for mild anisotropies like in the Ne + N_2 model case [32], while becoming better for higher collision energies, i.e. for larger J values contributing to the inelastic scattering process. Thus, a general conclusion from the cases for which the method has been tested seems to indicate the small anisotropies, high-ℓ values and high collision energies as being favourable factors towards realistic integral cross sections to be obtained from EP calculations. Finally, because of the arbitrariness of the definition of the G-functions and the f_λ^{eff} coefficients of eq.(4.108), other choices can be made outside the one originally suggested with this method, and alternative formulations have been recently suggested and begun to be tested [100]. They are all however, essentially within the physical implications of the same model, the EP model, which assumes a potential-dominant situation with weak m_j-dependence of the dynamics, which then allows some sort of prescription to yield m_j- (or Ω)-independent channel wavefunctions out of which the wanted S-matrix is then constructed.

4.5 The BSA and SDA treatments of purely vibrational inelasticity

In the previous Sections we have shown in some detail how the potential-dominant situation allows one to introduce simpler formulations of the angular momentum couplings essentially by modifying the structure of the centrifugal interaction and hence producing a substantial reduction of dimensionality for the vibro-rotational excitation processes.

In many cases of practical interest, however, one needs to consider many vibrational levels at several collision energies and the previous decoupling techniques still require a substantial amount of computational effort.

This is particularly true when relaxation experiments need to be interpreted over wide temperature ranges and therefore require repeated application of a given computational model.

Since in the above processes a single vibrational relaxation time is often obtained and discussed (see below), several authors [102-105] have utilized a model that extends the vibrationaltranslational inelastic process calculations to three dimensions and where the oscillating molecule is treated as a spherical body that is capable of pulsations (changes in r_{eq}) while its spherical shape is rigidly maintained. The incident particle, for the atom-molecule case, is represent by a point mass and the intermolecular potential is assumed to be a function of the distance of that particle from the surface of the sphere. This is called the Breathing Sphere (BSA) approximation and is expacted to be most appropriate for very rapidly rotating molecules with respect to the interaction time, since in that case it approximates the average potential that an incoming particle would encounter.

Starting from the SF representation of the rigorous CC equation, as given in (3.58) and (4.61), if one puts j=j'=0, then the total angular momentum representation yields: $J = \ell = \ell'$ and the corresponding equations are now given by the following expression:

$$\left\{ \frac{d^2}{dR^2} + k^2_{v'j'} - \frac{J(J+1)}{R^2} \right\} u^{J,oJv}_{oJv'}(R) = \sum_{v''} 2\mathcal{M} \langle o\ J\ v' | V | o\ J\ v'' \rangle\ u^{J,oJv}_{oJv''}(R) \tag{4.119}$$

where the potential matrix element has become:

$$\langle o\ J\ v' | V_{BSA} | o\ J\ v'' \rangle = \sum_{\lambda} V^{\lambda}_{ov'\to ov''}(R) \cdot f_{\lambda}(o\ J;\ o\ J;\ J) = V^{o}_{ov'\to ov''}(R) \tag{4.120}$$

Since the angular coefficient f_λ is = 1 for $\lambda = 0$ and zero otherwise, i.e.: $f_\lambda\ (j\ell;\ j'\ \ell';\ J) = \delta_{\lambda o} \cdot \delta_{jj'} \cdot \delta_{\ell\ell'}$ in the present situation and as required by the 3-j and 6-j triangular conditions in eq. (4.101). The eq.s (4.119) are therefore coupled only via the expansion over vibrational states with matrix elements (4.120) and only those that contain the spherically symmetric part of the full potential multipolar expansion. Within the hierarchy of the approximations discussed in the previous Sections, the is obviously the most drastic one, since both m_j and j-information are wiped out and the rotation-vibration coupling and flux redistribution is disregarded.

The treatment of this approximation, within the BF formalism of eq. (4.9), proceeds along very similar lines, since the $j = j' = o$ assumption also implies $m_j = o$, hence only one element remains in the SF→ BF rotation matrix, namely $\mathcal{D}^{o}_{oo}$

which is unity. In addition, since $\Omega = \Omega' = 0$, the two representations become identical. In eq. (4.9) the off diagonal, intermultiplet coupling terms $h^{Jj'}_{\Omega,\Omega\pm1}(R)$ given by (4.12a) also go to zero and the coupled equations reduce to:

$$\left\{ \frac{d^2}{dR^2} - \frac{J(J+1)}{R^2} + k^2_{v'j'} \right\} u^{J,oov}_{oov'}(R) = 2\mathcal{M} \sum_{v''} \langle v'oo|V|v''oo \rangle \, u^{J,oov}_{oov''}(R) \tag{4.121}$$

where the coupling matrix elements have been simplified from (4.12b) to the decoupled expansion in (4.121) and the vibrational indeces have been trivially added to the (4.9). The corresponding BF angular coefficients $g_\lambda(j\Omega;\, j'\Omega')$ given by eq. (4.10) now have to satisfy the following conditions: $g_\lambda = \delta_{\lambda o} \cdot \delta_{jj'} \cdot \delta_{\Omega\Omega'}$, henca they reduce to unity for the $\lambda = 0$ index and go to zero for the other anisotropic components.

The coupling vibrational matrix elements of the BF formulation of the BSA treatment therefore reduce exactly to the same expression as that given by eq. (4.120). Thus, for this special case, the SF and BF formulations are completely equivalent and both the coupling potential and the dynamical part of the coupled equations (4.119) and (4.121) are identical [106]. This is an interesting conclusion, since the Corolis terms (4.12a) turn out to be rigorously zero within the BSA scheme and the diagonal terms assume the form used in the original CS approximation assumptions (4.13). Within the CS treatment, however, the coupling matrix elements are all rigorously included on the r.h.s. of (4.121), as one can easily see by comparing it with the previous equation (4.14), where only $\Delta\Omega = o$ contributions are required in the coupling terms summation. The BSA approximation can therefore be regarded as a limit case where the molecular target is either forced to be in its $j = o$ rotor state or the interaction potential multipolar expansion only contains the $\lambda = 0$ contribution. Moreover, it is also worth noting that the EP treatment of the vibro-rotational coupling matrix elements of eq.(4.106) also reduces to the simpler form of eq.(4.120) whenever $j = j' = 0$ or $\lambda = 0$, as immediately seen from the expression of (4.108). In this case, not only is the m_j-average performed but the j-states are further decoupled, all with weighting factors equal to unity.

This approximation for rotational levels is derived essentially from the Massey criterion already discussed in Chapter 3 [104]. It assumes, in fact, a physical situation where the collision time is much longer than the rotational time and therefore no inelasticity of rotational degrees of freedom is activated during the encounters. If one remembers that, say for the H_2 molecule, the rotational time τ_{tot} is of the order of $\sim 10^{-13}$ sec. ($j = o$), the BSA dynamical requirement, for potential ranges of ~ 10 a.u., is satisfied for collision energies of the order of 10^{-4} e V. Unfortunately, this is just energy region for which the same adiabatic criterion predicts an even lower vibrational inelasticity, hence suggesting that the BSA approach should not be expected to be a good description of 3-dimensional anisotropic effects in energy transfer calculations. Results are, however, likely to be more realistic for weakly anisotropic potentials, since the $\Delta j \neq 0$ transitions are activated by $\lambda \neq o$ potential terms that are just the ones to be very small in the latter cases, thus indicating that the $\Delta v \neq o$ processes, which are controlled by the short-range nearly isotropic forces, in the latter instance become the dominant ones, relatively speaking.

The corresponding N coupled equations, as easily seen from (4.119) are however drastically reduced in number since the N_{BSA} value is given only by $(v_{max} + 1)$, hence the computational effort is only controlled by the number of partial waves contributing to the (4.116) expressions, with T-matrix elements that are here independent of j [107]. The vibrational states expansion is usually converging with three-to-four levels included above the one considered in the specific transition [108]. In many cases of physical interest, on the other hand, several hundred values of ℓ may have to be included to compute inelastic cross sections and therefore several hundred sets of perhaps 10 or 20 coupled differential equations must be solved at each incident energy. A further approximation was therefore suggested by Takayanagi [104] for solving large numbers of equations within the BS approximation. This is usually called the **M**odified **W**ave **N**umber (**MWN**) BS approximation or **MWN**-BSA.

Since at very large ℓ the $T^{\ell}_{v \to v'}$ contributions fall off rapidly with increasing ℓ, the MWN-BSA assumes that the major contributions to the scattering process are limited to a small region of R' over which $V(\underline{r}', \underline{R}')$ varies rapidly with R' while $\ell(\ell + 1)/R'^2$ is relatively constant. Therefore, for purposes of solving eq.s(4.119) or (4.121) the terms $J(J + 1)/R'^2$ or $J(J + 1)/R^2$ may be replaced by a constant $\ell(\ell + 1)/R^{*2}$, where R^* is the classical distance of closest approach. If one now defines a modified wave number $K^*_{v'}$ as given by:

$$k^{*2}_{v'} = k^2_{v'} - \ell(\ell + 1)/R^{*2} \tag{4.122}$$

its substitution into eq. (4.119) yields the simpler expression:

$$\left\{ \frac{d^2}{dR'^2} + k^{*2}_{v'} \right\} g^{\ell v}_{v'}(R') = \sum_{v''} < v''|V|v' >^{\ell} g^{\ell v}_{v''}(R') \tag{4.123}$$

which corresponds to the one-dimensional ($\ell = 0$) treatment of spherical interaction between vibrator states and a projectile at a collision energy of $k^{*2}_{v'}$. By further defining the dimensionless quantities:

$$\varepsilon = \frac{k^2_v/2\mathbf{M}}{h\nu} \tag{4.124a}$$

and:

$$\varepsilon^*_\ell = \frac{\ell(\ell + 1)}{2\mathbf{M}R^{*2} \cdot h\nu} \tag{4.124b}$$

with $h\nu$ being the vibrational quantum of energy involved in the $|v> \to |v'>$ transition, one can relate the cross section obtained from solving (4.123) [104], $P_{v\to v'}$, to be three dimensional cross section at the given ℓ value:

$$\sigma^{\ell}_{v\to v'}(\varepsilon) = \frac{\pi}{k^2_v}(2\ell + 1)|\,T^{\ell}_{v\to v'}(\varepsilon)|^2 = \frac{\pi(2\ell + 1)}{k^2_v} P_{v\to v'}(\varepsilon - \varepsilon^*_\ell) \tag{4.125}$$

Since: $\sigma^{BSA}_{v\to v'}(\varepsilon) = \sum_\ell \sigma^{\ell}_{v\to v'}(\varepsilon)$ and in view of the large number of angular momenta involved, the sum may be changed to an integral by multiplying each term in the sum by $\Delta\ell = 1$. The result is:

$$\sigma^{MWN-BSA}_{v\to v'}(\varepsilon) = \int_0^{\ell_{max}} \frac{\pi(2\ell + 1)}{k^2_v} P_{v\to v'}(\varepsilon - \varepsilon^*_\ell)\,d\ell \tag{4.126}$$

The upper limit is that partial wave value ℓ_{max} for which one can write: $\varepsilon^{*}_{\ell max} = \varepsilon$. By defining the new variable $x = \varepsilon - \varepsilon^{*}_{\ell}$ the eq.(4.126) can be rewritten as follows:

$$\sigma^{MWN-BSA}_{v\to v'}(\varepsilon) = \frac{\pi R^{*2}}{\varepsilon} \int_0^{\varepsilon} P_{v\to v'}(x)\, dx$$
$$= \pi R^{*2}\ \overline{P}_{v\to v'}(\varepsilon) \qquad (4.127)$$

where the average one-dimensional probability is given by:

$$\overline{P}_{v\to v'}(\varepsilon) = \frac{1}{\varepsilon} \int_0^{\varepsilon} P_{v\to v'}(x)\, dx \qquad (4.128)$$

Since this last integral is performed over inelastic probabilities for a wide energy range, it is essential to expect a smooth behaviour for $P_{v\to v'}(x)$ as a function of energy in order to yield stable and realistic results by numerical quadrature over relatively few points for computational saving. This will obviously depend on the system under study. It is, in fact interesting to note that for the He + H_2 case an extensive comparison has made between BSA ($0 \to 1$) excitation cross sections and rotationally summed IOSA integral cross sections of eq.(4.96) [79] by using the Krauss-Mies potential surface discussed in Chapter 3.

The results for this nearly isotropic system indicate that the BSA cross sections come out to be typically less than 20% smaller than the IOSA cross sections, a difference which is almost within the error limits of both calculations.

Moreover, the fact that BSA cross sections are consistently below the summed IOSA results is in agreement with the previous observation on this system that potential anisotropy enhances vibrational inelasticity in the collision [6].

The major drawback of the BSA calculations, as pointed out by their recent applications to relaxation times evaluations in simple systems [93, 101, 109], is the lack of proper accounting of flux redistribution among the open channels $|v_j m_j>$ of the target system.

This generally means that the prevented interplay of vibrational and rotational channels can force the inelastic cross sections for vibrational excitation to distort their flux intake because of unrealistic coupling and then to come out small at low collision energy. The effect could be reversed at higher collision energies, where the sudden dynamical conditions begin to get fulfilled and thus should yield high rotational inelasticity, while the BSA treatment channels excitation only into vibrational states and excludes flux redistribution into rotational channels. A correcting averaging procedure has therefore been suggested that maintains the simplicity of the BSA calculations while including the above considerations in the form of a modified potential interaction [110, 111, 106]. The approach was called the statistical averaging (SA) or statistical decoupling (SDA). The main idea is to provide a physically reasonable prescription for generating energy-dependent changes in a BS form of spherical interaction, so that the direct vibrational coupling results altered, as would hoperfully occur with the correct inclusion of the full anisotropic potential.

Thus, if the interaction potential is first written in the usual form:

$$\begin{aligned} V(\underline{r}, \underline{R}) &= \sum_{\lambda} V_{\lambda}(r, R)\, P_{\lambda}(\hat{r} \cdot \hat{R}) \\ &= V_{o}\,(r, R) + \sum_{\lambda \neq o}{}' \; V_{\lambda} \cdot P_{\lambda} \end{aligned} \tag{4.129}$$

one defines now a new potential, V_{SA}, that dependes also on the collision energy E_{coll} and is spherically symmetric:

$$V_{SA}(r, R, E_{coll}) = \mathbf{V}_{o}(r, R) + \bar{V}_{asym}(r, \mathbf{R}, E_{coll}) \tag{4.130}$$

where the averaged form of the anisotropic contributions is obtained by first rewriting each multipolar coefficient in (4.129) in the following general form $\lambda \neq o$:

$$V_{\lambda}(r, R) = a^{\lambda}_{SR}(r, R)\, V_{o,SR}(r, R) + a^{\lambda}_{LR}(r, R)\, V_{o,LR}(r, R) \tag{4.131}$$

One is here explicitely separating short-range (SR) and long-range (LR) contributions, with the new coefficients a^{λ}'s representing some parameters that measure the relative strengh of the asymmetric components that may also depend on (r, **R**).

The spherical contribution and the various anisotropic coefficients of (4.129) can be viewed as affecting the various possible processes caused by each collisions in different ways, since only the $V_{\lambda \neq o}$'s will act on rotational inelasticity whereas the spherically repulsive (SR) regions of interaction are usually recognized as those mainly responsable for vibrational inelasticity. Moreover the general Massey crite ion on adiabaticity also allows one to expect that those trajectories which go through the attractive regions of the PES should increase the anisotropic inelasticity since they correspond, for a given collision energy, to collision times which are shorter than those attained for purely repulsive interaction.

Thus, if one further defines an energy dependent multipolar expansion:

$$V_{asym}(r, \mathbf{R}, E_{coll}) = \sum_{\lambda \neq o} V'_{\lambda} \cdot P_{\lambda} = \sum_{\lambda \neq o} \frac{V_{\lambda}(r, R)}{E_{coll}} \cdot P_{\lambda}(\hat{r} \cdot \hat{R}) \quad (4.132)$$

and an ad hoc weighting factor $\mathbf{W}$:

$$\mathbf{W}_T = \prod_{\lambda \neq o} \mathbf{W}_{\lambda}(\underline{r}, \underline{R}, E_{coll}) \quad (4.133a)$$

where each contributing W_{λ} is given by:

$$\mathbf{W}_{\lambda}(\underline{r}, \underline{\mathbf{R}}, E_{coll}) = \exp[- V'_{\lambda} \cdot P_{\lambda}(\hat{r} \cdot \hat{R})] \quad (4.133b)$$

It is then possible to perform a weighted average of each V_{λ} coefficient of (4.129):

$$\begin{aligned} \bar{V}_{\lambda}(r, R, E_{coll}) &= \bar{V}_{\lambda,SR} + \bar{V}_{\lambda,LR} = \\ &= V_{o,SR}(r, R) < a^{\lambda}_{SR} \mathbf{W}_T P_{\lambda} >_{\gamma} + \\ &+ V_{o,LR}(r, R) < a^{\lambda}_{LR} \mathbf{W}_T P_{\lambda} >_{\gamma} . \end{aligned} \quad (4.134)$$

where the bracket symbols indicate integration over the internal variable $\gamma = \arccos \hat{r} \cdot \hat{R}$. The new multipolar coefficients are now independent of the relative orientation of $\underline{r}$ and $\underline{R}$ and the total anisotropic contribution of (4.132) can be written in a weighted, averaged form as follows:

$$\bar{V}_{asym}(r, R, E_{coll}) = \mathcal{N} \cdot \sum_{\lambda \neq 0} [\bar{V}_{\lambda,SR} + \bar{V}_{\lambda,LR}] \qquad (4.135)$$

where the normalization factor $\mathcal{N}$ is needed to preserve unitarity:

$$\mathcal{N}^{-1} = \int_{-1}^{1} W_T(\underline{r}, \underline{R}, E_{coll})\, d(\cos\gamma) \qquad (4.136)$$

This procedure thus completely defines the V_{SA} potential of (4.130).

The simplification of the original CC equations proceeds now very straightforwardly [106] and yields the SDA coupled equations (SF representation):

$$\left\{ \frac{d^2}{dR^2} - \frac{\ell(\ell+1)}{R'^2} + K^2_{v'} \right\} u^{\ell v}_{v'}(R') = 2\mathcal{M} \sum_{v''} V^{\ell v'v''}_{SA}(R'; E_{coll})\, u^{\ell v}_{v''}(R') \qquad (4.137)$$

where the wavevector $K^2_{v'}$ si related to the total energy and the vibrational eigenvalues:

$$k^2_{v'} = 2\mathcal{M}[\varepsilon_v - \varepsilon_{v'} + k^2_v/2\mathcal{M}] \qquad (4.138)$$

and the vibrational coupling matrix elements vary with the collision energy and are obtained by a further integration:

$$V^{\ell v'v''}_{SA}(R', E_{coll}) = \int dr'\, \chi^{j=0}_{v''}(r')\, V_{SA}(r', R', E_{coll})\, \chi^{j=0}_{v'}(r') \qquad (4.139)$$

This expression reduces to the BSA expansion only if the first term of the r.h.s. of (4.130) is used in the coupling. The weighted averaging of the full potential via a Boltzmann-like factor therefore modifies the strength of the coupling according to the magnitude of the $V_{\lambda \neq 0}$ coefficients and of the chosen collision energy. The latter in fact, through eq. (4.133b), weighs more importantly those regions of the PES that are strongly attractive and anisotropic and that contribute more heavily to rotational inelasticity, while excluding the weakly attractive or strongly repulsive parts of the V_λ's as less effective in channeling flux into rotationally excited states. Their net effect will be one generating spherical potential with energy dependent r-gradients, hence with different contributions to vibrational couplings via (4.139). Because of the ad hoc nature of the W_T form in

(4.133), the ultimate justification of the SDA lies in its agreeing with CC results or experiments. Its statistical form, however, suggests that in general it should hold for mildly anisotropic systems where no strong propensity rules for rotational excitation/deexcitation accompanying vibrational inelasticity are detected [94]. For such cases, in fact, the rotational degrees of freedom would be populated in collisions that do not strongly favour particular trajectories as is instead the case in systems with large anisotropy.

Thus, not very good results were obtained for $Li^+ + H_2$ [111], whereas the $He + H_2$ system showed reasonable agreement with CS and CC results [106] and indicated a correct increase in vibrational inelasticity when the full V_{SA} was used as opposed to the BSA treatmen with V_o only. These results have also been found to be in agreement with the rotationally summed IOSA cross sections recently obtained for the same system [79], where a rather slow numerical convergence on the vibrational states expansion was observed, indicating lack of convergence for the previous CC and CS results at the same E_{coll} [79].

Obviously the SDA needs to be tested on a wider range of cases and for more of those systems which have been also analyzed via the CC, CS, EP approaches before establishing more exactly its range of validity, but appears thus far to indicate a computational approach that is sufficiently simple to warrant further study.

4.6 - The LD simplifications

All the methods discussed in the previous Sections, and especially the EP and CS methods, exploit formal, analytic simplifications of the atom (molecule) molecule angular momentum coupling in order to reduce the number of coupled equations to be solved.

Comparison of their results for integral and different cross sections of vibro-rotational inelastic processes in simple systems, as those shown in Tables 1 and 2, when made with respect to the CC results has indicated that these approximations are most accurate for the study of collisions in which the adiabatic PES is dominated by isotropic and anisotropic short-range forces. Both the EP and CS approaches, in fact, achieve a reduction of dimensionality by transformations that simplify the coupling from the $\lambda \neq o$ terms of the multipolar expansion (4.129). In the process, the exact centrigugal barrier is altered.

In cases of short-range coupling the corresponding total cross sections at low collision energies are dominated by those encounters that take place at small impact parameters. The corresponding centrifugal barriers are therefore small and relatively unimportant with respect to the anisotropic potential terms, and thus the alterations introduced by the simpler coupling schemes have relatively little effect.

When the partners' coupling is long range in nature, on the other hand, the degree of agreemet with rigorous CC result deteriorates. The corresponding cross sections will, in fact, contain major contributions from collisions at large impact parameters and the centrifugal potential will play an important, if not dominant, role which will be significantly distorted or destroyed by the decoupling techniques discussed in previous Sections. Unfortunately, precisely these conditions characterize many collision problems of interest, especially in ion-molecule encounters with polar tergets, where the charge-dipole anisotropic interaction exhibits a very long range R^{-2} dependence that needs to be balanced by the correct inclusion of centrifugal cut-off (see Chapter 1) in order to obtain the potential-free asymptotic matching.

In order to deal with this centrifugal-dominant situation, an alternative approximate scheme has recently been suggested [2, 112 - 115]. This was called the 'ℓ-dominant' (LD) and 'decoupled ℓ-dominant' (DLD) approximation. The main observation by the authors regarded the J-dependence of the T-matrix elements magnitude as cotributing to the degeneracy-averaged integral cross sections [e.g. see eq. (4.116)].

It was, in fact, found by CC calculations and confirmed by CS calculations [5], that the element magnitude for elastic processes makes its largest contribution when the orbital angular momenta undergo transitions ($\ell = \ell'$) while for inelastic processes ($j \neq j'$) tha largest T-matrix element is associated with situations where the orbital angular momenta take their minumum allowed values, J-j and J-j' [2]. The T-matrix elements are all small whenever the ℓ values span the allowed region of $(J + j)$, i.e. they become greater than J.

A simple explanation of these effects can be provided by the coupling potential changes with R. Since its magnitude is generally decreasing with R, it is reasonable to find that the largest transition probabilities will be associated with encounters that exhibit the smallest classical turning points. In other words, the centrifugal barrier shields the collision partners from the coupling potential, as already seen in the IOSA results of Fig. 4.5, and thus the largest transition contributions of the T-matrix will come from the centrifugal minimum value of $\ell = J - j$[5, 112].

Physically, this implies that the preferred orientation for rotational inelasticity in collisions will come from the constructive addition of ℓ and j, i.e. from a nearly coplanar geometry with the molecular and orbital rotational motions in the same direction .

The LD method therefore basically involves neglecting all equations for $\ell > J$, since the above discussion was based on the observation that the T-matrix elements for rotational inelasticity become small when $\ell, \ell' > J$. The number of equations to solve is thus substantially reduced, since N_{LD} becomes equal to $N_{CC/2}$ for odd parity equations (i.e. for $(-)^{\ell+j+J}$ odd) and equal to $N_{CC}(1/2 + j^{-1}_{max})$ for even parity equations. The corresponding saving in computational time is somewhat intermediate between CS and CC, and markedly closer to CS times depending on the treatment reserved for the neglected matrix elements.

In the earlier attempts on Ar + N_2[2], a set of prescriptions was suggested for computing the $T^J_{\ell j,\ell' j'}$ values (elastic and inelastic) from the ones already obtained within the LD reduced range of the ℓ-index, and the obtained integral cross sections exhibited greater agreement with CC results than the EP and CS results with the same basis expansion. They departed from CC calculations in the small-J region where the potential-dominated situation is incorrectly represented by LD technique that eliminates, with the arbitrary cut-ooff, some of the important potential coupling terms.

A further application to Li^+- H_2 collisions [112] tested the LD reduction without any replacement prescription for T-matrix elements at $\ell, \ell' > J$, and similar results were found regarding a better agreement with CC calculation in the high-J region. The rotationally inelastic (o → 2,4) cross sections appeared to be in qualitative and quantitative agreement with exact CC cross sections, while the (o → 6) transitions were far less satisfactory. This last result was attributed to the dominance of small J values in obtaining this inelastic process, and thus to the greater error introduced by the LD approach of neglecting the corresponding

$\ell > J$ subspace. A considerable increase in accuracy with respect to CS and EP calculations was also found in this case.

It is worth pointing out that one expects this approximation to be most appropriate for integral cross sections (partial and total) that are most sensitive to the large-J contributions in the presence of long-range potentials. On the other hand, whenever one is interested in elastic and inelastic differential cross sections at moderate and large angles, where most rainbow effects are apparent and which will reflect collisions at small impact parameters, the LD approach will not be appropriate.

Its replacement or neglect approximates in fact the T-matrix elements, and while possibly capable of yielding reasonable values of total crass sections magnitudes, would certainly destroy or alter drastically the phase information needed for DCS calculations. Moreover, the dynamical importance of the $\ell, \ell' > J$ channels decreases as the centrifugal barrier becomes the dominant factor, hence if a significant contribution to integral inelastic cross sections (partial and total) arises in this centrifugal-dominated condition, the LD method will yield reasonable results for the relevant inelastic transitions.

Since, on the other hand, a general inelastic process $|vjm_j\rangle \rightarrow |v'j'm_{j'}\rangle$ can occur with reasonable probability whenener the coupling matrix elements $V^J_{vj,v'j'}(R)$ are larger than the energy transferred in the encounter, one sees clearly that the above condition of ℓ-dominance will usually be satisfied in the long range regions of interaction where the small tails of the potential coupling will only be comparable with rotational energy spacings.

This means that for vibrationally inelastic processes, where there is a large energy change, the potential coupling will be activating transitions only for those values of R corresponding to the steeply repulsive inner wall of the spherical component in (4.129). Here, since the potential dominates the dynamics, the LD approach would of course fail. One can therefore expect that the present approximation will only be valid for rotationally inelastic transitions.

The LD simplification essentially halves the number of coupled equations to be solved, but a further reduction was introduced [113] by examining the alignment of the angular momentum vectors and decoupling those relative orientations already shown by the LD numerical tests to be relatively unimportant. The dimensionality becomes equivalent to the one yielded by the CS equations and the method was called 'Decoupled ℓ-dominant' (DLD). In the rigorous CC equations (4.38) the summation on the r.h.s. (for rotations only) runs over j" and ℓ" and is reduced in the LD approach by limiting ℓ', $\ell'' \leqslant J$. Thus, the corresponding partial integral inelastic cross sections are defined by:

$$\sigma^{J}_{j\to j'} = \frac{\pi}{(2j+1)\,k_j^2} \sum_{\ell,\ell'} (2J+1)\,|S^{J}_{j\ell,j'\ell'}|^2 \qquad (4.140)$$

One can alternatively introduce the index μ, where:

$$\mu = \ell - J + j \qquad ; \qquad o \leqslant \mu \leqslant 2j \qquad (4.141)$$

Classically, this new index is the measure of the angle of alignment between $\underset{\sim}{j}$ and $\underset{\sim}{J}$, since: $\hat{j} \cdot \hat{J} \sim \mu/j$. The minimum of the centrifugal barrier contribution for a given J comes from $\ell = J - j$ which corresponds to $\mu = o$ in (4.141). Thus, the dynamical ℓ-dominance situation suggests that the major contributions to the inelastic cross sections at large -J values will originate from $\mu \sim o$ values. Moreover, for small μ values, the corresponding angular coupling coefficients f_λ of (4.101), which are given as $f_\lambda(j\mu, j'\mu')$ [113], acquire a particularly simple form that is independent of ℓ, ℓ' J and for a given (μ, μ') becomes constant as $J \to \infty$ [114]. From their asymptotic form one can see that strength the angular coupling for small μ values, and large J values, is maximized when $\Delta\mu = o$, thus suggesting a decoupling in the alignment index μ by restricting the collisions to the same initial and final values of μ.

This decoupling becomes obviously less valid as μ approaches j but the corresponding dynamical situation, as well as the one where $\mu > j$, has been shown by the LD calculations to contribute little to inelasticity in the large-J regions [113].

Thus, the coupling matrix elements in eq.s(3.58) become (rotational excitations only):

$$V^{DLD}_{J,j\mu,j'\mu'}(R') = \sum_{\lambda} V_\lambda(R')\, f_\lambda(j,\, J-j+\mu,\, j',\, J-j'+\mu;\, J) \qquad (4.142)$$

The index μ is restricted to the range $o \leqslant \mu \leqslant j_{max}$, where j_{max} has the usual maximum-index meaning, and j' assumes all the values $\geqslant \mu$. The corresponding partial integral inelastic cross sections are further simplified with respect to (4.140), i.e. one obtains:

$$\sigma^{J,DLD}_{j\to j'} = \frac{\pi}{(2j+1)\,k_j^2} \sum_{\mu=o}^{j_{min}} (2J+1)\,|S^J_{j\mu,j'\mu'}|^2 \qquad (4.143)$$

where j_{min} is the smaller of j and j' respectively. The dimensionality of the coupled equations has thus become equal to the one derived in Section 2 for the CS method, since it is now equal to j_{max} for each J value.

The numerical tests performed on Ar - N_2, Li^+ - H_2 [113], H^+ - CN[115] and HF - HF [99] indicate the rather complementary accuracy of the DLD approach with respect to the CS calculations. The latter method is in fact best suited for short-range anisotropic interactions and for weak anisotropy in general, while the former produces better agreement with CC computations for the study of those collisions in which strong, long-range forces play a major role. Whenever the collisions at large impact parameters do not contribute significantly to a specific inelastic process the DLD approach, however, fails to be realistic and physically acceptable. As said before this is always the case for vibrationally inelastic process and for the threshould behaviour of rotationally inelastic cross sections where the amount of energy transferred is nearly equal to the total energy, thus requiring a large value for the classical distance of closest approach. This in turn maps the very tail of the potential, already beyond the significant DLD contributions to inelasticity [116].

It is however interesting to see how this method consistently improves over BA calculations [99] and complements the regions of accuracy of the CS approach, thus suggesting that their combined used for ion-molecule or molecule-molecule collisions might provide a quantitative and yet less costly agreement with benchmark CC calculations.

4.7 - The Distorted Wave Approximations

Two of the best known methods adopted to provide approximations to the exact radial wavefunctions, and hence to the S-matrix, result from earlier quantum mechanical studies of inelastic collisions [117] and are the Born Approximation

(BA) and the Distorted Wave Born Approximation (DW BA). The BA, in which the whole potential is treated as a perturbation, is valid for large ℓ, or equivalently for large J, when the scattering is only affected by the tail of the interaction and there is little distortion of the zeroth order wavefunction or a small phase shift, as seen from eq.(1.71).

In the multichannel, differential formulation (SF representation) this means that the CC equations (3.58) can be written in an uncoupled form:

$$\left\{ \frac{d^2}{dR^2} + k^2_{j'v'} - \frac{\ell'(\ell'+1)}{R'^2} \right\} u^{Jj\ell v}_{j'\ell'v'}(R') = 2\mathcal{M} < j'\ell'v'J|V|j'\ell'v'> u^{Jj\ell v}_{j'\ell'v'}(R') \tag{4.144}$$

the solution of which can be taken as the zeroth order approximation in an iteration procedure. If one in fact writes it down as a solution which behaves asymptotically as [118]:

$$w^{Jj\ell v}_{j'\ell'v'}(R') \sim \sin [k_{j'v'}R' - \frac{1}{2}\ell'\pi + \eta^{Jjv}_{j'\ell'v'}] \tag{4.145}$$

where $\eta^{Jjv}_{j'\ell'v'}$ is a real phase shift, then the zeroth order approximation that solves the equation (4.144) with following index conditions:

$$\begin{aligned} u^{Jj\ell v}_{j'\ell'v'}(R') &= \delta_{jj'} \cdot \delta_{vv'} \cdot \delta_{\ell\ell'} \; w^{Jj\ell v}_{j'\ell'v'}(R') \\ &= \delta_{jj'} \cdot \delta_{vv'} \cdot \delta_{\ell\ell'} \; w^{(o)J}_{j\ell v}(R') \end{aligned} \tag{4.146}$$

can be taken as a starting solution [118].

The first order approximation $w^{(1)Jj\ell v}_{j'\ell'v'}(R')$, under the assumption that $|w^{(1)}| \leqslant |w^{(0)}|$ is then the solution of the following modified equation:

$$\left\{ \frac{d^2}{dR'^2} - \frac{\ell'(\ell'+1)}{R'^2} + k^2_{j'v'} - < j'\ell'v'J\;|U|\;j'\ell'v'J > \right\} w^{(1)Jj\ell v}_{j'\ell'v'}(R') = \tag{4.147}$$

$$= < j\ell vJ\;|U|\;j'\ell'v'J > w^{(o)J}_{j\ell v}(R')\;[1 - \delta_{jj'} \cdot \delta_{\ell\ell'} \cdot \delta_{vv'}]$$

and: $U = 2\mu V$ in the coupling matrix elements. Thus, the BA allows only for the distorsion caused by the centrifugal term in the above equations (cf.(1.71)), while here the spherical interaction potential is explicitely introduced to distort the $W^{(o)}(R')$ wavefunction.

The first order solution of (4.147) are given, following ref.s [117, 118], by:

$$w^{(1)Jj\ell v}_{j'\ell'v'}(R') = \delta_{jj'} \cdot \delta_{vv'} \cdot \delta_{\ell\ell'} \, w^{(o)J}_{j\ell v}(R') + [1 - \delta_{jj'} \cdot \delta_{vv'} \cdot \delta_{\ell\ell'}] \cdot \tag{4.148}$$

$$\left\{ Y^{Jj\ell v}_{j'\ell'v'}(R') \int_0^{R'} w^{Jj\ell v}_{j'\ell'v'}(R') < j\ell vJ|V|j'\ell'v'J > w^{(o)J}_{j\ell v}(R')dR' + \right.$$

$$\left. + \, w^{Jj\ell v}_{j'\ell'v'}(R') \int_{R'}^{\infty} Y^{Jj\ell v}_{j'\ell'v'}(R') < j\ell vJ|V|j'\ell'v'J > w^{(o)}_{j\ell v}(R')dR' \right\}$$

where Y is the solution of (4.144) which behaves asymptotically as:

$$Y^{Jj\ell v}_{j'\ell'v'}(R') \underset{R'\to\infty}{\sim} \exp[\, i(k_{j'v'}R' - \frac{1}{2}\ell'\pi + \eta^{Jjv}_{j'\ell'v'})] \tag{4.149}$$

By considering the asymptotic form of the first-order solution of(4.148) one can obtain the corresponding S-matrix element for the wanted transition $|jv\rangle \to |j'v'\rangle$:

$$S^{J,\ell,\ell'}_{jv\to j'v'} = (k_{jv} \cdot k_{j'v'})^{\frac{1}{2}} \exp[\, i(\eta'_{(o)} - \eta_{(o)})] \; 2i \, \beta^{Jj\ell v}_{j'\ell'v'} \tag{4.150a}$$

and the corresponding T-matrix element for the same transition:

$$|\, T^{J,\ell,\ell'}_{jv\to j'v'} \,|^2 = 4(k_{jv} \cdot k_{j'v'}) \cdot [\beta^{J,j\ell v}_{j'\ell'v'}]^2 \tag{4.150b}$$

where $\eta'_{(o)}$ and $\eta_{(o)}$ indicate compactly the real phase shifts appearing in (4.145) for the uncoupled solutions of the initial channel $|Jj\ell v\rangle$ and final channel $|Jj'\ell'v'\rangle$. The coupling term β describes the effect of the potential to first order:

$$\beta^{J,j\ell v}_{j'\ell'v'} = \int_0^{\infty} dR' \; w^{Jj\ell v}_{j'\ell'v'}(R') < j\ell vJ \,|V|\, j'\ell'v'J > w^{(o)J}_{j\ell v}(R') \tag{4.151}$$

The corresponding inelastic scattering cross sections are then given by:

$$\sigma^{DWA}_{jv\to j'v'} = \sum_J (2J + 1) \; \sigma^{DWA}_{jv\to j'v'}(J) \tag{4.152}$$

where the **DWA** opacities are given via the first order coupling terms:

$$\sigma^{DWA}_{jv\to j'v'}(J) = \left(\frac{4\pi}{2j + 1}\right) \left(\frac{k_{j'v'}}{k_{jv}}\right) \sum_{\ell,\ell'} [\beta^{J,j\ell v}_{j'\ell'v'}]^2 \tag{4.153a}$$

while the corresponding elastic cross sections are given by a different opacity expression:

$$\sigma^{DWA}_{jv\to j'v'}(J) = \frac{4\pi}{(2j + 1)k^2_{jv}} \left\{ \sum_{\ell} \sin^2 \eta^J_{j\ell v} + \sum_{\ell,\ell'}{}' \; [\beta^{J,\ell\ell'}_{jv}]^2 \right\} \tag{4.153b}$$

where the $\sum{}'$ indicates that the summation is limited by $\ell \neq \ell'$ with: $|J - j| \leqslant \ell \leqslant J + j$ and: $|J - j'| \leqslant \ell' \leqslant J + j'$. In the latter equation one sees then that the effect of coupling vanishes to first order [118].

The wavefunction of (4.148) can be iteratively improved by further adding higher order terms in the (4.147). The corresponding S-matrix can be obtained from the asymptotic form of the wavefunction at each higher order and the integral equation formulation acquires the form of a perturbation series for $\underline{\underline{S}}$ in powers of the potential [119]. The DWA approach corresponds to using just the zeroth and first order terms as in (4.150), while the Born Series can be written down as a direct analogy by disregarding the potential effect in the uncoupled equations (hence: $\eta_{(o)} = \eta'_{(o)} = o$) and using the free-motion eigenfunctions as zeroth order terms (see Chapter 1). Thus, one can write:

$$S^{J,\ell,\ell'}_{jv\to j'v'} = \delta_{jj'} \cdot \delta_{\ell\ell'} \cdot \delta_{vv'} - i\, B^{J,j\ell v}_{j'\ell'v'} \tag{4.150c}$$

where:

$$B^{J,j\ell v}_{j'\ell'v'} = 4 \sum_{\lambda} \left\{ (k_{jv} \cdot k_{j'v'})^{\frac{1}{2}} \int_0^\infty j_{\ell'}(k_{j'v'}R')\, V^{\lambda}_{j'v'}(\mathbf{R'})\, j_{\ell}(k_{jv}R')R'^2 dR' \right\} \cdot$$

(4.154)

$$\cdot \; f_{\lambda}(j'\ell', j\ell; J)$$

and where the interaction potential has been formally written in the multipolar expansion form, with $V^{\lambda}_{v,v'}(R')$ being the vibrational coupling matrix element for the λth multipolar coefficient.

Both the BA and DWA have been formulated in detail in both the SF representation [119] and BF representation [120]. The derivation of the DWA equations has also been given for polyatomic molecule-molecule collisions [122].

With analytic forms of the interactions potentials the B^J and B^J expressions acquire attractively compact forms, obviously simpler than the CC expression, especially for the BA approach [123] which however becomes valid at higher collision energies and for higher partial waves. The latter, however, are often contributing relatively little to the inelastic scattering process [115].

Both the DWA and BA treatments do not preserve the unitarity of the S-matrix [117] and therefore several methods have been suggested to rectify the situation which originates from indirect transitions begin forbidden in first order, thus violating the law of conservation of particles.

In a <u>ad hoc</u> renormalization procedure suggested by Takayanagi [124] and Bernstein [125], the conservation of particles requirement, but not the indirect coupling to higher orders, was incorporated into the DWA by arbitrarily modifying the coupling matrix elements of (4.154).

The exponential method [126, 127] expresses the S-matrix as a function of another matrix $\underline{\underline{A}}$ which is unitary and therefore, instead of expanding $\underline{\underline{S}}$ directly in the Born series, it is the new unitary matrix $\underline{\underline{\mathbf{A}}}$ that is expanded. Thus, one starts by writing [123]:

$$\underline{\underline{S}}^J = \exp(i\,\underline{\underline{A}}^J) \qquad (4.155)$$

where $\underline{\underline{A}}^J$ is required to be a hermitian matrix and which allows one to write the DWA matrix elements as:

$$S^{J,\ell,\ell'}_{jv\to j'v'} = \exp(i\,\eta'_{(o)})\,[\exp(i\,\underline{\underline{A}}^J)]^{\ell,\ell'}_{jv,j'v'}\,\exp(i\,\eta_{(o)}) \qquad (4.156)$$

The matrix $\underline{\underline{A}}^J$ can now be expanded in a perturbation series with the strength of the potential identified as the perturbation parameter α [123, 128]:

$$\underline{\underline{A}}^J = \sum_k \alpha^k \underline{\underline{A}}_k^J \tag{4.157}$$

and one obtains a unitarized $\underline{\underline{S}}$ as a power series of α, where the various terms correspond to the various order of the DWA scheme in which eq.(4.156) is used. One then finds that an element of the $\underline{\underline{A}}^J$ can be written as (4.151) or (4.154) i.e. generally as:

$$A^{J,\ell,\ell'}_{jv,j'v'} = 4 \sum_\lambda \left\{ (k_{jv}\, k_{j'v'})^{\frac{1}{2}} \int_0^\infty dR'\, R'^2\, w^{(o)}_{j'v'}(k_{j'v'}\, R')\, V^\lambda_{v,v'}(R') w^{(o)}_{jv}(k_{jv} R') \right\} \tag{4.158}$$

where with $w^{(o)}_{jv}$ we have compactly indicated the zeroth order solutions of the uncoupled equations for the initial and final channels as in (4.144). They are the radial part eigensolutions for the zeroth order Hamiltonian $\mathcal{H}^o$, which is the diagonal matrix operator in the basis of the molecular states [117]. In the exponential form of the Born treatment (EBA) the perturbation $\underline{\underline{V}}$ added to $\mathcal{H}^o$ to yield the total Hamiltonian is constituded by the entire molecular potential, so that the radial wavefunctions in (4.158) are spherical Bessel functions. Since $\mathcal{H}^o$ does not contain any part of the interaction, the phase shifts $\eta_{(o)}$ of eq. (4.156) are obviously zero.

The exponential distorted wave approach (EDWA) represents, as discussed before, a significant improvement over EBA since some portion of the intermolecular distorsion potential is added to $\mathcal{H}^o$. For general EPS, however, the integrals (4.158) now have to be performed numerically.

The Heitler method of unitarization [129, 130] starts by rewriting the S-matrix via the reactance matrix $\underline{\underline{K}}$ [117]:

$$\underline{\underline{S}} = \mathbb{1} - i\, \underline{\underline{T}} = (\mathbb{1} + i\, \underline{\underline{K}}) / (1 - i\, \underline{\underline{K}}) \tag{4.159}$$

For small matrix elements of $\underline{\underline{K}}$ one can write:

$$\underline{\underline{S}} \sim (\mathbb{1} + 2\, i\, \underline{\underline{K}}) \tag{4.160}$$

which yields the following relation with the T-matrix:

$$\underline{\underline{K}} = -1/2\, \underline{\underline{T}} \tag{4.161}$$

In the BA scheme the $\underline{\underline{T}}$ is real and symmetric, hence one can use (4.161) to rewrite (4.159):

$$\underline{\underline{S}} = (\mathbb{1} - \frac{1}{2} i \underline{\underline{T}}^{BA}) / (\mathbb{1} + \frac{1}{2} i \underline{\underline{T}}^{BA})$$

$$= \exp [i\, 2 \tan^{-1} (\underline{\underline{T}}^{BA}/2)] \qquad (4.162)$$

$$= \mathbb{1} + i \underline{\underline{T}}^{BA} - \frac{1}{2} (\underline{\underline{T}}^{BA})^2 - \frac{i}{4} (\underline{\underline{T}}^{BA})^3 + \ldots\ldots\ldots$$

It is clearly possible to use this form more generally in the same way as in the exponential form of before and thus, by expanding in eq.(4.162) the original correct T matrix (for small reactance matrices i.e. for small perturbing potentials), to obtain the Heitler Born Series as shown in (4.162) or the HDW series. It will be the number of terms taken in, and the chosen unperturbed Hamiltonian $\underline{\underline{\mathcal{H}}}^0$, that will decide which unitarized approximation one is using. The matrix elements however differ from the EDWA discussed before and become exact only in the sudden, high energy limit [128].

The numerical tests performed within these schemes seem to indicate that for low collision energies and large energy transfer values the EDWA produces better results, while the HDWA of above becomes preferable for higher collision energies and higher partial wave contributions, where the small K - matrix elements approximation that yields (4.161) becomes more realistic. Thus, any unitarized DWA will fail for strong, long-range potential coupling that will not satisfy (4.161) or will require higher terms in eq.(4.147). In general, however, one finds that the DWA is a reliable approximation when approaching the classical limit, away from threshold and for large J(or ℓ)values. The unitarization procedures allow for the extension of the range of validity of BA and DWA to smaller J (or ℓ) values [131]. A recent combination of a EDWA treatment with CS and EP methods has provided very promising indications for a substantial extention of the method to lower ℓ and lower collision energies [132].

An interesting comparison of IOSA, CS and EBA has been performed for the rotational inelasticity of Ar + CsF collision [72] at 87.7 mev of collision energy. Most of the scattering for the low-lying inelastic transitions was found there to result primarily from the intermediate and long-range parts of the EPS, and thus resulted in close agreement between the three computational models, with the EBA being essentially better and acceptable only in the high-J region. This indicates once more that the DWA method operates under physical conditions similar to the DLD-situation, i.e. at a centrifugal potential dominance over the strong potential coupling. The $\underset{\sim}{\ell} + \underset{\sim}{j}$ coupling is in fact correctly included between initial and final channels, but the potential coupling appears only to first order.

Analogously, the DLD approach reduces the total number of terms originating from the interaction potentials by excluding the corresponding ℓ, $\ell' > J$ contributions under the assumption that, in the high-J regime, most of the potential coupling would by wiped out by the larger centrifugal barriers and hence would not be importantly contributing to rotational inelasticity. On the other hand, it operates less drastically than the EDWA method, thus allowing for indirect coupling between all the included rotational channels.

An interesting extension of the DWA scheme has recently been tried with the use of IOSA conditions for rotational excitations and with the further use of the MWNA discussed in Section 4.5 [133].

The use of DWA methods was based on the observation that V-T processes have extremely low transition probabilities when compared either with elastic scattering or with purely rotational energy transfer processes. One can thus take as zeroth order wavefunction for (4.144) the solution that includes the correct rotational coupling and which yields the following modified T-matrix element (see eq. (4.150b) for comparison):

$$| T^{J,\ell,\ell'}_{vj\to v'j'} | = 2(k_{jv}\, k_{j'v'})^{\frac{1}{2}} \sum_{j''\ell''} \beta^{J,v\ell''j''}_{v'j'\ell'} \tag{4.163}$$

where:

$$\beta^{J,v\ell''j''}_{v'\ell'j'} = \int_0^\infty dR'\; w^{Jj\ell v}_{j''\ell''v}(R') < j''\ell''vJ|V|j'\ell'v'J > w^{(o)J}_{j'\ell'v'}(R') \tag{4.164}$$

whereas the vibrational transition can be treated in a DWA framework, the same is unlikely to be true for rotational inelasticity within the vibrationally elastic process, the effect of which is then included in the $w^{Jj\ell v}_{j''\ell''v}(R')$ but not in $w^{(o)J}_{j'\ell'v'}$ that only solves the uncoupled eq.(4.144). The further use of IOSA conditions and of realistic potential forms, leading to analytic expression for the interaction were then tested by R.B. Gerber via the added approximation of MWNA described before. The final, uncoupled equations then assumed the form of an s-wave equation for an exponential, the solution of which is known and allows one to obtain analytically the coupling matrix elements of (4.164). Simple numerical tests for the He - H_2 case provided encouraging agreement with previous CC calculations [133].

An improved version of the above combination of various approximate schemes has more recently been suggested by Balint-Kerti and coworkers [134, 135] who generalized the first order coupling term of (4.164) by including the correct rotational coupling between the rotational manifold of both initial and final vibrational states:

$$| T^{J,\ell,\ell'}_{jv\to j'v'} | = 2(k_{jv}\ k_{j'v'})^{\frac{1}{2}} \sum_{j''\ell''} \sum_{j'''\ell'''} \int_0^\infty w^{Jv'j'\ell'}_{v'j'''\ell'''}(R') \cdot$$

$$\cdot < j'''\ \ell'''\ v'\ J|V|\ j''\ell''\ v\ J > w^{Jvj\ell}_{vj''\ell''}(R') \qquad (4.165)$$

this approximation is therefore valid when the coupling between vibrational levels is small. They then proceeded to apply the IOSA reduction scheme to yield approximate expressions for the wavefunctions to insert into the (4.165) integrals. Those integrals were in turn obtained by a generalization of Mies's analytic formula [136].

The numerical test of this approximation, (DWIOSA), were again carried out for the standard He + H_2 system, for which various interactions and CC, CS, ET etc. calculations have been available for sometime. The agreement was found to be satisfactory in the intermediate energy range, while breaking down at low and high energies. The former failure is obviously due to the IOSA validity becoming questionable while the latter originates from the increased transition probabilities for vibrational inelasticity that therefore cannot any longer be treated within the DWA approximation.

Several approximate analytic methods for the calculations of the (4.165) integrals have been carefully compared [135] and have indicated that the added use of MWNA becomes rather poor for large values of the total angular momentum. This approximation, in fact, only takes into account tha value of the local kinetic energy term at the classical turning point [see eq. (4.122)] and therefore does not allow for the further energy correction due to the potential strength at the classical turning point. The definition of the free WNA [135], using both the kinetic energy and the slope of the effective potential for the definition of the modified wavenumber, then provided for the He + H_2 case an improved accuracy from the R-regions dominated by the centrifugal potential (high- J values) when distorded wave integrals were yielded within several approximate formulae.

4.8 General Conclusions

In the previous Sections of this Chapter we have concerned ourselves with the most recent time-independent, quantum-mechanical simplifications of the CC scattering equations that have been theoretically formulated and numerically applied to various atom-molecule, ion-molecule and molecule-molecule simple systems of diatomics.

The above survey, however, by no means exhausts the subject since the still rapidly growing field of molecular energy collisional transfers theory has produced several possible approaches in recent years. We feel, however, that those summarized in this Chapter are the ones that have come out of several previous years of general effort and that have been analyzed and tested by the larger number of scientists, thus affording a general picture of their performances that relies on several independent studies.

Classical-path methods [137], semiclassical approximations [138] and classical mechanics [139] have all been employed for treating vibro-rotational collisional excitations and have been discussed from the interesting viewpoint of analyzing more in details the classical limit of quantum-mechanical scattering theory [140]. Since they all belong to the general class of time-dependent treatments of the collision problem, though, they will not be dealt with in this study.

Another formal method that has received some attention in recent years, with the intention of reducing through it the number of channels needed in a scattering calculation, has involved the definition on an optical potential, since it is in fact possible to define such a potential rigorously so that the effect of those channels which are not explicitely included in a calculation are properly accounted for by the potential itself [117, 141].

The latter has then a complex, non-local and energy dependent form. Its main use has been in the phenomenological description of elastic scattering in the presence of coupling with reactive and inelastic channels although, for the purpose of treating vibrational inelasticity in the presence of rotational inelastic coupling, its form has often been approximated by a complex, local energy independent expression [142-145].

Recent implementation of the above method for physically realistic systems involved the vibro-rotational relaxation times and inelastic transitions calculations for He + H_2 , He + D_2, he + D_2 and Li + N_2 collisions by Gerber and coworkers [146, 147]. In their framework the multichannel scattering equations are replaced by a pair of coupled equations for each vib-rotational transition. The effect of the large manifold of the rotational levels that are open at the considered collision energies is thus introduced through an optical potential, in each of the coupled equations, and an approximate prescription is suggested to get the potential form. The general cross section trends and the corresponding relaxation times appear to be realistically given with a substantial saving of computational time.

It is therefore clear from the foregoing discussion that a large number of different methods are available today for the calculation of vibro-rotationally inelastic cross sections and it is only the specific nature of the physical problem, or the conditions under which the dynamic observables are required, that will suggest which is the most suitable method to use. The choice will have, of course,to strike a compromise between accuracy and computational feasibility.

In the absence of firm, quantitative general criteria only qualitative guidelines apply. Thus, approximations that yield non-unitary S-matrices can be accepted only when the off-diagonal matrix elements that contribute to inelasticity over tha whole of J values are small in magnitude when compared to unity. The BA and EBA always fail at small J values, although the latter has a wider range of validity.

The EP approach seems to provide only qualitative agreement with rigorous CC calculations, especially near-threshold and for strong anisotropic PES. On the other hand, it provides an efficient computational approach to relaxation time estimates for those more complex systems that are still out of range, for other, more rigorous methods. The more accurate CS scheme, in fact, fails for strongly anisotropic components of the $V(\underline{r}, \underline{R})$ with a long range behaviour and is also less suited for strongly forward scattering DCS behaviour.

The IOSA technique, on the other hand, still needs to be more accurately tested in terms of its low collision energy limits which in turn depend on the structure of the target internal bound states. Its application, combined with rigorous vibrational close coupling or first order DWA treatment of vibrational inelasticity, is only beginning to be implemented for realistic systems and has thus far produced rather encouraging results.

The correct treatment of centrifugal potential coupling combined with a dimensionality reduction in the high-J, high collision energy regions as suggested by the DLS approach has resulted in improved results for ion-molecule systems with long-range anisotropic contributions. Its validity near threshold of for vibrational inelasticity is however somewhat questionable and still requires more extensive numerical test and careful comparisons.

One can however say with enough confidence that the quantum mechanical approaches have made very rapid progress over the last few years. Considering the parallel development of sophisticated, state-selected experiments on inelastic collisions [148] between atoms, ions and molecules, it is certainly time to say that direct comparisons of observed and computed integral total cross sections and inelastic differential cross sections has become an increasingly more frequent reality [e.g. see ref. 38].

It is however still important to discuss in some detail the specific methods that have been suggested for numerically solving the relevant coupled equations and that have been optimized over the years to reduce computational efforts. Moreover, one also likes to see more specifically which of the many macroscopic phenomena affected by molecular collision dynamics have been directly studied from theoretical data and compared with the observed quantities. These points will be briefly discussed in the following Chapters.

REFERENCES

[1] A. E. De Pristo and M.H. Alexander, chem. Phys., 19 , 181 (1977).

[2] A.E. De Pristo and M.H. Alexander, J. Chem. Phys., 63 , 3552 (1975).

[3] P. Mc.Guire, Chem. Phys. Lett., 23 , 575 (1973).

[4] R.T. Pack, J. Chem. Phys., 60 , 633 (1974).

[5] P. McGuire and D.J. Kouri, J. Chem. Phys., 60 , 2488 (1974).

[6] P.McGuire, J. Chem. Phys., 62 , 525 (1975).

[7] P. McGuire, Chem. Phys., 13 , 81 (1976).

[8] G.A. Parker and R.T. Pack, J. Chem. Phys., 66 , 2850 (1977).

[9] D.J. Kouri and Y. Shimoni, J. Chem. Phys., 67 , 86 (1977).

[10] D. Secrest, J.Chem. Phys., 62 , 710 (1975).

[11] V. Khare, J. Chem. Phys., 67 , 3897 (1977).

[12] M.A. Brandt and D.G. Truhlar, Chem. Phys., 13 , 461 (1976).

[13] M. Tamir and M. Shapiro, Chem. Phys. Lett., 31 , 166 (1975).

[14] M. Jacob and G.C. Wick, Ann. Phys., 7 , 404 (1959).

[15] M. Shapiro and M. Tamir. Chem. Phys. 13 , 215 (1976).

[16] D.J. Kouri, Phys. Lett., 31 , 599 (1975).

[17] J.M. Launay, J. Phys. B., 9 , 1823 (1976).

[18] Y. Shimoni and D.J. Kouri, J. Chem. Phys., 66 , 2841 (1977).

[19] Y. Shimoni and D.J. Kouri, J. Chem. Phys., 65 , 3372 (1976).

[20] Y. Shimoni and D.J. Kouri, J. Chem. Phys., 65 , 3958 (1976).

[21] R. Schinke and P. McGuire, Chem. Phys., 28 , 129 (1978).

[22] V. Khare, D.J. Kouri and R.T. Pack. J. Chem. Phys. (1978).

[23] D.J. Kouri, T.G. Heil and Y. Shimoni, J. Chem. Phys., 65 , 226 (1976).

[24] V. Khare, D.J. Kouri, J. Chem. Phys. (1978).

[25] R.T. Pack, J.Chem. Phys., 66 , 1557 (1977).

[26] P. McGuire and H. Kruger, J. Chem. Phys., 63 , 1090 (1975).

[27] M.H. Alexander and P.J. McGuire, J. Chem. Phys., 64 , 452 (1976).

[28] P.McGuire and J.P. Toennics, J.Chem. Phys., 62 , 4623 (1975).

[29] M.H. Alexander and P. McGuire, Chem. Phys., 12 , 31 (1976).

[30] M. Alexander, Chem. Phys. Lett., 38 , 417 (1976).

[31] S.I. Chu and A. Dalgarno, Proc. Roy Soc. (London) A 342 , 191 (1975).

[32] S.I. Chu and A. Dalgarno, J. Chem. Phys., 63 , 2115 (1975).

[33] S. Green and P. Thaddeus, Astrophys. 205 , 266 (1976).

[34] S. Green, Chem. Phys. Lett. 38 , 293 (1976).

[35] B.J. Garrison et al. Astrophys. J.Lett., 200 , L 175 (1975).

[36] S. Green and L. Monchick, J. Chem. Phys., 63 , 4198 (1975).

[37] L. Monchick and S. Green, J. Chem. Phys., 63 , 2000 (1975).

[38] U. Buck and P. McGuire, (to be published).

[39] S. Green, J. Chem. Phys., 64 , 3463 (1976).

[40] S. Green and L. Monchick, J. Chem. Phys., 66 , 3085 (1977).

[41] L.D. Thomas et al., Chem. Phys., 27 , 237 (1978).

[42] D.J. Kouri and P.McGuire, Chem. Phys. Lett., 29 , 414 (1974).

[43] M.H. Alexander, J. Chem. Phys., 66 , 4608 (1977).

[44] P. McGuire, Chem. Phys., 8 , 231 (1975).

[45] L. Monchick, J. Chem. Phys., 67 , 4626 (1977).

[46] J.H. Ferziger and M.G. Kaper, Mathe-atical Theory of Transport processes in gases (N. Holland, Amsterdam, 1972).

[47] L. Monchick, Physics of Fluids, 7 , 882 (1964).

[48] W. Wielsen and R.G. Gordon, J. Chem. Phys., 58 , 4149 (1973).

[49] H. Rabitz, Ann. Rev. Phys. Chem., 25 , 155 (1974).

[50] S. Green, J. Chem. Phys., 62 , 3568 (1975).

[51] L. Monchick and D.J. Kouri, J. Chem. Phys. (1978).

[52] D.M. Chase, Phys. Rev. 104 , 838 (1956).

[53] K. Takayanagi, Progr. Theoret. Phys. Suppl. 25 40 (1963).

[54] K.M. Kramer and R.B. Bernstein, J. Chem. Phys., 40 , 200 (1964).

[55] R.D. Levine, Chem. Phys. Lett., 4 , 211 (1969).

[56] E.S. Chang and A. Temkin, Phys. Rev. Lett., 28 , 203 (1972).

[57] N. Chandra and F.A. Gianturco, Chem. Phys. Lett., 24 326 (1974).

[58] P.G. Burke, N. Chandra and F.A. Gianturco, J. Phys. B 5 , 2212 (1972).

[59] M. Shugard and A.U. Hazi, Phys. Rev. A12 , 1895 (1975).

[60] P.G. Burke and L.T. Sinfailarn, J. Phys B 3 , 641 (1970).

[61] R.J. Cross, J. Chem. Phys., 49 , 1753 (1968).

[62] U. Buck and V. Khare, Chem. Phys. 26 , 215 (1977).

[63] R.T. Pack, Chem. Phys. Lett., 14 , 393 (1972).

[64] T.P. Tsien and R.T. Pach, Chem. Phys. Lett., 6 , 54 (1970).

[65] C.F. Curtiss, J. Chem. Phys., 49 , 1952 (1958).

[66] L.W. Hunter, J. Chem. Phys., 62 , 2855 (1975).
[67] J.M. Bowman and S.C. Leasure, J. Chem. Phys., 66 , 288 (1977).
[68] D.J. Kouri, T.G. Heil and Y. Shimoni, J. Chem. Phys., 65 , 1462 (1976).
[69] M.A. Brandt and D.G. Truhlar, Chem. Phys. Lett., 23 , 48 (1973).
[70] R. Goldflam, S. Green and D.J. Kouri, J. Chem. Phys., 67 , 225 (1977).
[71] T.P.Tsien, G.A. Parker and R.J. Pack, J. Chem. Phys., 59 , 5373 (1973).
[72] D.E. Fitz and P.McGuire, Chem. Phys. Lett., 44 , 503 (1976).
[73] R. Goldflam, D.J. Kouri and S. Green, 67 , 5661 (1977).
[74] G.A. Parker and R.T. Pack, J. Chem. Phys., 68 , 1585 (1978).
[75] L. Monchick and E.A. Mason, J. Chem. Phys., 35 , 1676 (1961).
[76] L. Monchick and S. Green, J. Chem. Phys., 63 , 4198 (1975).
[77] R.T. Pack, Chem. Phys. Lett., 55 , 197 (1978).
[78] J.M. Bowman and S.C. Leasure, J. Chem. Phys., 66 , 288 (1977).
[79] S.C. Leasure and J.M. Bowman, Chem. Phys. Lett., 48 , 179 (1977).
[80] P. McGuire, J. Chem. Phys., 62 , 525 (1975).
[81] F.A. Gianturco and U. Lamanna, Chem. Phys. Lett., 6 , 326 (1970).
[82] R. Schinke and P. McGuire, Chem. Phys. 28 , 129 (1978).
[83] A. Karo et al., Status Report, Argonne Nat. Laboratory (1974).
[84] F.A. Gianturco and U.T. Lamanna, Chem. Phys. (1978).
[85] M. Rabitz, J. Chem. Phys., 57 , 1718 (1972).
[86] G. Zarur and H. Rabitz, J. Chem. Phys., 59 , 943 (1973).
[87] H. Rabitz Modern Theoretical Chemistry, Vol. 1 pp. 33 (Plenum Press, N.Y. 1976).
[88] H. Rabitz and G. Zarur, J. Chem. Phys. 61 , 5076 (1974).
[89] H. Rabitz and G. Zarur, J. Chem. Phys. 62 , 1425 (1975).
[90] M.H. Alexander, J. Chem. Phys., 61 , 5167 (1974).
[91] M.H. Alexander, J. Chem. Phys., 8 , 86 (1975).
[92] S.I. Chu and A. Dalgarno; Asptrophys. J. 199 , 637 (1975).
[93] M. Verter and H. Rabitz, J. Chem. Phys., 64 , 2939 (1975).
[94] S. Green, J. Chem. Phys., 62 , 3568 (1975).
[95] G. Zarur and H. Rabitz, J. Chem. Phys., 60 , 2057 (1974).
[96] S.I. Chu, J. Chem. Phys., 62 , 4089 (1975).
[97] B.H. Choi and K.T. Tang, J. Chem. Phys., 64 , 942 (1976).
[98] Y. Itikawa, J. Phys. Soc., 39 , 1059 (1975).
[99] A.E. De Pristo and M.H. Alexander, J. Chem. Phys., 66 , 1334 (1977).

[100] S. Augustin and H. Rabitz, J. Chem. Phys., 64 , 4821 (1976).

[101] R. Ramaswamy and H. Rabitz, J. Chem. Phys., 66 , 152 (1977).

[102] H.K. Shin, J. Chem.Phys., 46 , 3688 (1967).

[103] F.M. Mies and K. Shuler, J. Chem. Phys., 37 , 177 (1962).

[104] K. Takayanagi, Progr. Theor. Phys., (Kyoto), 8 , 497 (1952).

[105] R. Marriott, Proc. Phys. Soc., 83 , 159 (1964).

[106] F.A. Gianturco and U.T. Lamanna, Chem. Phys., 17 , 255 (1976).

[107] F.A. Gianturco and R. Marriott, J. Phys. B 2 , 1332 (1969).

[108] F.A. Gianturco and U.T. Lamanna, Chem. Phys. (1978).

[109] F.A. Gianturco, U.T. Lamanna and I. Baraldi, J. Chim; Phys., 74 , 437 (1977).

[110] F.A. Gianturco and U.T. Lamanna, Chem. Phys. 3 , 110 (1974).

[111] F;A; Gianturco and U.T. Lamanna, Chem. Phys., 25 , 401 (1977).

[112] A.E. De Pristo and M.H. Alexander, J. Chem. Phys., 63 , 5327 (1975).

[113] A.E. De Pristo and M.H. Alexander, J. Chem. Phys., 64 , 3009 (1976).

[114] A.E. De Pristo and M.H. Alexander, J. Phys B, 9 , L 39 (1976).

[115] A.E. De Pristo and M.H. Alexander, J. Phys. B 9, 2713 (1976).

[116] A.E. De Pristo and M.H. Alexander, Chem. Phys. Lett., 44 , 214 (1976).

[117] N.F. Mott and M.S.W. Massey, The Theory of atomic collisions, (Claredon Press Oxford, U.K. 1965).

[118] A.M. Arthurs and A. Dalgarno, Proc. R. Soc. (London Ser.) A 256, 540 (1960).

[119] K. Takayanagi, Adv. At. Mol. Phys., 1 , 149 (1965).

[120] C.F. Curtiss and A. Hardisson, J. Chem. Phys., 46 , 2618 (1967).

[121] L.W. Hunter and C.F. Curtiss, J. Chem. Phys., 58 , 3897 (1973).

[122] D. Russell and C.F. Curtiss, J. Chem. Phys., 59 , 1974 (1973).

[123] G.C. Balint-Kurti, Int. Review of Science , Phys. Chem. Series 2, Vol. 1, pag. 283 (Butterworth, London, 1975).

[124] K. Takayanagi, Sc. Rept. Saitama Univ. A3 , 87 (1959).

[125] R.B. Bernstein et al., Proc. Roy Soc. London Ser. A 274 , 427 (1963).

[126] I.C. Percival, Proc. Phys. Soc. (London) 76, 206 (1960).

[127] R.D. Levine, Mol. Phys., 22 , 497 (1971).

[128] R.D. Levine and G.G. Balint-Kurti, Chem. Phys. Lett., 6 , 101 (1970).

[129] M.J. Seaton, Proc. Phys. Soc. (London), 77 , 174 (1961).

[130] M. Van Seggern and P.J. Toennies, Chem. Phys. Lett., 6 , 101 (1970).

[131] G.G. Balint-Kurti and R.D. Levine, Chem. Phys. Lett., 7 , 107 (1970).

[132] S.M. Tarr and H. Rabitz, J. Chem. Phys., 68 , 642 (1978).

[133] R.B. Gerber, Chem. Phys., 16, 19 (1976).

[134] L.Eno and G.G. Balint-Kurti, Chem. Phys., 23 , 295 (1977).

[135] L.Eno, G.G. Balint-Kurti, and R. Saktreger, Chem. Phys., 29 , 453 (1978).

[136] F.H. Mies, J. Chem. Phys., 40 , 523 (1964).

[137] J.B. Delas, W.R. Thorson and S.K. Khudson Phys. Rev. A6 , 709 (1972).

[138] W.M. Miller, Adv. Chem. Phys., 30 , 77 (1975).

[139] D.L. Bunker, Methods Comp. Phys., 10 , 287 (1971).

[140] W.H. Miller, Adv. Chem. Phys., 25 , 69 (1974).

[141] D.A Micha, in Modern Theoret. Chem. Vol. 1 (W.M. Miller Ed., Plenum Press, N.Y. 1976).

[142] D.J. Kouri, J. Chem. Phys., 48 , 448 (1968).

[143] .G. Wolken Jr., J. Chem. Phys., 56 , 259 (1972).

[144] E. Ficocelli-Varracchio, Chem. Phys., 18 , 37 (1976).

[145] E. Ficocelli-Varracchio, Mol. Phys., 30 , 1117 (1975).

[146] R.B. Gerber, N.C. Zaritsky and U. Minglegrin, Mol. Phys. 35 , 1247 (1978).

[147] N.C. Zaritsky and U. Minglegrin and R.B. Gerber, Mol. Phys. 35 , 1269 (1978).

[148] e.g. see the recent review by M. Faubel and J.P. Toennies, Adv. At. Mol. Phys., 13 , 229 (1977).

5. NUMERICAL METHODS FOR THE COUPLED EQUATIONS: A SURVEY

5.1 - Introduction

In the previous Chapters we indicated and discussed in some detail the theoretical models which have been worked out from first principles and employed in treating the various collisional processes that preside over the transfer of transitional and rotovibrational energies in simple diatomic gases. Also underlined was how the need to know many cross sections at several collision energies, and for increasingly more sophisticated systems, has in recent years spurred the development of dimensionality reduction schemes which could hopefully allow the actual computation of the required cross sections with the necessary level of accuracy but with a much less expensive computational effort.

It is, in fact, of considerable scientific interest as well as practical interest to have reliable information on the detailed kinetics of chemical reactions in the gas phase and this therefore implies that, among other things, one would like to be able to predict with relative ease and reliability such quantities as the many cross sections for formation of products in specific internal quantum states and, in general, the energy and angular distributions of the products of gas phase reactions.

Ultimately, then, the problem is reduced to the formulation of numerical procedures that could efficiently yield the relevant cross sections by solving a varying number of coupled equations which originate say, from the static expansions discussed in Chapter 3. It is the purpose of this present Chapter to survey briefly some of the methods that have been implemented and tested in recent years; more extensive reviews which have appeared in the recent literature are listed in the bibliography [1].

It might be useful as an introduction to compare the numerical methods which may be used for the bound-state problem that provides B.O. potential hypersurfaces, as discussed in Chapter 2, with those used for the continuum state problem. The electronic bound states appear of first as being more complicated in that the relevant equantions to be solved are non-linear, but they become however a simpler problem with regard to the form of the boundary conditions to be imposed in the limit of large electronic coordinates, since these conditions require that each M.O. function φ_i should tend to zero exponentially as $\underline{r}_i$ goes to ∞ . In order to obtain solutions which exhibit the correct asymptotic form, and which are also normalized, it is the necessary to solve an eigenvalue problem at each iterative step. The imposition of a normalization condition at each stage complicates the relevant numerical work but helps in guinding the system through a speedy convergence of the iterations.

For the continuum problem, the corresponding radial functions have an oscillatory asymptotic form so that, once this form is known, one may obtain the relevant cross sections. It is however much more difficult to find suitable basis functions for the continuum, as opposed to the various, fairly successful, basis sets that have been applied to atomic and molecular bound state problems and that have proven to be a good representation of the low-lying electronic states of many simple systems (e.g., see the discussion carried out in Chapter 2). The corresponding attempts for obtaining continuum functions were based on algebraic expansions of the scattering wavefunction in terms of a set containing functions of appropriate long range behaviour plus functions designed to describe short range behaviour, and have mainly been directed to electron-atom scattering problems [2, 3, 4]. In this last instance variational criteria for determining the expansion coefficients need to be established, and along lines similar to those used for the bound state cases.

The resulting basis set, however, is usually constituted by rather complicated functions which do not allow for the analytical evaluation of the necessary integrals, this being even more true for the molecular collisions which interest us here. Moreover, the convergence of iterative methods may also be less statisfactory for the continuum problems and spurious resonances may appear in the computed cross sections.

The great advantage of the presently discussed problem comes however from the simplification which states that the CC equations are linear in the functions u_i (R) and that these linear differential (or integro-differential) equations may be solved by non-iterative numerical methods. The set of equations that need to be solved can be written compactly in a matrix form as:

$$\left\{ \frac{d^2}{d\,\mathbf{R}^2}\mathbb{1} + \mathbb{K}^2 - \mathbb{V}(\mathbf{R}) \right\} \mathbb{U}(\mathbf{R}) = 0 \qquad (5.1.)$$

where $\mathbb{V}(\mathbf{R})$ is the Hermitian matrix of the coupling potential plus the centrifugal barrier and $\mathbb{K}^2$ is the diagonal wavevector matrix. Even more succinctly one can define a new Hermitian matrix $\mathbb{W}(\mathbf{R})$:

$$\mathbb{W}(R) = -\mathbb{K}^2 + \mathbb{V}(R) \qquad (5.2.)$$

and therefore write down the following set of linear equations:

$$\left(\frac{d^2}{d\,R^2} \right) \mathbb{U}(R) = \mathbb{W}(R)\,\mathbb{U}(R) \qquad (5.1')$$

which has three interesting properties that can be exploied in finding efficient computational methods [1]:

(i) the collision energy E_{coll} appears in (5.2) as a scalar which multiplies the unit matrix added to the coupling potential matrix, hence a change of collision energy, ΔE, causes the eigenvalues of $\mathbb{W}$ to all be changed by the same amount.

(ii) there are no first derivatives and there is no matrix coupling the first derivatives of the unknown set of continuum functions to be determined.

(iii) the coupling potential matrix is symmetric and may often be a slowly varying function of internuclear distance R, thus $\mathbb{W}$ can be diagonalized rather easily since it is also symmetric.

The partial or total exploitation of the above points has led to the testing and perfecting of various computational procedures for heavy particle collisions and the following Sections will briefly survey the respective formulations of some of them. Before doing that, however, let us look more closely at the detailed form of the coupled equations.

If the dimensions of the matrices in (5.1') is N x N, the problem consists in finding N linearly independent solutions for each collisions energy which satisfy the following boundary conditions (in complex form):

$$u^S_{ii'}(R) \underset{\mathbf{R}\to\infty}{\sim} \delta_{ii'}\, k_i^{-\frac{1}{2}} \exp\left[-i(k_i R - \ell_i\, \pi/2)\right] - \qquad (5.3)$$

$$- S_{ii'} k_i^{-\frac{1}{2}} \exp [i(k_i R - \ell_i \pi/2)] \quad (5.3)$$

where the superscript refers to the incoming and outgoing waves form and the nature of the interaction $V(R)$ is presumed to be such that:

$$\lim_{R \to \infty} RV(R) = 0 \quad (5.4)$$

and S_{ii} in the corresponding S-matrix element.

Moreover, since the second order matrix equation admits 2N solutions that are linearly independent, N of them are eliminated by further imposing another condition for each of the sought solutions:

$$u_{ii}(o) = 0 \quad (5.5)$$

This in the usual requirement that solutions be regular at the origin. It originates from the physical fact that the interaction potentials are always repulsive when $R \to 0$ and therefore the corresponding probability of particle being present at the origin is zero.

Because of the real nature of the coupling matrix elements in $\mathbb{W}(R)$ one can look for a real form of the required solutions by rewriting their asymptotic form as given eq.(5.6) for real, non zero k_i:

$$\begin{aligned} u^K_{ii'}(R) \underset{R \to \infty}{\sim} & \delta_{ii'} k_i^{-\frac{1}{2}} \sin(k_i R - \ell_i \pi/2) \\ & + k_i^{1/2} K_{ii} \cos(k_i R - \ell_i \pi/2) \end{aligned} \quad (5.6)$$

which defines the real, symmetric K-matrix that relates the two forms of the solutions [5]:

$$\mathbb{U}^S(R) = -2i \; \mathbb{U}^K(R) \, (-\mathbb{1} - i\mathbb{K}) \quad (5.7)$$

hence:

$$\mathbb{S} = (\mathbb{1} + i\mathbb{K} \, (\mathbb{1} - i\mathbb{K})^{-1} \quad (5.8)$$

In the case where closed channels are included ($k_i^2 < 0$), the asymptotic form of the required solution is not given by eq.(5.6) but rather by a linear combination of a growing exponential, $e^{|k_i|R}$ and of a decaying exponential $e^{-|k_i|R}$. Since the flux conservation requirements need the incoming flux to be distributed only within the open channels, the $\exp[|k_i|R]$ solution is also physically unacceptable and has to be eliminated. This causes, in turn, the imposing of N_c additional conditions, where N_c is the number of closed channels. If $N_o = N - N_c$ is the number of open channels, one can then write down the following set of general asymptotic conditions:

$$u^k_{ii'}(R) \underset{R \to \infty}{\sim} \delta_{ii} \, k_i^{-\frac{1}{2}} \sin(k_i R - \ell_i \, \pi/2) + k_i^{-\frac{1}{2}} \, K_{ii'} \, \cos(k_i R - \ell_i \, \pi/2) \; ; \quad 1 \leqslant i \leqslant N_o \tag{5.9}$$

$$u^k_{ii'}(R) \underset{R \to \infty}{\sim} \mathbf{A}_{ii'} \, \exp[-|k_i|R] \; ; \qquad N_o + 1 \leqslant i \leqslant N$$

Since the eq.s(5.1') are homogeneous in $u_{ii'}(R)$, the multiplication of a solution by an arbitrary factor is also a solution. This implies that the asymptotic amplitudes produced by the K-matrix are arbitrary, and thus they are commonly fixed in scattering problems by setting:

$$\frac{d}{dR} \mathcal{U} \, (o) = \mathbb{D}_o \tag{5.10}$$

where $\mathbb{D}_o$ is an arbitrary non-singular matrix, thereby reducing the apparent two-points boundary value problem of (5.3) and (5.5) to an initial-value problem [6]. Retaining only the N regular solutions, one can thus write them as follows:

$$\begin{array}{ll} u_{11}, u_{21}, \ldots \ldots \ldots \ldots u_{N1} & \text{first solution,} \\ u_{12}, u_{22}, \ldots \ldots \ldots \ldots u_{N2} & \text{second solution,} \\ u_{1N}, u_{2N}, \ldots \ldots \ldots \ldots u_{NN} & N\textbf{th} \text{ solution.} \end{array} \tag{5.11}$$

where the first subscript indicates the component of a given solution and the second subscript numbers the ordering sequence of solutions. A general solution of (5.1') is thus given by the set of N functions:

$$u_i = a_1 u_{i1} + a_2 u_{i2} + \ldots + a_N u_{iN} \qquad i = 1, 2, \ldots N \qquad (5.12)$$

where the a's arbitrary constants.

For intermolecolar potentials containing terms proportional to R^{-n} $(n > 2)$, the singularity of the interaction at the origin prevents the starting of the integration at R = 0. Usually this difficulty is bypassed by the introduction of an hard-core in the potential ($V = \infty$) for $R < R_S$ and by using the potential's correct form for $R > R_S$. The R_S value should be chosen sufficiently far into the classically inaccessible region so as to make the final phase shifts and amplitudes independent of this choice. If, however, it comes too close to the origin, the linear independence may be lost by exponential growth of the solutions in this region. The determination of R_S is thus a matter of trial and error, and with an appropriate choice for its value the linear independence of the N solutions may be obtained by further choosing a linearly independent set of initial derivate vectors that provide the $\mathbb{D}_0$-matrix of (5.10) at $R = R_S$ [7].

Since the solutions, that are linearly independent at the beginning of the integration, tend to become numerically linearly dependent because of the loss of accuracy along the propagation, it is common practice to stabilize the solutions periodically by taking their linear combinations as a new starting point [8] for proceeding towards the asymptotic region.

5.2. - The De Vogelaere's method

One of the numerical methods which has been used rather extensively to propagate the continuum functions of the initial-value problem formulated in the previous section has been a generalized form of a fourth order, step-by-step method based on difference formulas [9]. This method allows the solution of equations of the following type for the uncoupled case:

$$y''(R) = f[R, y(R)] \qquad (5.13)$$

with the first derivative missing. The corresponding interval of $R(0, \infty)$ is conveniently discretized in a finite number of points R_o, R_1, ... R_n, ... R_{fin} for which the f functions values are tabulated, together with the additional values at $R_{n+1/2} = 1/2\,(R_n + R_{n+1})$. The forward difference formulas for the first and second derivatives required for the derivation of the De Vogelaere's formulas are given by:

$$
\begin{aligned}
h\,y'_n &= (\Delta - \tfrac{1}{2}\Delta^2 + \tfrac{1}{3}\Delta^3 - \tfrac{1}{4}\Delta^4 + \ldots..)\,y_n \\
&= (\Delta + \tfrac{1}{2}\Delta^2 - \tfrac{1}{6}\Delta^3 + \tfrac{1}{12}\Delta^4 + \ldots..)\,y_{n-1} \\
&= (\Delta - \tfrac{3}{2}\Delta^2 + \tfrac{1}{3}\Delta^3 - \tfrac{1}{12}\Delta^4 + \ldots..)\,y_{n-2}
\end{aligned}
\qquad (5.14)
$$

and:

$$
\begin{aligned}
h^2 y''_n &= (\Delta^2 - \Delta^3 + \tfrac{11}{12}\Delta^4 - \ldots..)\,y_n \\
&= (\Delta^2 - \tfrac{1}{12}\Delta^4 + \ldots..)\,y_{n-1} \\
&= (\Delta^2 + \Delta^3 - \tfrac{1}{12}\Delta^4 + \ldots..)\,y_{n-2}
\end{aligned}
\qquad (5.15)
$$

where h is the interval of the R points and:

$$\Delta y_n = y_{n+1} - y_n \qquad (5.16)$$

Neglecting terms higher than Δ^3 we obtain:

$$hy'_n + \frac{h^2}{6}(4y''_n - y''_{n-1}) = (\Delta + \Delta^2)\,y_{n-1} \qquad (5.17a)$$

and:

$$y_{n+1} = y_n + hy'_n + \frac{h^2}{6}(2y''_n + y''_{n-1}) \qquad (5.17b)$$

Similarly, neglecting terms higher than Δ^4:

$$
\begin{aligned}
2hy'_n + \frac{1}{3}h^2(2y''_n + 4y''_{n+1}) &= (2\Delta + \Delta^2)\,y_n \\
&= y_{n+2} - y_n
\end{aligned}
\qquad (5.18a)
$$

and:

$$y_{n+2} = y_n + 2hy'_n + \frac{h^2}{3}(2y''_n + 4y''_{n+1}) \tag{5.18b}$$

$$\frac{h}{3}(y''_n + 4y''_{n+1} + y''_{n+2}) = \frac{1}{4}(2\Delta^2 + \frac{1}{6}\Delta^4)y_n \tag{5.18c}$$

$$y'_{n+2} - y'_n = \frac{1}{h}(2\Delta^2 + \frac{1}{6}\Delta^4)\, y_n \tag{5.19a}$$

$$y'_{n+2} = y'_n + \frac{h}{3}(y''_n + 4y'_{n+1} + y''_{n+2}) \tag{5.19b}$$

On substituting $f_n = y''_n$ and changing the interval size to (1/2 h) in eq.s (5.17), (5.18) and (5.19) one obtains the De Vogelaere formulas for the single channel case:

$$y_{n+\frac{1}{2}} = y_n + \frac{1}{2}hy'_n + \frac{1}{24}h^2(4\, f_n - f_{n-\frac{1}{2}}) \tag{5.20a}$$

$$y_{n+1} = y_n + hy'_n + \frac{1}{6}h^2\,(f_n + 2f_{n+\frac{1}{2}}) \tag{5.20b}$$

$$y'_{n+1} = y'_n + \frac{1}{6}h\,(f_n + 4f_{n+\frac{1}{2}} + f_{n+1}) \tag{5.20c}$$

The errors in these expressions are of the order of h^4, h^5 and h^5 respectively and so the accuracy is comparable to that of a fourth order Runge-Kutta process [10]. Here, however, only one intermediate point is required in carrying out the integration from R_n to R_{n+1}. Moreover the main advantages of this method are the ease of starting the integration and the simple way in which the interval value can be changed.

To start the integration it is necessary to know not only y_0 and y'_0 but also $f_{-1/2} = f_{-1/2}[R_{-1/2}, y_{-1/2}]$. An estimate of $y_{-1/2}$ can be made by using backward differences. Thus, if:

$$\nabla y_n = y_n - y_{n-1} \tag{5.21}$$

then:

$$y_{-\frac{1}{2}} = (1 - \frac{1}{2}\nabla - \frac{1}{8}\nabla^2 - \ldots\ldots)\, y_0 \tag{5.22a}$$

$$hy'_0 = (\nabla + \frac{1}{2}\nabla^2 + \ldots\ldots)\, y_0 \tag{5.22b}$$

$$h^2 y_0'' = (\nabla^2 + \ldots\ldots)\, y_0 = h^2 f_0 \qquad (5.22c)$$

Hence:

$$y_0 - \frac{1}{2}\, hy_0' + \frac{1}{8}\, h^2 y_0'' = (1 - \frac{1}{2}\,\nabla - \frac{1}{8}\,\nabla^2)\, y_0 \qquad (5.23)$$

which yields:

$$y_{-\frac{1}{2}} = y_0 - \frac{1}{2}\, hy_0' + \frac{1}{8}\, h^2 f_0 \qquad (5.24)$$

The interval h may be changed conveniently by replacing eq. (5.20a) with:

$$y_{n+\frac{1}{2}} = y_n + \frac{h_2}{2}\, y_n' + \frac{1}{24}\, h_2^2 \left[\left(3 + \frac{h_2}{h_1} \right) f_n - \frac{h_2}{h_1}\, \overline{f}_{n-\frac{1}{2}} \right] \qquad (5.25)$$

where h_1 is the previous value of the interval, h_2 is its new value, and $\overline{f}_{n-1/2}$ denotes the value of the function at $R = R_0 - h_{1/2}$.

In the general case of multichannel coupled equations of size N, the previous notation now becomes a matrix notation, where each matrix element is obtained via the same equations of above. They need however to carry another label referring to the solution under study. The function f obviously depends now on all the ordered solutions $y_\beta(R)$ with $\beta = 1, 2, \ldots\ldots N$.

The continuum solutions are then propagated out to the large-R region, up to where the coupling matrix elements generated by the interaction potential among the expansion functions becomes negligible and essentially potential-free scattering situation is achieved.

In this instance one can then write, in matrix form:

$$\mathbb{Y}(R) = \mathbb{J}(R)\mathbb{A} - \mathbb{N}(R)\mathbb{B} \qquad (5.26a)$$

and:

$$\mathbb{Y}'(R) = \mathbb{J}'(R)\mathbb{A} - \mathbb{N}'(R)\mathbb{B} \qquad (5.26b)$$

where:

$$J_{ij}(R) = \delta_{ij}\, k_j^{\frac{1}{2}}\, R\, j_{\ell_j}(k_j R) \quad ; \quad \text{for } k_j^2 > 0. \qquad (5.27a)$$

$$= \delta_{ij} (-)^{\ell_j + 2} (\frac{1}{2}\, k_j)^{\frac{1}{2}}\, R\, h_{\ell_j}^{(2)}(k_j R) \;; \text{if } k_j^2 < 0. \qquad (5.27b)$$

and:

$$N_{ij}(R) = \delta_{ij}\, k_j^{\frac{1}{2}}\, R\, n_{\ell_j}(k_j R) \quad ; \qquad \text{if} \quad k_j^2 > 0\ . \tag{5.28a}$$

$$= \delta_{ij}\, i^{\ell_j+2} \left(\frac{1}{2} k_j\right)^{\frac{1}{2}} R\, h_{\ell_j}^{(1)}(k_j R)\ ;\ \text{if}\ k_j^2 < 0. \tag{5.28b}$$

where j_ℓ and n_ℓ indicate spherical Bessel and Neumann functions (see Chapter 1) and $h_\ell^{(1,2)}$ the corresponding Hankel functions [11]. The J_{ii} matrix element behaves as $k_i^{-1/2} \sin(k_i R - \ell_i \pi/2)$ whenever the ith channel is open and as $\exp(k_i R)$ if the same ith channel is closed. Correspondingly, the N_{ii} element behaves as $-k^{-1/2} \cdot \cos(k_i R - \ell_i \pi/2)$ for the open channel case and as $\exp(-k_i R)$ for the closed channel case.

The matrices defined in eq.s (5.26) need to verify the Wronskian relationship:

$$\mathbb{J}\mathbb{N}' - \mathbb{J}'\mathbb{N} = \mathbb{1} \tag{5.29}$$

hence the $\mathbb{A}$ and $\mathbb{B}$ matrices are easily obtained through the previous equations and are given by:

$$\mathbb{A} = \mathbb{N}'\mathbb{Y} - \mathbb{N}\mathbb{Y}' \tag{5.30a}$$

$$\mathbb{B} = \mathbb{J}'\mathbb{Y} - \mathbb{J}\mathbb{Y}' \tag{5.30b}$$

One usually makes the assumption that if the rows, components of any ith solution as in eq. (5.12), are linearly independent then the relevant asymptotic forms will also be linearly independent. The asymptotic conditions (5.9) can then be written as:

$$\mathbb{Y}^k \sim \mathbb{J} - \mathbb{N}\mathbb{K} \tag{5.31}$$

where $\mathbb{K}$ is the (N + N) $\mathbb{K}$-matrix of the problem under study. This implies that the coefficients of (5.9) are obtained from matching the initial solutions, propagated up to the asymptotic region, together with the solution of eqs (5.26) and under the condition (5.31):

$$\mathbb{Y}^{in}\mathbb{a} = [\mathbb{J}\mathbb{A} - \mathbb{N}\mathbb{B}]\,\mathbb{a} = \mathbb{J} - \mathbb{N}\mathbb{K} = \mathbb{Y}^K \tag{5.32}$$

hence:

$$\mathbb{a} = \mathbb{A}^{-1} \tag{5.33}$$

and:

$$\mathbb{K} = \mathbb{B}\,\mathbb{A}^{-1} \qquad (5.34)$$

which lead directly to the formulation of the K-matrix.

The initial conditions (5.5) and (5.10) to be used in eq.(5.9) are specified, in the present method, as a column of zeros and successive columns of non singular matrices of initial derivatives. The integration over the radial range must be performed N times for each set of N coupled equations. Since the De Vogelaere method is based on finite difference formulae that are exact for solutions of a polynominal form, their application to functions with oscillatory character is acceptable provided that the truncation error is made to be negligible. This implies that a sufficient number of integration points should be taken for each oscillation [7,12].

More precisely, the eigenvalues of the coupling matrix $\mathbb{W}(R)$ of eq.(5.2) will have to correspond to eigensolutions which are given locally by linear combinations of sinusoidal functions with period:

$$\lambda_\alpha(R) = 2\pi\,(\sqrt{w_\alpha(R)})^{-1} \qquad (5.35)$$

where $w_\alpha(R)$ is the positive eigenvalue at the point R.

The corresponding h interval of (5.14) will therefore be chosen small enough to approximate the oscillating solutions sufficiently. The minimum value of λ in (5.35) will be chosen from the wavelength associated with the largest wavenumber channel [13]:

$$\lambda_{min}(R) = 2\pi\,[k^2_{max} - 2\,\mathcal{M}\,V_0(R)]^{-\frac{1}{2}} \qquad (5.36)$$

where $V_0(R)$ corresponds to the strongest coupling contribution from the multipolar expansion of the potential, this usually being the spherical term with an attractive well. Eq.(5.36) is written for s-wave scattering, but it obviously holds even better in the presence of a centrifugal barrier.

The integration step is then usually chosen as equal to $\lambda_{min}(R_0)/n$ where R_0 is the position of the potential minimum and n is an integer between 10 and 20, for the region of integration around the well, while a larger steplength can be chosen for the attractive asymptotic region [14].

The importance of a proper choice for the starting integration point R_S also needs a few words of comment. If its value is chosen to lie too close to the classical turning point the calculations will not lead to the correct scattering matrices with the required symmetry and unitarity properties. Moreover, the region of convergence for many-channel calculations may not be tested correctly via only the spherical term of interaction if ionic systems are considered [13], as the latter are dominated by long-range anisotropic contributions. This implies therefore that complete, coupled-channel calculations need to be used for convergence tests whenever charge-dipole or charge-quadrupole effects are present [13].

Finally, it should be stressed that the determination of the accuracy of the S-matrix obtained from a step-by-step integration such as we have in the present method, involves the balanced choice of different integration meshes, which in turn have to be adjusted for different energies. The latter is obviously a time-consuming stage of the computations, especially when several collision energies are needed and which thus require the repeating of the entire calculation for each of them. In the following Sections we we will make a brief comparison with other approaches to the coupled differential equations of (5.1').

5.3. - The Numerov methods

An alternative, step-by-step integration of the single-channel form of eq. (5.1') consists in applying the method of Numerov [10, 15], based on finite difference expressions which do not use the first derivative of the unknown function y(R) of eq. (5.13):

$$(1 + \frac{1}{12} h^2 f_{n-1})\, y_{n-1} - (2 - \frac{10}{12} h^2 f_n)\, y_n + (1 + \frac{1}{12} h^2 f_{n+1})\, y_{n+1} = 0 \quad (5.37)$$

where, as before, h is the interval and the subscript n refers to the value of the function at the nth mesh point.

The truncation error is $1/240\ \delta^6\ y_n$, which is the smallest error for a three-point method [12], and the method becomes unstable when h is chosen so that $k^2h^2 > 6$;

this is practically avoided, however, by choosing a small enough truncation error [16].

In many problems of physical interest the coupling potential function f(R) is a rapidly varying function of the radial variable R for small R values, it grandually becomes smoother at larger R and finally it tends to zero in the asymptotic region.

The efficient integration of such a function therefore requires a steady increase in interval sizes which can easily be chosen to be multiples of their previous values.

For the coupled-channel case the previous (5.13) can be rewritten in matrix form as:

$$\mathbb{Y}'' + \mathbb{F}\mathbb{Y} = 0 \qquad (5.38)$$

hence the previous three-point formula is easily generalized to the following expression:

$$\mathbb{Y}_{n+1} = [\mathbb{1} - \tfrac{\mathbb{1}}{12}\mathbb{h}^2\mathbb{F}_{n+1}]^{-1} \cdot [(2\mathbb{1} - \tfrac{10}{12}\mathbb{h}^2\mathbb{F}_n)\mathbb{Y}_n - (\mathbb{1} + \tfrac{\mathbb{1}}{12}\mathbb{h}^2\mathbb{F}_{n-1})\mathbb{Y}_{n-1}] \qquad (5.39)$$

Using a matrix of constants as in (5.10) and a zero matrix as in (5.5) for the starting conditions, one can integrate as before into the asymptotic region and there perform the matching with the boundary conditions at two points [17, 18]. Each step obviously requires the inversion of the matrix, as shown in the square brackets on the r.h.s. of eq. (5.39), although the integration over the entire radial range is performed only once for all the N coupled equations. An interative procedure applied to the Numerov's method has been suggested [12] in order to avoid the matrix inversion at each step, but the integration needs then to be repeated N times, with a numerical accuracy that can be predetermined within the quickly converging iterative process itself.

Another variant of the Numerov's method, suggested for electron-atom scattering [19] and for heavy-particle collinear collisions [20], utilizes the Fox-Goodwin method [21]. Starting from eq. (5.39), one can multiply it to the right by $\mathbb{Y}_n$ and define $\mathbb{R}_n = \mathbb{Y}_{n-1} \cdot \mathbb{Y}_n^{-1}$. This allows one to rewrite the equation as follows:

$$\mathbb{R}_{n+1} = [\,\mathbb{1} - \frac{\mathbb{1}}{12}\mathbb{h}^2\,\mathbb{F}_{n+1}]^{} \times \{(2\cdot\mathbb{1} - \frac{10}{12}\mathbb{h}^2\,\mathbb{F}_{n}) - (\mathbb{1} + \frac{\mathbb{1}}{12}\mathbb{h}^2\,\mathbb{F}_{n-1})\,\mathbb{R}_{n}\}^{-1} \qquad (5.40)$$

This result provides a recurrence relation for the radial functions, provided that the constraint at the origin be assured via the position: $\mathbb{R}_1 = 0$.

From the point of view of the speed of convergence, it has been found that the Numerov is slightly slower than the De Vogelaere method, while its iterative version improves speed considerably. The Fox integrator, on the other hand, is comparable with the De Vegelaere approach, but always requires the inversion of a full matrix [20], i.e. no advantage can be taken from a simplified structure of the $\mathbb{F}$ matrices.

Generally speaking, finite difference methods are easier to program, although suffering from the drawback of requiring a large number of integration points for collisional systems with a short associated De Broglie wavelength (large reduced masses and or large collision energies).

Moreover, the step-by-step methods produce a more controllable and higher numerical accuracy, which becomes important for those cases where the structure of the coupling potential is reliably known. For more empirical choices of the potentials, however, it might become more interesting to examine those methods which can afford coarser meshes for the one-dimensional integration and more approximate interpolations along it. This aspect will be thus discussed in the following Section.

5.4. - The method of piecewise analytic solutions

The general idea of the method is one of generating approximate fits to the known, or roughly known, potential functions over various sections of the range of interactions and of subsequently obtaining correspondingly simpler analytic solutions [22, 23]. These approximate analytic solutions are then propagated from one interval to the next by imposing on them the usual continuity conditions up to the asymptotic region.

Let us start by considering the case of a single equation:

$$\frac{d^2}{dR^2} \varphi(R) - [\varepsilon - U(R)] \varphi(R) = 0 \tag{5.41}$$

with the usual meaning of the ε and $U(R)$ symbols.

In order to find its solution within a given interval centred at a given $\overline{R}$, one divides the interval into smaller ranges ΔR within which an analytic form of $U(R)$, be it $U_0(R)$, can be found to solve the problem. The corresponding solutions are called $A(R)$ and $B(R)$.

If $U_0(R)$ is given by a linear interpolation:

$$U_0(R) = U(\overline{R}) + (R - \overline{R}) \left[\frac{dU(R)}{dR} \right]_{R=\overline{R}} \tag{5.42}$$

then the eigensolution of the linear Hamiltonian are given by Airy functions [11]. For a constant $U_0(R)$ the solutions will be trigonometric functions, while a quadratic interpolations yields cylindric parabolic functions.

The Wronskian of the problem can be defined as:

$$W = A(R) \cdot B'(R) - A'(R) \cdot B(R) \tag{5.43}$$

hence, one can also write that:

$$W^{-1} = AB'' - A''B = - A[\varepsilon - U_0(R)] B - [\varepsilon - U_0(R)] AB = 0 \tag{5.44}$$

Let us define a new function $b(R)$ as:

$$b(R) = W^{-1}(A \varphi' - A'\varphi) \tag{5.45}$$

and its derivative, via (5.41), is then given by:

$$\begin{aligned} b'(R) &= W^{-1}\{A[- \varepsilon + U(R)] \varphi - [- \varepsilon + U_0(R)] A\varphi\} \\ &= W^{-1} A[U(R) - U_0(R)]\varphi \end{aligned} \tag{5.46}$$

The closer the chosen $U_0(R)$ comes to the true U(R) within the examined interval, the smaller b' becomes, i.e. $b' \to 0$ and $b(R) \sim$ const.

Similarly:

$$a(R) = W^{-1}[B'\varphi - B\,\varphi'] \tag{5.47}$$

and:

$$a'(R) = -W^{-1}\mathbf{B}[U - U_0]\varphi \tag{5.48}$$

Since eq.s (5.45 - 5.48) are linear in the sought eigenfunction and its first derivative, one can combine them to yield:

$$\varphi(R) = a(R) \cdot A(R) + b(R)\cdot B(R) \tag{5.49a}$$

$$\varphi'(R) = a(R) \cdot A'(R) + b(R)\cdot B'(R) \tag{5.49b}$$

but since, from (5.49a):

$$\varphi'(R) = A' \cdot a + B' \cdot b + A \cdot a' + B \cdot b' \tag{5.50a}$$

then it must also be that:

$$A \cdot a' + B \cdot b' = 0 \tag{5.50b}$$

Using now eq. (5.49a) in eq.s (5.46) and (5.48) one also obtains the derivatives of a (**R**) and b(R):

$$a'(R) = -\,W^{-1}B\,[U - U_0]\;(A \cdot a + b \cdot B) \tag{5.51a}$$

$$b'(R) = W^{-1}\,A[U - U_0]\;(A \cdot a + b \cdot B) \tag{5.51b}$$

These last equations are now first order equations, a fact that therefore reduces the difficulty of the problem of finding the A's and B's functions. Moreover, for an expansion centred around $\bar{R}$, the mid-point interval, it must hold that $U_0 \sim U$.

Hence, from (5.46) and (5.48), one also that: $a' \sim 0$, $b' \sim 0$; a(R) and b(R) are thus slowly-changing functions of the radial variable within the chosen interval ΔR. One can therefore rewrite eq.s (5.51) by the approximate (1st order) expression:

$$\bar{a} = a(R_\ell) \qquad ; \qquad \bar{b} = b(R_\ell) \tag{5.52}$$

where R_ℓ is the radial value at the left-side extreme of ΔR. If R_r is its value at the end (righ-side) of ΔR, the average values of the functions a(R) and b(R) can be obtained from the following expressions:

$$< \frac{da}{dR} > \sim - \frac{1}{W \cdot \Delta R} \int_{R_\ell}^{R_r} B(R) \, [U - U_0] \cdot [A \cdot \bar{a} + B \cdot \bar{b}] \, dR \tag{5.53a}$$

$$< \frac{db}{dR} > \sim \frac{1}{W \cdot \Delta R} \int_{R_\ell}^{R_r} A(R) \, [U - U_0] \cdot [A \cdot \bar{a} + B \cdot \bar{b}] \, dR \tag{5.53b}$$

It then becomes possible to write:

$$a(R_r) \sim \bar{a} + \Delta R < \frac{da}{dR} > \tag{5.54}$$

$$b(R_r) \sim \bar{b} + \Delta R < \frac{db}{dR} > \tag{5.54b}$$

If the values of the functions φ and φ' are already known up to R_ℓ, then $\bar{a}$ and $\bar{b}$ are also known and it is easy to compute a (R_r) and $b(R_r)$, since the linear expansion of U allows one to solve the integrals in eq.s (5.53). Substitution into eq.s (5.49) yields the new φ and φ' values at $R = R_r$, thus producing the new $\bar{a}$ and $\bar{b}$ for the next interval. This procedure outlines a perturbative computation of a(R) and b(R) within each interval ΔR, which can be chosen to be small enough to give the required numerical accuracy. For a linear form of U_0, the correction is of the order of $(\Delta R)^4$ and it has been shown [22] that, if δ is the required relative accuracy of $\varphi(R)$, the width of the interval is related to the next interval by:

$$(\Delta R)_{n+1} = (\Delta R)_n \cdot \left\{ \delta \left[|a| + |b| \right] \Big/ \max(|\Delta a|, |\Delta b|) \right\}^{1/3} \tag{5.55}$$

In the many-channel case, the previous treatment remains unchanged, although is now given via a more cumbersome notation. In matrix form, one has in fact that:

$$\left[\mathbb{1}\left(\frac{d^2}{dR^2}\right) + \mathbb{1}\cdot\varepsilon - \mathbb{U} \right] \boldsymbol{\varphi} = 0 \tag{5.56}$$

If the approximate coupling matrix is chosen to be diagonal, by neglecting off-diagonal first derivatives of $\mathbb{U}_0$, as well as its higher derivatives, the eq.s (5.56) become uncoupled and the solutions are expressed in terms of Airy functions:

$$(U_0)_{mn} = \delta_{mn} \cdot \left\{ U_{mn}(\bar{R}) + \Delta R \left[\frac{dU_{mm}(R)}{dR}\right]_{R=\bar{R}} \right\} \tag{5.57}$$

The corresponding 2N solutions are given by the diagonal matrices $\mathbb{A}$ and $\mathbb{B}$ which produce matrix forms of the Wronkian of (5.43) and of the radial function of (5.49). Since the $\mathbb{U}$ -matrix also has to be diagonal, a basis set for the expansion of $\boldsymbol{\varphi}$ in each interval needs to be chosen in order to obtain a matrix which performs the transformation:

$$\mathbb{U}(R_n) = \mathbb{M}_n^{\dagger}\, \tilde{\mathbb{U}}\, (\bar{R}_n)\, \mathbb{M}_n \tag{5.58}$$

and $\tilde{\mathbb{U}}$ is a diagonal matrix at the midpoint of the nth interval. The propagation of $\boldsymbol{\varphi}$ to the next interval requires the matching at the boundary and the definition of the (matrix) operator $\mathbb{T}_n$:

$$\boldsymbol{\varphi}_{n+1} = \mathbb{M}_{n+1}^{\dagger}\, \mathbb{M}_n\, \boldsymbol{\varphi}_n = \mathbb{T}_n\, \boldsymbol{\varphi}_n \tag{5.59}$$

the propagation is thus initialized in the first chosen interval by the perturbative solutions of (5.57) which provide the starting values of $\boldsymbol{\varphi}_0$ and $\boldsymbol{\varphi}_0'$ within the basis expansion of the zeroth interval. The next step then involves the calculations of the a(R), b(R) trial function increments from (5.53a) and (5.53b) which in turn allow one to propagate the eigensolution components to the end of the interval. The largest gradient of the a(R), b(R) components is used to chose the ΔR value of the following interval and eq.s (5.58), (5.59) can be employed to diazonalize the coupling matrix at midpoint and to obtain the propagator matrix $\mathbb{T}_n$ for moving the radial solutions onto the new basis for the next interval.

This treatment of the uncoupled piecewise solutions has been formulated in a widely used computer program for the efficient numerical solution of these equations [24] and several calculations of the systems discussed in the previous Chapters have been performed via the present integrator technique. In this approach, as discussed above, the first-order corrections to the uncoupling approximation are related to several integrals which involve the various uncoupled solutions [22]. Since general analytic expressions do not exist for such integrals, the program then performs averaging simplification that replace the true solutions with those obtained introducing a constant coupling $\bar{\alpha}$ which is an average value for all channels included in the expansion. A comparison between the uncoupled solution and the 'average coupling' solution then provides the necessary thresholds for the neglect of small correction terms to the uncoupled solutions. Possible modifications which are needed when the square magnitude of the relevant S-matrix elements becomes too small have also been the subject of discussion [25].

The correct asymptotic solution of eq. (5.56) are given by the familiar expression (in matrix form):

$$\mathbb{\Phi} \underset{R\to\infty}{\sim} \mathbb{J}(R) - \mathbb{N}(R)\cdot\mathbb{R}^{-\frac{1}{2}}\cdot\mathbb{K}\cdot\mathbb{R}^{\frac{1}{2}} \tag{5.60}$$

where the $\mathbb{J}(R)$ and $\mathbb{N}(R)$ matrix elements are given by eq.s (5.27a) and (5.28a), the $\mathbb{K}$ matrix is the one defined by eq.s (5.7) and (5.8) and:

$$R_{ij} = \delta_{ij}\,[\,\varepsilon - U_{ii}\,(R\to\infty)]^{\frac{1}{2}} \tag{5.61}$$

The outward integration of (5.56) yields, on the other hand, an asymptotic solution of the form:

$$\mathbb{\Phi} \underset{R\to\infty}{\sim} \mathbb{J}\cdot\mathbb{X} - \mathbb{N}\cdot\mathbb{Y} \tag{5.62}$$

where $\mathbb{X}$ and $\mathbb{Y}$ are constant matrices. If one now multiplies (5.62) to the right by $\mathbb{X}^{-1}$ one gets:

$$\mathbb{\Phi}\cdot\mathbb{X}^{-1} = \mathbb{J} - \mathbb{N}\cdot\mathbb{Y}\cdot\mathbb{X}^{-1} \tag{5.63}$$

hence the need to obtain the correct solutions (5.60) imposes the following relationships with eq.(5.62):

$$\mathbb{R}^{-\frac{1}{2}} \cdot \mathbb{K} \cdot \mathbb{R}^{\frac{1}{2}} = \mathbb{Y} \cdot \mathbb{X}^{-1} \tag{5.64a}$$

$$\mathbb{X}^{-1} \cdot \mathbb{R}^{-\frac{1}{2}} \mathbb{K} = \mathbb{Y} \cdot \mathbb{R}^{-\frac{1}{2}} \tag{5.64b}$$

$$(\mathbb{R}^{\frac{1}{2}} \cdot \mathbb{X})^{\dagger} \mathbb{K} = (\mathbb{R}^{\frac{1}{2}} \mathbb{Y})^{\dagger} \tag{5.64c}$$

The matrix elements X_{ij} and Y_{ij} are found by matching the function values and their radial derivatives to the Bessel function solutions, hence the solutions of the linear equations (5.64c) allows one to obtain the required Reactance matrix.

The relevant physical quantities for the scattering process are finally written in terms of the complex S-matrix of eq. (5.8) and one can find its imaginary part by solving the linear equation:

$$\mathbb{S}_i = \mathrm{Im} \left\{ \frac{\mathbb{1} + i\mathbb{K}}{\mathbb{1} - i\mathbb{K}} \cdot \frac{\mathbb{1} + i\mathbb{K}}{\mathbb{1} - i\mathbb{K}} \right\} = 2\,\mathbb{K}\,(\mathbb{1} + \mathbb{K}^2)^{-1} \tag{5.65}$$

hence its real part follows from one matrix multiplication:

$$\begin{aligned} \mathbb{S}_r &= (\mathbb{1} - 2\,\mathbb{K}^2 + \mathbb{K}^2)\,(\mathbb{1} + \mathbb{K}^2)^{-1} \\ &= \mathbb{1} - \mathbb{K} \cdot \mathbb{S}_i \end{aligned} \tag{5.66}$$

The great advantage of the present piecewise method of solution, with respect to the step-to-step methods analysed in the previous Sections, is that once an approximate fit has been established for a particular effective potential characterized by, say, a given value of the total angular momentum J, then the following calculation for some chosen energy value can be performed.

Any further calculation for different collision energies, and for that same J value, can now be repeated very quickly and much faster than in the previous cases.

Moreover, in the previous methods, that are based upon implicit polynominal approximations to the radial functions, the step sizes must be kept quite small in order for numerical accuracy and stability to be acceptable. Thus, since the λ values decrease with increasing collision energies, the corresponding number of step goes up for higher energies or light collision partners.

In the present approach, on the other hand, a polynomial approximation instead is searched for the potential function and thus the step size is not directly related to the energy. The analytical solutions of the simplified problem can, in fact, have any number of oscillations in a single step and one can use a coarser radial grid to obtain an approximation to the true potential as opposed to that needed for the step-by-step treatment. This obviously provides a considerable saving in computational time with regard to the potential itself, since fewer values of the multidimensional potential surface are needed.

Wiht the increase in accuracy of the knowledge of molecular interactions, hawever, the use of higher order methods of solution might be preferable, even if a 0th order method like the present one is clearly most efficient in providing the relevant answers when many collision energy values are needed [12].

5.5. - The solutions via integral equations

The solution of the time-independent scattering problems discussed in this notes consists, most generally, of finding the eigenstates of the equation:

$$\mathcal{H} |\alpha> = (\mathcal{H}_0 + \hat{V}) |\alpha> = E |\alpha> \tag{5.67}$$

subject to the conditions imposed by the physics of the problem:

$$|\alpha> \underset{\hat{V} \to 0}{\sim} |\alpha_i> \tag{5.68}$$

where:

$$\mathcal{H}_0 |\alpha_i> = E |\alpha_i> \tag{5.69}$$

and the state $|\alpha_i>$ is the prepared incoming state of the non-interacting partners. This simply means that, for vanishing interaction, the scattered wave would just be the incoming wave prepared for non interacting particles and $\mathcal{H}_0$ is the Hamiltonian operator for the eigenstates of such particles.

The condition (5.68) is usually assured by writing down a solution that reads as follows:

$$(E - \mathcal{H}_0 + i\,\varepsilon)(|\alpha_i^\varepsilon> - |\alpha_i>) = \hat{V}\,|\alpha_i^\varepsilon> \tag{5.70}$$

where ε is a real number and the l.h.s. operator of (5.70) is a nonsingular operators for any finite value of ε. Formally, a solution of (5.70) can then be written as:

$$|\alpha_i^\varepsilon> = |\alpha_i> + (E - \mathcal{H}_0 + i\varepsilon)^{-1}\,\hat{V}\,|\alpha_i^\varepsilon> \tag{5.71}$$

and, in the limit of ε approaching zero, the state $|\alpha_i^\varepsilon>$ approaches a solution of (5.67) that also satisfies eq. (5.68). Since one may take both positive and negative values for ε, there will be in general two different solutions. Hence they shall be labelled as:

$$\lim_{\varepsilon\to 0} |\alpha_i^\varepsilon> = |\alpha_i^+> \quad \text{or} \quad |\alpha_i^->\tag{5.72}$$

Equation (5.71) is usually known as the Lippman-Schwinger equation [26] and can be more compactly rewritten by defining an operator $\tau\,(E + i\varepsilon)$ such that:

$$\tau(E + i\varepsilon)\,|\alpha_n> = \hat{V}\,|\alpha_n^\varepsilon> \tag{5.73}$$

for any of the eigenstates $|\alpha_n>$ of $\mathcal{H}_0$ with eigenvalue E. Substitution into (5.71) and multiplication by $\hat{V}$ leads to the following expression:

$$\begin{aligned}\tau(E + i\varepsilon)\,|\alpha_i> &= \hat{V}\,|\alpha_i> + \\ &+ \hat{V}\,(E - \mathcal{H}_0 + i\varepsilon)^{-1}\,\tau(E + i\varepsilon)\,|\alpha_i>\end{aligned} \tag{5.74}$$

which is now an operator equation for the operator τ:

$$\tau = \hat{V} + \hat{V}(E - \mathcal{H}_0 + i\varepsilon)^{-1}\,\tau \tag{5.75}$$

The matrix elements of which are given by:

$$T_{mn} = < \alpha_m|\tau|\alpha_n > = < \alpha_m|\widehat{V}|\alpha_n > + + < \alpha_m|\widehat{V}(E - \mathcal{H}_0 + i\varepsilon)^{-1}\tau|\alpha_n > \tag{5.76}$$

By choosing the usual coordinate representation of the states $|\alpha_m>$, $|\alpha_n>$ of the total system and by expanding them over complete eigenfunctions of $\mathcal{H}_0$, of k^2 and of the total angular momenta which are required by the problem under study, the wavefunction of the final state $|\alpha_n>$ can then be obtained in terms of a sum of integrals over amplitude density contributions [27, 28]. Each of the contributions is then defined as:

$$F^+_{jn}(\underline{R}) = \iint d\underline{r}_A\, d\underline{r}_B\, \varphi^*_j(\underline{r}_A)\, \chi^*_j(\underline{r}_B)\, V(\underline{r}_A, \underline{r}_B, \underline{R})\, \psi^+_n(\underline{r}_A, \underline{r}_B, \underline{R}) \tag{5.77}$$

where the functions $\varphi_j(\underline{r}_A)$ and $\chi_j(\underline{r}_B)$ define the separate A and B partners at positions $\underline{r}_A$, $\underline{r}_B$ respectively and in their states as required by the total state $|\alpha_j>$ and ψ^+_n is the solution in the state $|\alpha_n>$. $\underline{R}$ is the usual relative coordinate in the chosen frame of reference. In case of non-local potential a further non-local form of F^+_{jn} can be defined without additional complexity.

Since from a knowledge of the amplitudes one may compute the complete scattering wavefunction, one can use the former quantity rather than with the latter wavefunction [29]. It then turns out that the asymptotic form of the T-matrix elements of (5.76) can be computed from the corresponding asymptotic form of F^+_{jn}, $F^{\pm}_{jn}(R)$, which can be written in the standard spherical harmonics expansion as:

$$F^{\pm}_{jn}(\underline{R}) = \sum_{\ell,m} Y_{\ell m}(\hat{R})\, F^{\pm}_{jn,\ell m}(R) \tag{5.78}$$

The T-matrix element (asymptotically) is then given by:

$$T^{\pm}_{jn,\ell m} = i\, k_j\, 2 \mathcal{M} \int_0^\infty j_\ell(k_j R')\, F^{\pm}_{jn,\ell m}(R')\, R'^2\, dR' \tag{5.79}$$

which is thus yielded without computing the complete wavefunction and can be obtained via extremely stable numerical quadratures [29].

The above method has been implemented in several calculations [27, 28] and found to by accurately given by simple quadrature formulae [29].

It has also been formulated via the use of Volterra equations by Sams and Kouri [30, 31], who showed that a complete solution of the coupled channel case can be written (in matrix form):

$$\mathbb{U}(R) = \int_{R_0}^{R} \{ \mathbb{n}(R)\cdot \mathbb{j}(x) - \mathbb{j}(R)\cdot \mathbb{n}(x) \} \mathbb{V}(x)\, \mathbb{U}(x)\, dx \qquad (5.80)$$

where the matrix $\mathbb{j}$ contains regular spherical Bessel functions and the matrix $\mathbb{n}$ spherical Newmann functions (see before). The potential matrix $\mathbb{V}$ contains all the information about the interaction between partners. In practice one must take $\mathbb{U}(R_0) = 0$, where $R_0 > 0$, as was discussed in the previous Sections for the starting point of the outward integrations.

The advantage of eq.(5.80) is also that when the integration is replaced by a quadrature with pivots $\{x_t\}$ the value of the integral at $R = x_t$ does not depend on $\mathbb{U}(x_T)$ [30]: thus the correct solution can be built up step by step. If one recasts eq.(5.80) in the equivalent form:

$$\mathbb{G} = \mathbb{V}(\mathbb{n}\cdot\mathbb{Y}_1 - \mathbb{j}\cdot\mathbb{Y}_2) \qquad (5.81)$$

where:

$$\mathbb{Y}_1(R) = \int_{R_0}^{R} \mathbb{j}(x)\cdot\mathbb{G}(x)\, dx \qquad (5.82a)$$

and:

$$\mathbb{Y}_2(R) = \int_{R_0}^{R} \mathbb{n}(x)\cdot\mathbb{G}(x)\, dx \qquad (5.82b)$$

Usually the simplest quadrature formula, namely the trapezoidal rule, is sufficient. A more complicated formula permits, in fact, a smaller step size but the increase in computational time is seldom balanced by a worthwhile increase in accuracy [29]. If one defines $\mathbb{G}_T$ as: $\mathbb{G}(R_0 + th)$, t = 1, 2, ... N and h is the chosen interval, the one-step advancing procedure is then given by:

$$\mathbb{G}_t = \mathbb{V}_t(\mathbb{n}_t\cdot\mathbb{Y}_{1,t-1} - \mathbb{j}_t\cdot\mathbb{Y}_{2,t-1}) \qquad (5.83a)$$

and:

$$\mathbb{Y}_{1,t} = \mathbb{Y}_{1,t-1} + h \cdot \mathbb{j}_t \cdot \mathbb{G}_{T_t} \qquad (5.83b)$$

$$\mathbb{Y}_{2,t} = \mathbb{Y}_{2,t-1} + h \cdot \mathbb{n}_t \cdot \mathbb{G}_{T_t} \qquad (5.83c)$$

The starting conditions are given by the matrix values: $\mathbb{Y}_{1,0} = 0$ and $\mathbb{Y}_{2,0} = 1$ and R_0 must be chosen inside the classically forbidden region but not too far inside to avoid the excessive growth of the elements of $\mathbb{G}_T$ for large potential couplings. As shown in the previous treatments, for large enough values of the number of points N, the K-matrix in the asymptotic region is given by:

$$\mathbb{K} = \mathbb{Y}_{1,N} \cdot \mathbb{Y}_{2,N}^{-1} \qquad (5.84)$$

and thus the T-matrix can be easily constructed.

In the integral equation formalism , both in the method of amplitude densities [27] and in the method of Volterra-type coupled equations [31], the T-matrix elements are obtained directly, once the coupled equations solution is known, and no matrix inversions are required to extract the required elements that yield cross sections. Moreover, the stability problems of the differential equations approach are avoided by defining amplitude densities, while numerical quadrature techniques for the Volterra equations (5.80) might allow time saving by reducing the number of matrix inversion operations needed to yields asymptotic solutions.

5.6. - The coupled channel R-matrix method.

As we have seen in the previous Sections, the recovery of dynamical information from the coupled equations that need to be solved is usually accomplished in a two-step process.

Firstly, by exploiting the linearity of the relevant differential equations, one solves an initial value problem and then, as a second stage, one constructs those linear combinations of primitive solutions that satisfy the desired boundary conditions. A successful attainnement of such conditions requires therefore linear independence of the solutions to that initial value problem.

As the number of terms in the linear basis expansion is increased, however, the exponential growth of the closed channels of that expansion tends to build a spurious linear dependence into the primitive solutions [32, 33], a problem that is inherent to all the close coupling calculations performed via the initial value methods of Sections 5.2 to 5.4.

The presentation of the correct dynamical information is then usually performed through a series of periodical and time-consuming, stabilization transformations, each of which reselects initial values to preserve the required linear independence.

In the present and in the following Sections in will then briefly discuss, in the spirit of the survey-like nature of this Chapter, two recent approaches to the numerical solutions of the coupled equations that strive for the most complete possible elimination of the instability problems, while also attaining a reduction of computational times.

The R-matrix theory, recently proposed for heavy particle collision problems, has a long history in the theory of nuclear reactions [34, 35] and has also been successfully revived in electrom-atom collision problems [36, 37]. Its generalization and implementation has also dealt, even more recently, with electron-molecule collisions [38, 39, 40]. The essential idea of the theory is to realize and emphasize that the dynamics of a system of two interacting particles with internal structure differs depending on the relative distance R between them.

According to the nature of the problem under study, one can then use different theoretical approaches and different wavefunction representations of the various regions into which the whole space of the acting interaction can be divided. The resulting piecewise wavefunctions need obviously to be matched on the boundary and the coupling is carried over from one region to the next by the matching process of the relevant R-matrix elements.

In the actual applications to electron scattering from atoms and molecules only 2 main regions have been used, the first being the internal region where the particles interect strongly and in a complicated manner. It is the region where the potential is non-local and it is impossible to distinguish between the two subsystems: so far it was studied using the configuration-interaction (CI) approach similar to the one discussed for bound-state problems in Chapter 2.

The second, outer or external region, describes the R values for which the two partners can be regarded essentially as independent in the sense that the interaction can be described by a multichannel, local potential that is a slow function of the intersystem distance. In the simplest case of s-wave, structureless partners, the Schrödinger equation in the internal region assumes the following familiar form:

$$\left\{ - \frac{1}{2\mathcal{M}} \cdot \frac{d^2}{dR^2} + V(R) - \frac{k^2}{2\mathcal{M}} \right\} \psi(R) = 0 \qquad (0 \leqslant R \leqslant a) \tag{5.85}$$

with the usual initial-value boundary condition:

$$\psi(R = 0) \; = \; 0 \tag{5.86}$$

The behaviour of the dimensionless quantity:

$$\frac{a}{\psi(a)} \cdot \frac{d}{dr} \psi(a) \,\Big|_{R=a} = \mathcal{R}_a \tag{5.87}$$

shows that, by varying the collision energy, there is no essential difference between positive-energy and negative solutions [40] and that the distinction between bound and continuum states is introduced only by considering the external region with the relevant asymptotic boundary conditions. If $\mathcal{R}_a$ decreases steeply within a small energy region, then the wavefunction is much larger in the internal region than on the boundary. With the introduction of proper boundary conditions, such a type of $\psi(R)$ will correspond to a bound state or to a resonance state depending on whether the energy is negative or positive.

If the required boundary condition is:

$$\mathcal{R}_a = \beta \tag{5.88}$$

an infinite number of energy eigenvalues and eigenfunctions are introduced. These eigenfunctions form a complete set and therefore they can be used as a basis expansion for the function $\psi(R)$ within the internal region. If the single-channel R-matrix is defined in a more general way as being given by:

$$\mathcal{R}(a, b) = \left\{ \frac{a}{\psi(a)} \cdot \frac{d}{dr} \psi(a) \,\Big|_{R=a} - b \right\}^{-1} \tag{5.89}$$

whenever b is chosen to be equal to β, then $\mathcal{R}(a, \beta)$ has a pole at each E_k generated by (5.88) and the expansion of $\psi(R)$ leads to the following expression [41]:

$$\mathcal{R}(a, \beta) = \frac{1}{2a} \sum_k \frac{[\psi(a)]^2}{E_k - E} \tag{5.90}$$

and an approximate form of it straightforwardly leads to the one-level resonance formula discussed in Chapter 1.

For heavy-particle collisions, reactive and sub-reactive, the definition of the R-matrix elements similar to eq. (5.87) can then be performed in the coupled channel case for several sectors spanning the whole region of interaction. A new general technique has been recently presented [42], in which one does not compute the scattering wavefunction but propagates insted a generalization of the inverse logarithmic derivative of (5.87) from some point R_i to R_f, until a final R value is reached where such a matrix is matched to known asymptotic forms.

This approach has shown the merit of being quite stable numerically, even for cases involving strong coupling between open and closed channels. The coupled equations to be treated in this case may be written in the form:

$$\frac{d^2}{dR^2} F_k(R) = \sum_{k'}^{N} V_{kk'}(R) F_{k'}(R) \tag{5.91}$$

where the coupling matrix elements include all the usual contributions. One now divides the configuration space in which subsystems interact into a set of boxes where the potential within each box is approximated via a numerically convenient form that is simple enough to allow the solution of (5.91) in an easy accurate way. The most usual approach has been to expand the matrix elements $V_{KK'}(R)$ in a power series about the center of the box R_m:

$$V_{kk'}(R) = V_{kk'}(R_m^i) + \frac{\partial}{\partial R} V_{kk'} \Big|_{R=R_m} (R - R_m^i) + \ldots.. \tag{5.92}$$

where the superscript i labels the specific box under exam.

When only the first constant term is retained, the above equation may be solved in terms of trigonometric functions, while when some of the linear terms are retained the Airy functions provide another set of solutions as discussed in the previous Section 5.4.

To ensure the continuity of the wavefunction and its derivative throughout the whole space, one requires various matching procedures at the boundary of each box and the introduction of solution vectors $F_{\ell}(R^{i})$ and $F_{r}(R^{i})$ obtained via the above couplings and corresponding to translational solutions at the inner <u>left</u> and <u>right</u> boundaries of the ith box. Matching the wavefunction at the interval boundaries requires the knowledge of an overlap matrix between the expansion functions $\phi_K(R)$ of each neighbouring box, the element of which are given by:

$$\mathcal{O}_{kk'}(i, i-1) = \langle \phi_k(i-1; R) | \phi_{k'}(i; R) \rangle \tag{5.93}$$

Another need is now the knowledge of its energy-dependent counterpart, this one being obtained after transforming the primitive basis of each ith box into a (energy dependent) basis which diagonalizes the full interaction matrix in that same box [42]:

$$\{\mathbb{T}^{(i)}\}^{\dagger}\mathbb{V}^{(i)}\ \mathbb{T}^{(i)} = \lambda_i^2\ (E) \tag{5.94}$$

Hence, after this transformation one obtains (in matrix form) the following uncoupled expression:

$$\mathbb{O}(i, i-1; E) = \{\mathbb{T}^{(i-1)}\}^{\dagger}\ \mathbb{O}(i, i-1)\ \mathbb{T}^{(i)} \tag{5.95}$$

The matching procedure between neighbouring sections is then accomplished by requiring that:

$$\mathbb{F}_r(R^{i-1}) = \mathbb{O}(i, i-1; E) \cdot \mathbb{F}_{\ell}(\mathbf{R}^{i}) \tag{5.96a}$$

and:

$$\frac{d}{dr}\mathbb{F}_r(\mathbf{R}^{i-1}) = \mathbb{O}(i, i-1; E)\frac{d}{dr}\mathbb{F}_{\ell}(\mathbf{R}^{i}) = \mathbb{F}'_r(R^{i-1}) \tag{5.96b}$$

Within each box one can now write the solutions is terms of the step propagator $\mathbb{P}^{(i)}$ [42] and in a supermatrix notation:

$$\begin{pmatrix} \mathbb{F}_{\ell}(R^{i}) \\ \mathbb{F}'_{\ell}(R^{i}) \end{pmatrix} = \begin{pmatrix} \mathbb{P}_1^{(i)} & \mathbb{P}_2^{(i)} \\ \mathbb{P}_3^{(i)} & \mathbb{P}_4^{(i)} \end{pmatrix} \begin{pmatrix} \mathbb{F}_{r}(R^{i}) \\ \mathbb{F}'_{r}(R^{i}) \end{pmatrix} \tag{5.97}$$

where the propagators matrix elements are given by:

$$[\mathcal{P}_1^{(i)}]_{jk} = [\mathcal{P}_4^{(i)}]_{jk} = \begin{cases} \delta_{jk} \cosh d\,|\lambda_j| & ; \quad \lambda_j^2 > 0\,. \\ \delta_{jk} \cos d\,|\lambda_j| & ; \quad \lambda_j^2 \leqslant 0\,. \end{cases} \tag{5.98a}$$

$$[\mathcal{P}_2^{(i)}]_{jk} = \begin{cases} \delta_{ij}\,|\lambda_j^{-1}| \sinh d\,|\lambda_j| & ; \quad \lambda_j^2 > 0\,. \\ \delta_{jk}\,|\lambda_j^{-1}| \sin d\,|\lambda_j| & ; \quad \lambda_j^2 \leqslant 0\,. \end{cases} \tag{5.98b}$$

$$[\mathcal{P}_3^{(i)}]_{jk} = \begin{cases} \delta_{jk}\,|\lambda_j| \sinh d\,|\lambda_j| & ; \quad \lambda_j^2 > 0\,. \\ -\delta_{jk}\,|\lambda_j| \sin d\,|\lambda_j| & ; \quad \lambda_j^2 \leqslant 0\,. \end{cases} \tag{5.98c}$$

and where one defines $d = R_\ell^i - R_r^i < 0$ and also: $R_\ell^i = R_r^{i-1}$ and $R_r^i = R_\ell^{i+1}$. The d value then defines the radial size of each given box.

The requirements of eq. (5.96) and (5.96b) can be used in the matrix equation (5.97) to obtain the R-matrix recursion relations.:

$$\begin{pmatrix} \mathbb{F}_r(R^{i-1}) \\ \mathbb{F}_r(R^i) \end{pmatrix} = \begin{pmatrix} \mathcal{R}_1^{(i)} & \mathcal{R}_2^{(i)} \\ \mathcal{R}_3^{(i)} & \mathcal{R}_4^{(i)} \end{pmatrix} \begin{pmatrix} \mathbb{F}_r'(\mathbf{R}^{i-1}) \\ -\mathbb{F}_r'(\mathbf{R}^i) \end{pmatrix} \tag{5.99}$$

where the new matrices $\mathcal{R}_m^{(i)}$ are obtained in terms of step propagators of the translational solutions for the constant potential and of energy-dependent overlap matrices between the beighbouring boxes basis set expansions of (5.95) [42]. In this procedure one never computes value of wavefunctions or of their derivatives since only the R-matrix elements are evalued. The negative sign on the r.h.s. of (5.99) points out the symmetry properties of the R-matrix in each box: $\mathcal{R}_1^{(i)}$ and $\mathcal{R}_4^{(i)}$ are symmetric while $\mathcal{R}_3^{(i)} = (\mathcal{R}_2^{(i)})^\dagger$, provided that the overlap matrices $\mathcal{O}$ are orthogonal matrices.

A more general derivation of eq.s (5.99) has been recently formulated in terms of Green's functions [43] and without resorting to the simplified form of the interaction as in (5.92). It was also shown there that even the traslational wavefunction for each channel could be expanded in a basis set by taking advantage of the fact that the configuration space to be spanned in the present treatment only involves a single box and therefore such an expansion should converge fairly rapidily and even for considerably larger values of d, the box size , an those needed by the numerical solutions of (5.97).

5.7. - The Variable Phase Method

As a final point, we shall discuss now another method which has been recently applied to electron-atom collision and electron-molecule collisions, and that appears to be a powerful technique far solving secon-order ordinary differential equations, that is the Variable Phase Method (VPM).

In recognizing, for the physical problems under study, that different radial regions sample different forms of interaction one is again dividing the whole space in different sectors, as previously discussed in Section 5.6. In the case of electron-neutral atom scattering, the usual procedure therefore consists of solving directly the system of linear, second-order differential equations by using the asymptotic expansion technique [44] for very large R values in the near-threshold region. As mentioned before, however, the presence of closed channels can give solutions via the previous method that show rather poor linear independence of the relevant vectors at the boundary matching points. The necessary stabilization procedures that are called for, however, require that the scattering wavefunctions must be integrated numerically from a point at which the satisfactory convergence in the expansion has been already achieved for all channels, a requirement which ofter pushes the required R value must further out than what is needed for most channels.

The VPM approach has been based on the well known fact [45] that a linear homogeneous equation of second order, such as the scattering equation, can be reduced to a Riccati equation wich is satisfied by the K-matrix. This matrix function is a monotonic, non oscillating function which can be obtained with great accuracy and carries a clear physical meaning: at each R value it yields the reactance matrix for the corresponding part of the potential.

As was already discussed in Section 5.2, the matrix solution of the differential equations in the outer region can be written as in eq.s (5.26a) and (5.26b).

The condition on the $\mathbb{Y}$ (R) derivative, on the other hand, is equivalent to the supplementary condition:

$$\mathbb{J}(R)\,\mathbb{A}' = \mathbb{N}\cdot\mathbb{B}' = 0 \tag{5.100}$$

by remembering that the matrix $\mathbb{K}$ is given by eq. (5.34), the further differentiation of (5.26b) yields:

$$\begin{aligned}\mathbb{Y}''(R) &= \mathbb{J}''\cdot\mathbb{A} - \mathbb{N}''\cdot\mathbb{B} - \mathbb{W}(\mathbb{J},\mathbb{N})\,\mathbb{A}'\cdot\mathbb{N}^{-1}\\ &= \mathbb{J}''\cdot\mathbb{A} - \mathbb{N}''\cdot\mathbb{B} - \mathbb{A}'\,\mathbb{N}^{-1}\end{aligned} \tag{5.101}$$

where $\mathbb{W}$ is the wronskian matrix of eq. (5.29). Using now the coupling potential matrix of (5.1) into (5.101) one has that:

$$\begin{aligned}\mathbb{A}' &= -\mathbb{N}\cdot\mathbb{V}\cdot\mathbb{Y}\\ \mathbb{B}' &= -\mathbb{J}\cdot\mathbb{V}\cdot\mathbb{Y}\end{aligned} \tag{5.102}$$

Then the K-matrix of (5.34) can be given as:

$$\begin{aligned}\mathbb{K}' &= (\mathbb{B}' - \mathbb{K}\cdot\mathbb{A}')\,\mathbb{A}^{-1}\\ &= -(\mathbb{J} - \mathbb{K}\cdot\mathbb{N})\,\mathbb{V}\cdot\mathbb{Y}\cdot\mathbb{A}^{-1}\end{aligned} \tag{5.103}$$

By using again eq.(5.26) one finally obtains the Riccati equation for the K-matrix [46]:

$$\mathbb{K}' = -(\mathbb{J} - \mathbb{K}\cdot\mathbb{N})\,\mathbb{V}\,(\mathbb{J} - \mathbb{N}\cdot\mathbb{K}) \tag{5.103a}$$

One now needs to determine the value of $\mathbb{K}$ at the initial boundary of the outer region in order to begin the integration of (5.103) up to the asymptotic form of the solution. Since one also knows, as in the R-matrix theory of the previous Section, the logarithmic derivative of $\mathbb{Y}$, it is then possible to write, from (5.26), that:

$$\mathbb{N}\cdot\mathbb{Y}'\cdot\mathbb{Y}^{-1}\mathbb{B} = \mathbb{J}\cdot\mathbb{Y}'\cdot\mathbb{Y}^{-1} - \mathbb{Y}' \tag{5.104}$$

now, via eq.(5.26b), the above result can be rewritten as:

$$(\mathbb{N}' - \mathbb{N}\cdot\mathbb{Y}^{-1}\cdot\mathbb{Y}')\mathbb{B} = (\mathbb{J}' - \mathbb{J}\cdot\mathbb{Y}^{-1}\mathbb{Y}')\mathbb{A} \tag{5.105}$$

From the K-matrix definition of (5.34) one then has that:

$$\mathbb{K} = (\mathbb{J}' - \mathbb{J}\mathbb{Y}^{-1}\cdot\mathbb{Y}')(\mathbb{N}' - \mathbb{N}\cdot\mathbb{Y}^{-1}\mathbb{Y}')^{-1} \tag{5.106}$$

and remembering the definition of $\mathbb{R}$ as the inverse logarithmic derivative of $\mathbb{Y}$ as in (5.89), one can write:

$$\begin{aligned}\mathbb{K}(\mathbf{R} = a) = \{[\mathbb{J}(\mathbb{1} + b\,\mathbb{R}(a, b) - R\,\mathbb{J}'\cdot\mathbb{R}(a, b)] \,\cdot \\ \cdot\,[\mathbb{N}(\mathbb{1} + b\,\mathbb{R}(a, b) - \mathbf{R}\,\mathbb{N}'\,\mathbb{R}(a, b)]^{-1}\}_{R=a}\end{aligned} \tag{5.107}$$

The R-matrix method can then be used to produce the logarithmic derivative of the wavefunction at any given value of the radial variable.

For the particular case of R = 0, the specific choice of the ($\mathbb{J}$, $\mathbb{N}$) basis, i.e. $\mathbb{J}$ regular at the origin and $\mathbb{N}$ irregular, allows the direct selection of the physical solutions by imposing: $\mathbb{K}(\mathbf{R} = \mathbf{0}) = 0$.

The corresponding logarithmic derivative at the next boundary value R = a, and for the usual choice of the matrix $\underline{\underline{b}}$ to be a diagonal matrix with zeros as elements, is then given by:

$$\begin{aligned}\mathbb{Y}'\,\mathbb{Y}^{-1}(\mathbf{R} = a) &= \frac{1}{a}\;\mathbb{R}^{-1}(a, 0) \\ &= |(\mathbb{J}' - \mathbb{N}'\mathbb{K})(\mathbb{J} - \mathbb{N}\cdot\mathbb{K})^{-1}|_{R=a}\end{aligned} \tag{5.108}$$

It therefore follows that, once the K-matrix has been obtained within the inner region, the problem of connecting its value to the outer region is solved via the result of eq. (5.108). From there on the calculations therefore reduce to the direct determination of the reactance matrix from eq.(5.107) and to its propagation through a stable, monotonic procedure as in eq.(5.106).

Although the above method has only been tested so far electron scattering with atomic and with CO targets [47], it seems particularly promising for those heavy particle systems that exhibit long-range potential forms with weak coupling. They allow, in fact, choices of large steps for the propagation procedure without loss of accuracy and likely increase of computational speed.

REFERENCES

[1] Methods in Computational Physics, Vol. X (B. Alder, S. Fernbach and M. Rotenberg Ed.s, Academic Press. N.Y. 1971).

[2] L. Hulthen, Ark. Mat. Astron. Fys., 35A, N. 25 (1948).

[3] W.Köhn, Phys. Rev., 74, 1763 (1948).

[4] S.I.Rubinow, Phys. Rev., 98, 183 (1955).

[5] L.S.Rodberg and R.M.Thaler, Introduction to the Quantum Theory of Scattering (Academic Press, N.Y. 1967).

[6] W.J. Cody and K.Smith, Argonne Nat. Lab. Rept., ANL-6121 (1960)

[7] W.A.Lester Jr. and R.B. Bernstein, J. Chem. Phys., 48, 4896 (1968).

[8] A.F. Wagner and V. McKoy, J. Chem. Phys., 58 2604 (1973).

[9] R.De Vogelaere, Jour. Res. N.B.S. 54, 119 (1955).

[10] L. Fox and D.F. Mayers, Computing Methods for Scientists and Eigineers (Clarendon Press. Oxford, 1968).

[11] M. Abramowitz and I.A. Stegun, Handbook of Mathematical Functions (Dover, New York 1968).

[12] A.C. Allision, J.Comput. Physics, 6, 378 (1970).

[13] W.A. Lester Jr., Method Comp. Phys., 10, 211 (1971).

[14] P.McGuire and D.A.Micha, Int.J. Quantum Chem., 6,111 (1972).

[15] J.M. Blatt, J. Comput. Physics, 1, 382 (1967).

[16] P.G. Burke and M.J. Seaton, Method. Comp. Phys., 10, 1 (1971).

[17] L.L. Barnes et al., Phys. Rev., A 137, 388 (1965).

[18] K. Smith et al., Phys. Rev. A 147, 21 (1966).

[19] D.W. Norcross and M.J. seaton, J. Phys. B 6, 614 (1973).

[20] G. Bergeron, X. Chapuisat, J. M. Launay, Chem. Phys. Lett., 38, 349 (1976).

[21] L. Fox, The numerical solution of two-point boundary value problems in ordinary differential equations (Oxford Univ. Press, London 1957).

[22] R.G.Gordon, J.Chem.Phys., 51., 14 (1969).

[23] R.G.Gordon, Meth.Comput.Phys., 10, 81 (1971).

[24] R.G.Gordon, Program 187, Quantum Chemistry Program Exchange (QCPE) Indiana University, Bloomington, Ind. (U.S.A.).

[25] M.U.Alexander, J.Comp. Phys., 20, 248 (1976).

[26] T.Y.Wu and T. Ohmura, Quantum theory of scattering (Prentice Hall Inc., Engle-

wood Cliffs, N.J., 1962).

[27] B.R. Johnson and D. Secrest, J. Math. Phys., 7, 2187 (1966).

[28] B.R. Johnson and D.Secrest, J.Chem. Phys., 48, 4682 (1968).

[29] D.Secrest, Meth. Comp. Phys., 10, 243 (1971).

[30] N.W. Sams and D.J. Kouri, J. Chem. Phys., 51, 4809 (1969).

[31] N.W. Sams and D.J. Kouri, J. Chem. Phys., 51, 4815 (1969).

[32] W. Eastes and D. Secrest, J. Chem. Phys., 56, 640 (1972).

[33] B.H. Choi and K.T. Tang, J. Chem. Phys., 63, 1775 (1975).

[34] E.P. Wigner, Phys. Rev., 70, 15 (1946).

[35] E.P. Wigner, Phys. Rev., 70, 606 (1946).

[36] P.G. BURKE, A. Hibbert and W.D.Robb. J. Phys. B., 4, 153 (1971).

[37] P.G. Burke and W.D. Robb, Adv. At. Mol. Phys., 11 143 (1975).

[38] N. Chandra and F.M. Gianturco, Chem. Phys. Lett.

[39] P.G. Burke, I. Mackey and I. Shimamura, J. Phys. B, 10, 2497 (1977-).

[40] B.I. Schneider, Chem. Phys. Lett., 31, 237 (1975).

[41] E.P. Wigner and L. Eisenbud, Phys. Rev., 72, 29 (1947).

[42] J.C. Light and R.B. Walker, J. Chem. Phys., 65, 4272 (1976).

[43] B.I. Schneider and R.B. Walker, J. Chem. Phys., (1979).

[44] P.G. Burke, D.D. McVicar and H.M. Shey, Proc. Phys. Soc., 83, 397 (1964).

[45] M. Le Dorneouf and Vo ky Lan, J. Phys. B, 10, L 35 (1977).

[46] I. Calogero, Variable Phase approach to potential scattering (Academic Press, N.. Y. 1967).

[47] M. Le Dorneouf and Vo ky Lan, private communication (1978).

6. ROTOVIBRATIONAL RELAXATION MODELS IN SIMPLE GASES

6.1 - Introduction

In the previous Chapters we analyzed in some detail the 'forward' progression of the theoretical treatments that allow one to obtain dynamic observables from first principles. This meant that we had to define an interaction Hamiltonian between colliding partners, compute the corresponding potential forms at various levels of accuracy and then go on to evaluate detailed, state-to-state total and differential cross sections that involved transitions between levels of the rotational and vibrational structures of the colliding partners in their lowest electronic states, usually closed- shell configurations.

A further step, one that could finally link the theoretical results with some bulk experimental findings, would be to examine the relationship between computed cross sections and the relaxation/excitation phenomena which take place in a realistic gas mixture at a given temperature. When a gas is at equilibrium, the distribution of energy between traslation, vibration and rotation is usually governed by Boltzmann distribution relations. This is to say that the translational, vibrational and rotational temperatures are equal: $T_{tr} = T_{vib} = T_{rot}$. Any small change induced in the system then causes molecules to gain and lose vibrational and rotational energy in collisions, and the nature of each specific scattering process is thus of fundamental importance in understanding the energy accumulation in a given mode or the energy deposition after the induced perturbation [1].

Just to cite a prime, current example in molecular processes, various studies of the performance characteristics of chemical lasers show that collisional deactivation of, say, the Hydrogen Halides by various partners is an important criterion in the choice of reagent mixtures. For examples, studies of CW transverse- flow lasers suggest that the large rate for self-deactivation and vibrational energy transfer in H(D)F molecules are among the reasons why CW devices operate under conditions of partial rather than the total inversion characterizing pulsed lasers [2].

The general process of collisional energy redistribution, rather than specific chemical reactions, has thus been postulated as the most important single limiting factor. It is theorefore of crucial importance to know the specific deactivation cross sections, together with their dependence on the $|vj\rangle$ state chosen as a starting lasing state [3].

Similar conclusions have been reached in the analysis of chemical transfer lasers, where the identification of a major pumping mechanism depends largely on the degree to which intermolecular V-V and V-R couplings are found responsible for the energy transfer rates, while also the performance of pulsed electrical lasers like CO seems to be strongly determined by the details of vibrational self- activation. All these examples demonstrate the importance in current molecular laser studies of several aspects of molecular collision dynamics, like V-R-T couplings, the interplay of the potential energy surface for each system examined, multiquantum transitions and anharmonicity effects.

Recently a careful analysis of the above elements has in fact appeared in the literature stressing the general need for better and more theoretical work on the collisional aspects of laser studies in Hydrogen Halides [4].

In order to examine more closely the relevant processes, one can say in general terms that the close connection between theoretical outcomes and experimental data can be seen at various levels of 'directness' , dependig on the amount of averaging over microscopic processes performed in the actual measurements. This aspect in turn forces the theory to reduce the degree of state-selection in the process studied, needing a more complex error analysis when determinig the important factors that cause its agreement or disagreement with measured data. Although the emphasis of this Chapter is again on theoretical developments, it seems however relevant at this point also to briefly review the appropriate experimental measurements which usually prompt the formulation of a theory in first place. It is obviously beyond our present scope to list in detail all possible experiments, but it will help to clarify the specific theoretical models of the following Sections to have them preceeded by a limited outline of possible observations and of the existing relations between the various ways of preparing the system or of performing its pertubation away from initial conditions.

6.2 - An outline of experiments.

The various methods that can be used to produce a disturbance of the equlibrium distribution of the molecular gas, or mixture of gases, broadly fall into two main classes. When a gas undergoes adiabatic compression, only the translational energy is directly influenced, while in a process of sudden compression the whole energy goes initially to increase the translational temperature. This step is then followed by the relaxation due to T-R and T-V energy transfer collisions until equilibrium between the three modes is re-established. Such rapid compressions are produced either by ultrasonic sound or by passage of shock- fronts and form the basis of many classical methods for determining relaxation times [1,5]. These experiments may be performed by observing some property of the system, such as temperature, refractive index, etc., that is sensitive to the energy tranfer process. The relaxation time here can also be obtained by measurig the rate of density change or by following the population of the excited level by absorption or emission spectroscopy. Direct measurements of vibrational temperature by atomic line reversal can be used. Shock tube experiments have provided an extensive set of data for many diatomics and several polyatomics over fairly large temperature rangers from 5000 to 300 °K.

The other, broad class of experiments can be defined as being constituted by partially state-selected processes under the general heading of laser fluorescence. In this instance, the gas is absorbing electromagnetic radiation and thus selected vibrational and rotational modes may be excited in its molecules. Use is made here of the large power densities in narrow spectral widths and the high degree of collimation of a laser operating on a given vibrational and or rotational transition to excite the same transition in a gas sample outside the active laser medium. When the radiative lifetimes of the excited vibrational states are long in comparison to the collisional deactivation time (as in the case of, say, HF where τ_{rad} is of the order of 5 msec), the measured fluorescence decay rates can be directly related to the collisional relaxation rates [6,7]. The radiative excitation results, in fact, in an increase of the vibrational temperature and/or the rotational temperature but has no affect on T_{tr}. Relaxation to equilibrium by V-T or R-T processes may then be followed either by monitoring the decrease in T_{vib} or T_{rot} by spectroscopic methods, or by monitoring the accompanying increase in T_{tr}. Gases with non-equlibrium values of T_{vib} and of T_{rot} are also sometimes produced in fast chemical reactions where the relaxation may then be followed by similar methods [8].

All these measurements are ultimately relared to the definition of a state-to-state rate $K_{i\to f}$ for energy transfer in a collision between two molecules M_1 and M_2, or between an atom A and a molecule BC, by a translational energy conversion into or from a given rotovibrational state of, say, BC $|i\rangle = |vjm_j\rangle$, which yields a final molecular state $|f\rangle = |v'j'm_j'\rangle$. The $K_{i\to f}$ is then defined by the number of transitions per second per molecule from state $|i\rangle$ to state $|f\rangle$ and can be obtained by averaging the total cross section for the process considered over a known velocity ditribution form at the given temperature T of the gas:

$$K_{i\to f}(T) = \int_0 d\underline{v}\, f(\underline{v}, T)\, \underline{v}\, \sigma_{i\to f}(E) \qquad (6.1)$$

These rates obviously enter into the kinetic mechanisms of the bulk media. They are obtained through a selective disturbance of the system, in the case of 'fluorescence' measurements, by monitoring the return to equilibrium [9] or by studying their orientation dependence in anisotropic conditions [10]. The latter treatments then constitute a more detailed way of obtaining experimental information; since the former methods perturb the original system in a non-selective manner and the obtained relaxation times are in general a further convolution of all state-to-state rates defined in equation (6.1).

Moreover, in the real cases of flowing gases, the appropiate fluid dynamics effects must be taken into account when relating experiments to computed observables [11]. This last point will be briefly reconsidered in the next Section.

A set of more selective, less averaged, data can be obtained from the study of pressure broadening of spectral lines, since the latter effect is generally due to collisions between gaseous partners that interrupt the normal radiative process and that therefore provide information on energy transfers between molecules which are, in some sense, in a labelled initial state caused by the radiative absorption [12]. In general terms one can therefore say that the convolution over the many possible final state $|f\rangle$ implicit in the integration of (6.1) can be selectively monitored over each chosen initial subset of molecular state $|i\rangle$, which in turn depends on the specific state preparation performed by the absorbed radiation.

According to elementary kinetic theory, the viscosity η of a gas and its conductivity k are related by relatively simple equations that, at a micrococoscopic level, are ultimately referred to bulk transport equations [13]. The dependence of the viscosity on the temperature and on the nature of the examined molecular gas can thus be related to the traslational energy transfers into the internal degree of freedom, via the specific interaction potential [14], and therefore can provide a further set of experiments for which the theoretical models of the previous Chapters can be tried and tested.

6.3 - The rate equations.

In order to relate the microscopic cross sections to the bulk measurements briefly discussed above, it is worthwile to remind ourselves of some of the main physical quantities involved in the process.

First of all, we are really considering the specific heat of a substance, defined as the rate of change of its energy with temperature, and its influence on the energy behaviour when a sound wave is passed through the gas and produces a sinusoidal variation of T for each element of volume. At low frequencies, the local thermodynamic equilibrium is mantained and the specific heat, C_V , calculated from measured velocities appears to be in good agreement with other measurements. At ultrasonic frequencies, however, not all the energy in the gas changes with the fast-varying temperature and therefore the corresponding specific heat comes out to be lower. The sound velocity is thus higher that at low frequencies. This behaviour gives rise to the known phenomenon of the dispersion.

The standard explanation for this behaviour is well established [15] by now, and is based on the suggestion that energy could be exchanged between the translational and vibrational degree of freedom of the gas molecules only at a finite rate and therefore, after an abrupt change in the translational energy, there is a delay before the vibrational and translational energies are again at equilibrium.

When the translational energy of the travelling sound wave has its period for a cycle comparable with the above delay, then no energy equilibrium is attained and the vibrational degree of freedom are not fully contributing to C_V . The changes in T_{vib} and T_{tr} are out of phase, the former lagging behind the latter with a smaller amplitude. Because of this lag, energy is absorbed as the sound wave travels through the gas and so its amplitude decreases. This energy absorption, and its velocity of absorption, vary with the sound frequency or, more generally, with the frequency of the disturbance in phase with the translational temperature.

The dispersion process is thus related to the presence of a physical quantity that fails to follow rapid changes in another one, which is in turn following the disturbance. Since these physical quantities are normally in equilibrium with each other, the former is said to be relaxed. The lag inherent in the relaxation process is characterized by a relaxation time (τ) which can be defined in various ways. A particularly simple form has been introduced in the original paper by Landau and Teller on the theory of sound dispersion [16] , (where its now popular form of dependence on $T^{-1/3}$ was also given):

$$- \frac{dE}{dt} = \frac{1}{\tau} \{E - E(T)\} \tag{6.2}$$

where E is the instantaneous value of the relaxation energy that lags behind the translational energy, while E (T) is the value that the same energy would have at the equilibrium temperature T . It therefore follows that the rate of attainment of equilibrium is proportional to the extent of the deviation from it, and the equation (6.2) integrates to give an exponential time function for the relaxation, as found in many experiments.

The microscopic origin of the time-lag in attaining equilibrium after a general perturbation of the gas energy, as briefly summarized above, is obviously residing in the nature of the transition probabilities for the collisional transfer, that energetically require kinetic energy values several times greater that the average thermal values. The transfer of a typical vibrational quantum, for instance, neeeds on the average from ten to several thousand collisions for polyatomic molecules; diatomic molecules can even require as many as 10^{10} collisions at room temperature. As a consequence, our theoretical models need to investigate the mathematical links between each individual (elastic and inelastic) binary collision and the more microscopic kinetic rates that are pressure dependent and require more information on the dynamics associated with each element of the fluid.

The most general aim of such a study is therefore constituted by the time and/or spatial dependence of the relaxation processes in molecular gases that have been pre prepared in a non-equilibrium state. It is thus necessary to combine the equations of fluid mechanics with those describing the kinetics of the molecular level populations and the most natural way to do so is to start from the classical formulation of the transport-relaxation equations of a non-equilibrium ensemble [17] . Since the quantum conditions obviously have to be also considered, thus making the problem of solving the Liouville equations even more complex, one naturally chooses a more approximate formulation that arrives at the Boltzmann equations [18] or to their generalization wich also includes the molecular internal state. The latter is called the Wang Chang- Uhlenbeck (WCU) equation [19]. A further lack of rigour in deriving the above steps of the formulation comes when in the presence of degenerate states, such as the rotational states with their associated degenerate m_j substates, so this has to be corrected by the Waldmann-Snider equation [20].

In most experiments, however, no direct measure is made of the m_j substate presence and under those circumstances the molecular distribution function is assumed to be independent of m_j and diagonal. The Waldmann-Snider equation is then reduced to the WCU equation.

The latter equation can be used then as a general starting point and can be written in the following way [20]:

$$\frac{\partial}{\partial t} f_i(\underline{r}, \underline{v}_1, t) + \underline{v}_1 \cdot \nabla_{\underline{\mathbf{r}}} f_i(\underline{r}, \underline{v}_1, t)$$

$$= \sum_{ljm_j} \iint d\Omega \, d\underline{v}_2 \, \{ f_\ell(\underline{r}, \underline{v}_2', t) f_m(\underline{r}, \underline{v}_1', t) \frac{(2\ell+1)(2m+1)}{(2i+1)(2j+1)} - \tag{6.3}$$

$$- f_j(\underline{r}, \underline{v}_2, t) \, f_i(\underline{r}, \underline{v}_1, t) \cdot (\underline{v}_2 - \underline{v}_1) \frac{d\sigma}{d\Omega} [ij \to lm; (\underline{v}_2 - \underline{v}_1); \Omega]$$

where only rotational degrees of freedom have been considered for the two molecules 1 and 2 involved in the binary collisions of the dilute-gas situation described in the present treatment. On the lhs of eq. (6.3) f_i represents the distribution function, at time t, for a molecule in a rotational state $|i\rangle$, and degeneracy $(2i+1)$, at the position $\underline{r}_1$ with velocity $\underline{v}_1$; the gradient $\nabla_{\mathbf{r}}$ only acts on the position coordinate $\underline{r}$. The differential cross section for the specific, degeneracy-averaged, process $|ij\rangle \to |lm\rangle$ depends explicitely on the orientation Ω and has been summed over final degenerate states and averaged over initial degenerate rotational states.

To further proceed in obtaining the WCU equation in a tractable form, it is often assumed that the distribution function is separable in its rotational and translational components, a decoupling based on the physical evidence that rotational relaxation is slower than the translational one, as briefly discussed before. It is therefore possible to write:

$$f_i(\underline{r}, \underline{v}, t) = \rho_i(\underline{r}, t) \cdot Q(\underline{r}, \underline{v}, t) \tag{6.4}$$

where the Q's are now normalized Maxwell-Boltzmann velocity distributions for particles at the translational temperature $T_{tr}(r,t)$ of the gas under study.

A practical procedure for going ahead in solving (6.3) is the so-called method of moments. The general idea is to expand the distribution function in some complete set of functions and then to consider each member of this basis as a 'normal mode', with which one can hope to associate a characteristic relaxation time and between which there will characteristic coupling coefficients. According to this picture the behaviour of the gas is to be described in terms of the amplitudes of the basis modes and one is then concentrating not so much on the distribution function but rather on its Fourier coefficients.

The entire relaxation process is then seen as an initial excitation of a single mode and a following response measured in terms of the rate of decay of the amplitude of this mode, its propagation together with other coupled modes to distant parts of the fluid, and the equilibrium state to which the system ultimately tends. The functions ρ_i of eq. (6.4) have now the meaning of a number density, or number of molecules per cm^3 of the ith state, with ρ being the total number density = $\sum_i \rho_i(\underline{r},t)$. The conservation of flux between each of the states (modes) produces, after a series of well documented passages [18] , the following new form of eq. (6.3) in terms of a Master-type equation :

$$\{\frac{\partial}{\partial t} + \underline{v}\cdot\nabla_{\underline{r}}\}\rho_i(\underline{r},t) + \rho_i(\underline{r},t)\nabla_{\underline{r}}\cdot\underline{v} = \tag{6.5}$$

$$= \rho^2(\underline{r},t) \sum_{j\ell m} \{-P_i\cdot P_j\ K_{ij\to\ell m} + P_\ell\cdot P_m\cdot K_{\ell m\to ij}\} = \rho \cdot F_i$$

where:

$$\rho_i(\underline{r},\ t) = P_i(\underline{r},\ t)\ \rho(\underline{r},t) \tag{6.6}$$

and $K_{ij\to lm}(T_{tr})$ is a bimolecular rate constant. The P's represent the probabilities of a molecule being in each given state, say the ith , at the time t and at the position $\underline{r}$. The differential operators on the lhs of eq.(6.5) contain non-stationary effects together with convective and spatial-dependent effects. The rate constants are thus providing the direct coupling between the internal (in this example , rotational) energy and the thermal bath. They should more specifically, be expressed as sums of terms indicating direct and exchange rates, in the case of only one type of molecular species being present in the gas and being allowed to undergo collisions with itself.

One simpler, limiting form of eq. (6.5) which is intuitively suggested by some of the experiments and wich has been often used, considers the gas as spatially uniform at constant density ρ , and confined in the finite volume of a box. The boundary conditions are thus assumed to be spatially invariant and the operators shown within the curly brackets on the lhs of eq. (6.5) simply reduce to $\partial/_{\partial t}$. The spatial velocity no longer appears since the bulk volume of the gas is assumed to be stationary. After some simple manipulations [11] , one finally comes up with a more familiar form the Master equation for each of the involved modes of the system :

$$\frac{\partial}{\partial t} P_i(t) = \rho(t) \sum_{j\ell m} \{-P_i P_j K_{ij\to\ell m} + P_\ell P_m K_{\ell m\to ij}\} \tag{6.7}$$

The above equation is also assumed to have been derived under energy conservation conditions between the thermal bath and the average rotational energy of the molecular gas , at each time t and for each given T_{tr} .

The simplest example of a vibrational rate equation , on the other hand, is provided by the case of a diatomic molecule as only component in the gas and by the special model - situation in which only one internal degree of freedom (set of moments) is interacting with the collisional partners treated as structureless projectiles. As is always the case , it is assumed that the pressure of the gas makes only the binary collisions relevant for the energy transfer mechanism, each molecule making Z of such collisions per second :

$$Z = 4\rho\, r_a^2\, (\pi KT/ 2\mu)^{\frac{1}{2}} \tag{6.8}$$

where the collision number Z is referred to the limiting form of the elastic cross section, πr_A^2 , with r_A being the collision radius [5]. Further , one can define :

$$a = \frac{h\nu}{KT} \quad ; \quad K_a = \rho .\ h\nu \tag{6.9}$$

and ν is the frequency of the oscillator, assumed to be an harmonic oscillator (HO). For the sake of simplicity , V-R coupling is not considered.

If $\overline{T}$ is the equilibrium temperature of the gas, T' its value after a sudden change due to the perturbation, then the corresponding vibrational energies are :

$$\varepsilon(\overline{T}) = \overline{\varepsilon} = \frac{K_a e^{-a}}{1 - e^{-a}} \tag{6.10a}$$

$$\varepsilon(T') = \varepsilon = K_a \cdot \sum_i i \cdot \rho_i/\rho \qquad (6.10b)$$

Another common assumption made in this simplified treatment is that of restricting transitions to Δ i= $\pm$ 1 , which is to be expected for vibrational relaxation processes which, from experimental evidence [21] , are mainly considered as step-wise deactivation processes with very little contributions from multiple-jump cross sections. One can therefore obtain the rate of energy change between the various modes of the only vibrational ladder considered as simply the algebraic sums over only index i in eq. (6.7) :

$$\frac{d\varepsilon}{dt} = Z \cdot K_a \cdot \sum_i \{P_{i\to i+1} \cdot \rho_i/\rho - P_{i\to i-1} \cdot \rho_i/\rho\} \qquad (6.11)$$

The simple structure of the HO levels, combined with the principle of detailed balance between the P_i's , allows one to rewrite eq. (6.11) in an even simpler form:

$$\frac{d\varepsilon}{dt} = Z \cdot K_a \cdot P_{1\to 0} \{\sum_i e^{-a}(i+1) \frac{\rho_i}{\rho} - \sum_i i \frac{\rho_i}{\rho}\} \qquad (6.12)$$

making use of (6.10b) and since : $\Sigma\rho i/\rho$ = 1 one can further write:

$$\frac{d\varepsilon}{dt} = Z\, P_{1\to 0}\, [\bar{\varepsilon} - \varepsilon] \cdot (1-e^{-a}) \qquad (6.13)$$

which is independent of the molecular density ρ (t).

This final expression for the energy relaxation equation can be related to a temperature relaxation equation in a rather simple way, when acoustical perturbations are considered [22] . In this case the value of T_{tr} varies periodically around the T value , and the instantaneous temperature T' lags behind via the smaller oscillations due to the single mode considered:

$$T_{tr} = \bar{T} \pm \Delta T_{tr} \qquad (6.14a)$$

$$T' = \bar{T} + \Delta T_a \qquad (6.14b)$$

One can therefore make use of these equations, and of the relationship between internal energy of the molecule and the molar specific heat that is related to the single oscillator considered in the model:

$$\Delta\varepsilon = \varepsilon - \bar{\varepsilon} = c_a\, \Delta T_a \qquad (6.15)$$

to rewrite eq. (6.13) in the following manner :

$$\frac{dT'}{dt} = Z\,P_{1\to o}\,[T_{tr} - T']\,(1-e^{-a}) \qquad ;(6.16)$$

By comparing this equation with the Landau-Teller result of eq. (6.2), one gets the definition of isothermal single relaxation time [5] for the single mode (moment) that contributes to the density function of the present model:

$$\frac{1}{\tau} = Z\,P_{1\to o}\,[1-e^{-h\nu/KT}] \qquad (6.17)$$

This derivation also assumes that at all times the distribution of the energy among the vibrator's levels corresponds to a definite vibrational temperature, i.e. that the distribution follows a Maxwell-Boltzmann behaviour.

The real life relaxation processes are, unfortunately, rather more complicated than the definition of eq. (6.17) seems to imply, and therefore one should take into consideration in the expansion of the density function a much larger number of moments or modes, in order to achieve acceptable levels of accuracy with respects se e al experimental findings, both for vibrational and rotational relaxation processes, especially for the low- temperature regimes of dilute gases. This simply means that single mode relaxation times (or pure mode relaxation behaviours) are very often prevented by numerous other effects not considered in the above model example of SSH theory, like anharmonicity contributions and rotovibrational coupling within each manifold, only to cite two factors. As a consequence of this, the uncoupled rate equations (6.12) should really be written via a more explicit inclusion of all the probalities involved, along each of the energy ladders for all the required normal modes, rather than only via the P_{1-0} probability as eq. (6.17).

For polyatomic systems, or for molecular mixtures, one should begin the increase in complexity by writing down, for each normal coordinate involved, a more general set of first order coupled equations that no longer require the single-step constraint of vibrational transitions. Hence, for each of the energy levels within each normal mode of vibration with its rotational manifold, one obtains a generalization of eq. (6.12):

$$\frac{d\rho_i}{dt} = \sum_j \gamma_{j\to i}\,\rho_i - \sum_i \gamma_{i\to j}\,\rho_i \qquad (6.18)$$

Where the coefficients γ's are simply equal to the ZP's defined in the previous equations. They represent, in the firt sum on the rhs of (6.18) , the transition probabilities from the jth state to the ith, increasing the population of the latter. In the second sum they indicate instead the transitions which deplete the ith level population. This result can be more compactly written via the following form of the coupled Master equations:

$$\frac{d\rho_i}{dt} = \sum_i \Gamma_{ij} \bullet \rho_i (t) \tag{6.19}$$

where $\Gamma_{ji} = \gamma_{j\to i}$ and $\Gamma_{ii} = - \sum_j \gamma_{i\to j}$, $j \neq i$. They are now the elements of a square matrix $\mathbf{\Gamma}$ that are obviously related to the specific cross sections, usually after convoluting them over Maxwellian velocity distributions of all the relevant molecules at the gas translational temperature T_{tr} under consideration:

$$\gamma_{i\to j} = 4\pi\rho(T) \left(\frac{\mu}{2\pi KT}\right)^{3/2} \int_0^\infty \sigma_{i\to j} (v) \exp \left[\frac{-\mu v^2}{2KT}\right] v^3 dv \tag{6.20}$$

and may be further multiplied by the corresponding statistical weight, g_j, of the jth vibrational level. It is in fact related to the opposite, de-excitation process via the detailed balance relation:

$$\gamma_{i\to j} \cdot g_j \cdot \exp \left[- \frac{\varepsilon_i}{kT} \right] = \gamma_{j\to i}\, g_i \exp [- \varepsilon_{j/KT}] \tag{6.21}$$

For a gas mixture in translational equilibrium at the temperature T, however, the rate coefficients (6.20) further involve another summation over the mixture components:

$$\gamma_{ij}(T) = \sum_{\alpha=1}^{c} \chi_\alpha\, \gamma^\alpha_{i\to j}(T) \tag{6.22}$$

for a given mixture with c components of fractional molecular concentration χ_α , with: $\sum_{\alpha=1}^{c} \chi_\alpha = 1$.

The above derivation still implies, however, that the relaxation process of the molecular gas, or of the gas mixture, does not depend on the internal structure of the projectiles. This will be valid, for instance, in the case of rare gases diluted into a molecular gas, but requires instead a careful examination of its validity for pure molecular species or for molecular mixtures, since complex transitions or near-resonant V-V exchanges acquire a great deal of relevance in many systems and require a more general formulation, like in eq. (6.7), to yield acceptable results.

When one is considering only the structures of the rotational manifolds within a given vibrational state, to examine mainly T-R relaxation phenamena, the form of the rate constants (6.20) needs to take into account the principle of detailed balance by summing each individual cross section over m_j-degeneracy, which simply provides a specific expression for the g's appearing in (6.21):

$$K_{j'\to j}(T) = \frac{(2j+1)}{(2j'+1)} \exp\left[(\varepsilon_{j'} - \varepsilon_j)/KT\right] K_{j\to j}(T) \qquad (6.23)$$

where the γ's are now written as K's to indicate the pure rotational relaxation situation within a vibrational state.

If one now considers the effect of complex relaxation, i.e. the R-R exchanges between molecules in the gas mixtures a low-T and the role of V-V exchanges (which are however much less important at low temperatures, especially for simple diatomics), then the above definition of the rates needs explicitely to contain indeces for the two rotational situations in the two (equal) partners. Hence:

$$K_{j_1 j_2 \to j_1' j_2'}(T) = K^d_{j_1 j_2 \to j_1' j_2'} + K^e_{j_1 j_2 \to j_1' j_2'} \qquad (6.23b)$$

where the small interference terms have been disregarded and where the labels d,e refer to the direct and exchange process respectively between molecules 1 and 2. The result of this symmetry requirement is that, in equation (6.7), the sums over ℓ and m need only be taken as $\ell \leq m$.

To solve now equation (6.19) becomes therefore the prime problem in the study of relaxation processes or, in general, in all the studies of temporal evolution of a density function constructed under the physical constraints of spatial isotropicity and energy conservation already discussed before. According to the specific meaning that one wants to give to the coupling matrix elements, and hence according to the nature of the corresponding cross sections that are used in eq. (6.20), the solution of (6.19) will correspond to rotational relaxation, vibrational relaxation or to vib-rotational relaxation. Moreover, a wide variety of initial conditions can be chosen for the ρ_i (0) appearing on of the rhs of the latter equation, thus allowing for the study of the temporal evolution and the testing of the single relaxation time conditions implied by the (6.17) [23,24,25].

Over the set of energy levels of a given rotovibrational structure i, the Master equation (6.19) can be rewritten in a matrix form :

$$\dot{\underset{\sim}{\rho}} = \mathbf{\Gamma} \cdot \underset{\sim}{\rho} \qquad (6.24)$$

where the vector $\underset{\sim}{\rho}$ (t) contains the individual populations of the relevant levels and $\mathbf{\Gamma}$ is the square matrix of the coupling matrix elements, i.e. the individual rate constants obtained from the specific cross sections. If it is further assumed that the structureless partner concentration in much larger than the one for the molecular species under consideration, i.e. if one ignores other possible modes present in the process as negligibly contributing to the density function time evolution, the above matrix equation will then carry all the information we need.

For diatomics, on the other hand, the equation (6.23) includes in principle the vibrotational states taking part in the process and constitutes an infinite set of first-order coupled equtions that need to be obviously truncated at some realistic level of convergence thereshold. The formal solution is then given by:

$$\underset{\sim}{\rho}(t) = \exp[\mathbf{\Gamma} t] \cdot \underset{\sim}{\rho}(o) \qquad (6.24b)$$

where $\underset{\sim}{\rho}(o)$ is the initial population distribution. The diagonalization of $\mathbf{\Gamma}$ can easily be done via the matrix $\mathbb{M}$ such that:

$$\mathbf{\Lambda} = \mathbb{M}^{-1} \mathbf{\Gamma} \mathbb{M} \qquad (6.25)$$

where $\mathbf{\Lambda}$ is now a diagonal matrix that reduces (6.24) to a set of uncoupled equations:

$$\omega_i(t) = \exp[\lambda_i\ t]\ \omega_i(o)\ , \qquad \text{all } i\ . \qquad (6.26)$$

where the ω's are elements of a transformed population vector related to the original one by the equation:

$$\underset{\sim}{\omega}(t) = \mathbb{M}^{-1}\ \underset{\sim}{\rho}(t) \qquad (6.27)$$

More specifically, one sees, from the form of (6.19), that $\mathbf{\Gamma}$ is not in general a symmetric matrix but is related to a symmetric matrix $\mathbb{K}$ by the similarity transformation:

$$\mathbb{K} = \mathbb{S}^{-1}\ \mathbf{\Gamma}\ \mathbb{S} \qquad (6.28)$$

where:

$$S_{ij} = \delta_{ij} \cdot \rho_i(\infty)^{1/2} \qquad (6.29)$$

and the $\rho(\infty)$'s are the equilibrium distribution at $t=\infty$:

$$\rho_i(\infty) = \frac{g_i \exp[-\varepsilon_i/KT]}{\sum_j g_j \exp[-\varepsilon_j/KT]} \qquad (6.30)$$

The diagonal matrix $\boldsymbol{\lambda}$ is then obtained via the standard orthogonal transformation:

$$\boldsymbol{\lambda} = \mathbb{U}^{-1} \mathbb{K} \mathbb{U} \qquad (6.31)$$

and its eigenvalues are usually assumed ordered and distinct. In terms of the individual populations the matrix elements are therefore given by:

$$\rho_i(t) = \sum_{m,j}^{M} (\mathbb{S}\mathbb{U})_{im} \exp(\lambda_m t) (\mathbb{S}\mathbb{U})^{-1}_{mj} \rho_j(o) \qquad (6.32)$$

The summation runs over the total number M of states included in the matrix $\mathbb{\Gamma}$. The jth columm of $\mathbb{S}\mathbb{U}$, or of $\mathbb{M}$, can be identified as the jth right eigenvector of Γ and similarly the jth row of $(\mathbb{S}\mathbb{U})^{-1}$ corresponds to the jth left eigenvector of $\mathbb{\Gamma}$. It can easily be shown [26] that λ_m is ≤ 0 for all m. An analysis of the population vector corresponding to λ=o shows that it corresponds just to the equilibrium distribition in the long-time region, which is therefore time-independent as expected. The first column of the $\mathbb{S}\mathbb{U}$ matrix therefore corresponds to the ρ_i (t =∞) values for all the involved M levels, as one can easily see from the definition of the S_{ij} matrix elements in (6.29). The limiting behaviour of eq.(6.32) is therefore given by the following expression:

$$\begin{aligned}\lim_{t\to\infty} \rho_i(t) &= \sum_j (\mathbb{S}\mathbb{U})_{i1} (\mathbb{S}\mathbb{U})^{-1}_{1j} \rho_j(o) \\ &= \rho_i(\infty) \sum_j S^{-1}_{jj} U_{j1} \rho_j(o) \qquad (6.33) \\ &= \rho_i(\infty) \sum_J \rho_j(o) = \rho_i(\infty)\end{aligned}$$

where use has been made of the fact that $U^{-1}_{1j} = U_{j1}$, and the populations are treated as normalized, i.e. the total number density is considered equal to the unity when generating the rate constants, which then represent rate/density usually given in units of cm^3/mole-sec.

The ordered eigenvectors λ_m that correspond to negative eigenvalues now describe the population imbalances caused by having perturbed the system away from equilibrium. Their concerned relaxation to equilibrium leads in turn to purely exponential time dependence, although one expects in a general case that the real situation should be give instead by a linear combination of such pure modes. The realistic time dependence will therefore be given by a sum of exponentials. In the long-time limit, however, the rate of relaxation to equilibrium will ultimately be governed by the mode with the slowest rate, so that the exponentially observed relaxation time will be given, within the theory, by λ_{min}^{-1} where λ_{min} is the eigenvalue of eq. (6.31) with the smallest, non-zero value of the modulus.

One can see better what has been discussed above by looking again at eq. (6.24) and rewriting it as follows:

$$\begin{aligned} \underset{\sim}{\rho}(t) - \underset{\sim}{\rho}(\infty) &= \exp[\mathbf{\Gamma} t]\, \underset{\sim}{\rho}(o) - \underset{\sim}{\rho}(\infty) \\ &= \exp[\mathbf{\Gamma} t] \cdot \{\underset{\sim}{\rho}(o) - \underset{\sim}{\rho}(\infty)\} \quad ; \qquad (6.34) \\ \Delta\, \underset{\sim}{\rho}(\tau) &= \Delta\, \underset{\sim}{\rho}(o)\, \exp[\mathbf{\Gamma} t] \end{aligned}$$

where the $\Delta\underset{\sim}{\rho}$ vectors represent the population departure from equilibrium at the time t and at starting time t=0 (the preparation of the system) ; the last step in eq. (6.34) follows from the fact that $\underset{\sim}{\rho}(\infty)$ is an eigenvector of (6.25) with zero eigevalue. One can now rewrite eq. (6.32) in the following way:

$$\Delta\, \underset{\sim}{\rho}(t) = (\mathfrak{SU})\, \exp[\boldsymbol{\Lambda} t]\, (\mathfrak{SU})^{-1}\, \Delta\, \underset{\sim}{\rho}(o) \qquad (6.35)$$

Suppose now that the system has been prepared in a set of states such that one can write the various elements of the $\Delta\, \underset{\sim}{\rho}(0)$ vector in the following way:

$$\Delta\, \rho_i(\theta) = \alpha \cdot (\mathfrak{SU})_{ij} \qquad (6.36)$$

with α being a constant. Then one can also rewrite eq. (6.35) in the following simpler form:

$$\Delta\, \underset{\sim}{\rho}(t) = \Delta\, \underset{\sim}{\rho}(0)\, \exp[\lambda_j t] \qquad (6.37)$$

which represents now a pure mode, an exponential relaxation with the jth mode. Since, in genral, the system preparation is not performed in a pure mode fashion, any arbitrary vector $\Delta\underset{\sim}{\rho}(o)$ will be given as a linear combination of the vectors that make up the $\mathfrak{SU}$, or the $\mathfrak{M}$, matrix. It however remains important to realize that an examination of the vectors' structure of the latter matrix can be used to interpret physically the relaxation process associated with a specific relaxation time τ_j.

Whenever it becomes again reasonable to assume that the crucial step in the relaxation process is constituted by the transition from the i=1 level (or from its manifold for rotational relaxation) to the i=0 level,while all other levels, including those within a rotational manifold, have relaxed already to a Boltzmann distribution, the above diagonalization is on longer cessary and one goes back to the evaluation of a two-level relaxation process in which the rotational populations have already reached equilibrium, while the slower vibrational relaxation is essentially a single mode process involving only the v=1 and v=0 levels of the molecular ith normal vibrational frequency. One of these two-level rate constants can, for instance, be written via the following simple formula :

$$\gamma^{i}_{1\to 0}(T) = \sum_{j_1, j_2} K^{i}_{1j_1\to 0j_2} \cdot \frac{(2j_1 + 1)\exp[-\varepsilon^{i}_{1j_1}/KT]}{\sum_{j_1} (2j_1 + 1)\exp[-\varepsilon^{i}_{1j_1}/KT]} \tag{6.38}$$

while the corresponding rate for the correlates process is very simply given by the application of the principle of detailed balance:

$$\gamma^{i}_{0\to 1}/\gamma^{i}_{1\to 0} = \frac{g_1}{g_0} \exp[-E^{i}_{1} - E^{i}_{0}/KT] \tag{6.39}$$

The corresponding eigenvalue equation (6.31) is now given by a 2x2 matrix with the obvious eigenvalues:

$$\begin{aligned} \lambda_1 &= 0 \\ -\lambda_2 &= \gamma^{i}_{0\to 1} + \gamma^{i}_{1\to 0} \\ &= \gamma^{i}_{1\to 0}\{1 + \exp[-E^{i}_{1} - E^{i}_{o}/KT]\} \end{aligned} \tag{6.40}$$

where it as been assumed for simplicity that $g_1 = g_0 = 1$. This is exactly the same result as the one suggested by the SSH treatment and briefly outlined in the paragraphs which lead to eq. (6.17) , the latter being the expression for the single relaxation time, the inverse of the corresponding rate of equation (6.40) , and valid for energy separations between the relevant vibrational levels much larger than KT. For most simple molecular gases this is certainly the case, especially for the lower-lying vibrational levels and for temperature ranges below 1000 °K.

6.4 - The He - H_2 relaxation and other examples.

Any theoretical determination of relaxation rates, as it should have become clear thoughout the present Notes, may be decomposed into three different but closely linked tasks: (i) the construction of the potential energy hypersurface for the many-body system examined and for a range of relative geometries that allow for the treating of both rotational and vibrational inelasticities; (ii) the solution of the scattering problem on this surface in order to determine ro-vibrationally inelastic cross sections as a function energies; (iii) the formation of thermal averages of the inelastic cross section over the desired ranges of temperature to obtain the corresponding rate constants, out of which the relaxation times are computed. Each clearly requires quite a large amountof work and frequently becomes an indipendet problem in its own right, a reality that explains why most of the studies performed on molecular systems have usully concentrated on the last step and treated the previous two points via some sort of simple modelling or through ad hoc choices, to proceed further in producing relaxation times that could then be compared with experiments [4,27,28]. This type of approach has been accompanied by rather limited success, especially when confronted with the increasingly higher level of accuracy attained in experiments. More recently, however, the widened range of computational knowledge on the evaluation of potential surfaces, and on the various decoupling techniques that strongly reduce the dimensions of step (ii), have produced various attempts at treating all the steps involved with ab initio methods and within a quantum mechanical framework. Because of the lack of many details on the structure of potential surfaces for anything more complicated than a three-atom neutral system, even the above attempts have been, however, rather few in number and had to apply often strongly reductive schemes in calculating the numerous cross sections needed over the relevant ranges of collision energies.

For molecule-molecule studies, as an example, the case of molecular Hydrogen has been examined by various authors from the point of view of its rotational relaxation mechanisms and behaviour [29], rotovibrational relaxation processes [30] and mainly vibrational relaxation rates [31], all studies which have adopted a generally a priori approach within quantum treatments.

To our knowledge, no other molecular cases have been carried through all the steps listed above without either employing semiclasical methods for the dynamics or using simple empirical prescriptions for the region of molecular interaction (and of molecular orientations) relevant to the coupling that gives rise to inelastic transition probabilities. Only recently, an accurate fit to the ground state HF - HF potential energy hypersurface [32] has been directly used, through its spherical component in a multipolar expansion, to compute vibrationally inelastic cross sections, and from them the corresponding transition rates and low-to -high temperature relaxation times without the use of classical mechanics or of strongly decoupling the dynamical channels [33]. In both the above molecular cases, however, no definitive answer can be given on the ultimate success of the theory since discrepancies with experiments have indeed been found and attributed to either the need of a better knowledge of the surface in specific interaction regions or to the necessity for extending the size of the coupled equations yielding the relevant inelastic cross sections. Both these points have recently been examined in a detailed and stringent comparison of low-energy computed cross sections the rotationally inelastic processes and the differential cross section experiments, for the case of H_2 - H_2 collisions [34].

Another, this time atom-diatom, neutral system that has received attention for the calculation of its vibrational transition rates via ab initio treatments through quantal schemes is the CO - He mixture at low and intermediate temperatures. Using a computed form of the potential hypersurface that had been obtained via the EGM approach discussed in the previous Chapter 2 [35], Verter and Rabitz [36] examined an extensive set of vibrationally inelastic transition rates over temperature ranges between 100 °K and 3000 °K. In studying the time evolution of the corresponding Master equations discussed in (6.23), they employed different preparations (initial conditions) for the vibrating target occupied levels and found marked departure from the Landau - Teller behaviour surmised by equation (6.17). The measured rates seemed, however, to be qualitatively given by the single relaxation time (pure mode)behaviour. This last result was also confirmed by other authors [37] who employed a slightly different potential surface form of the spherical component of the interaction (disregarding, as in [36] the anisotropic part) and also examined the effect of the changes in such interaction on the calculated relaxation times. Their results thus exhibited a marked dependence on the repulsive branch of such interactions, while recalculations of them performed recently in a preliminar way [38] indicate that one is still far from being able to know the correct potential hypersurface with the necessary accuracy.

On the other hand, the neutral system which has been studied most extensively, both theoretical and the experimental viewpoints, is problably the He - H_2 mixture in a dilute situation for its molecular component. Because of this, therefore, it will be used here as an interesting test case for the various levels of accuracy with which all the three steps listed above have been approached, thus providing a sort of direct comparison for the various merits and performance of different models vis à vis the experimental findings.

The theoretical calculations of Rabitz and Zarur [39] used their EP decoupling tecnique to obtain cross sections via the SCF potential surface produced by Gordon and Secrest (GS) [40] and then went on to obtain rate constants for pure rotational, as well as rotovibrational, relaxation behaviour. They could then be used to solve the corresponding Master equations and to observe in a computed fashion the time evolution of this simple system [39]. M.H. Alexander employed the same potential surface of the previous authors and computed again rotational and rotovibrational inelastic cross sections within the EP decoupling scheme [41,42], but also performed a study of differences in the chosen potential surfaces on the finally obtained rates, by using the earlier calculations of Krauss and Mies (KM) [43] and a different numerical fit to the GS potential surface calculations [44].

MC Guire and Toennies [45] employed in turn three different potential energy surface forms to first obtain the elastic and inelastic cross sections and then the corresponding relaxation rates: the KM surface, the GS surface and a modification of the latter which also included the long range term of the dispersion interaction. The decoupling technique for solving the quantum mechanical formulation of the scattering process that they used was the CS approximation alredy presented in detail in Chapter 4. They found that the presence of the attractive, long range term becomes extremely important in providing agreement with the available experiments at low temperatures [46].

Gianturco and Lamanna [47] solved the coupled scattering equations via the SDA approach alredy introduced in Chapter 4 and obtained relaxation rates by using the GS, the KM and the alternative fit to the GS as given in reference [44]. They also found that differences among the potential surfaces employed affected the final computed cross sections even more that the use of the various decoupling techniques adopted to treat the dynamics. Alexander and McGiure provided a further test [48] for the same problem by obtaining the interaction as a combination of the GS calculations for its anisotropic part and of an experimantallly derived potential (SG) [49] for the spherical component of the multipolar expansion of the total potential energy surface. The CS approximation was again used by them to treat the dynamics. To further improve the agreement of the calculations with the available very-low T experiments ($\leqslant$ 300 °K), M.H. Alexander went on to include in the theoretical treatment the effect of mixing o.H_2 and p.H_2 in the normal Hydrogen gas sample actually considered, to be in keeping with the experiments [50]. A later application of the same potential energy surface, followed by the CS approximation for the dynamics, in order to obtain vibrational relaxation rates for D_2 in collision with He was also per formed by the same author [51].

The quality of the ab initio potential energy surfaces employed in the studies listed above has recently been improved by CI calculations of Tsalpine and Kutzelnigg (TK) [52]. These calculations exhibit the shallow Van der Waals minimum, a necessary feature because of the dispersion interaction between the inert (chemically) species of this system but that was missing in the SCF calculations of the GS potential surface. The semiempirical alternative form of the interaction produced by Shafer and Gordon [49] also exhibited the long range attractive well, but the latter was quantitatively different in position and steepness of the repulsive wall, besides exhibiting a minimum in the perpendicular approach while the results of [52] presented it for the collinear configuration. The TK calculations have recently been fitted to an analytic form [53].

The effect of this, improved, form of the interaction potential on the computed cross sections, vibrationally inelastic between the lower-lying levels of the target molecule, has been examined by using both the CS approximation [54] and the rigorous CC treatement of the dynamics [55]. The latter calculations showed that the TK interaction produces a much smaller inelasticity at low collision anergies, its cross sections being about five times smaller than those produced via the GS potential surface. The former surface, however, yields cross sections that rise much more rapidly with collision energy and also produces smaller rotational inelasticity within the manifold associated with the vibrational levels of the primary transition, a fact related to the smaller value of its anisotropic component with respect to the GS value [55] of the same.

It is interesting to note at this point that, when the corresponding rate constants were computed for the vibro-rotational relaxation process [55] both the full diagonalization of eq. (6.23) and the 'pseudo two-state' formula of eq. (6.37) were used One of the results was then that, up to temperatures of $\sim$ 2000 °K, the rotational relaxation within a vibrational manifold is order of magnitude faster than the vibrational relaxation proper and therefore it becomes possible to employ a sort of 'speudo two-state' approach with a reasonable degree of confidence. The process in question, in fact, essentially follows the mode with the smallest, non-zero λ eigenvalue.

A comparison of computed relaxation rates which have been obtained by using the vāriuos forms of computed potential surfaces discussed above, is shown in Figure 6.1 and was taken from the recent computation of Raczkowski et al. [55]: the agreement of their results with experimantal findings is only reasonable and the effect of the dif ferent employed surfaces appears to be clearly larger than that of decoupling the correct CC dynamical equations via some approximate scheme like the EP approach plotted in the Figure. The same useful comparison of various calculations, taken from the same work [55], is shown for the low-temperature intervals in the following Fig. 6.2 and in the form of a classical Landau - Teller plot.

The marked change of the temperature dependence shown by the experiments [46], which level off at about 50 °K, is not presented by the computed values of the $K_{1\rightarrow0}$: the TK potential surface produces rate values which are one or two orders of magnitude too low, while failing to go up again in the very-low temperature region, as shown by the experiments. The use of the GS potential form, on the other hand, yields larger rates than those given by measurements. The latter are in fact a factor of three or four smaller in the interval of temperatures shown in Figure 6.2 On the hand, the use of the CS decoupling approximation fares rather well with both the potential surfaces used to compute the rates shown in the latter plots: they are both closest to experiments, exhibit the expected low-T curvature and contain within them the experimental positions of the rates. However the quality of the computed cross sections that originate such rates, i.e. to perfom their direct calculation for all the needed collision energies instead of using some extrapolation formula for their low-E_{coll}, or for their high-E_{coll}, behaviour as given by a fitting to already known valus of the $\sigma_{i\rightarrow j}$'s appears to be an important factor in controlling the accuracy of the final computed rates. It also appears to be mainly responsible for the more marked disagreement shown by the CC calculations of Fig. 6.2 [55].

The detailed analysis of one of the most popular systems studied theoretically thus far to generate vibrotational relaxation rates, has then been summarized in the previous pages. It certainly appears as a very instructive case in indicating some general conclusions about the various theoretical and computational steps involved in any fully ab initio, quantal treatment of systems exhibiting weak chemical interactions and therefore presenting rather small inelastic cross section, all very sensitive to the details of the adopted procedure.

First of all, the total number of cross section values that need to be computed within the examined range of collision energies strongly depends on the structural properties of the studied partners : for the He-H_2 case the power-law behaviour often suggested for the $\sigma_{i\rightarrow j}$ vs. E_{coll} plots [39], seems to hold rather well. Other, more anisotropic systems, on the other hand, often exhibit resonances and orbiting effects that actually require very small energy grids in order to provide numerically stable

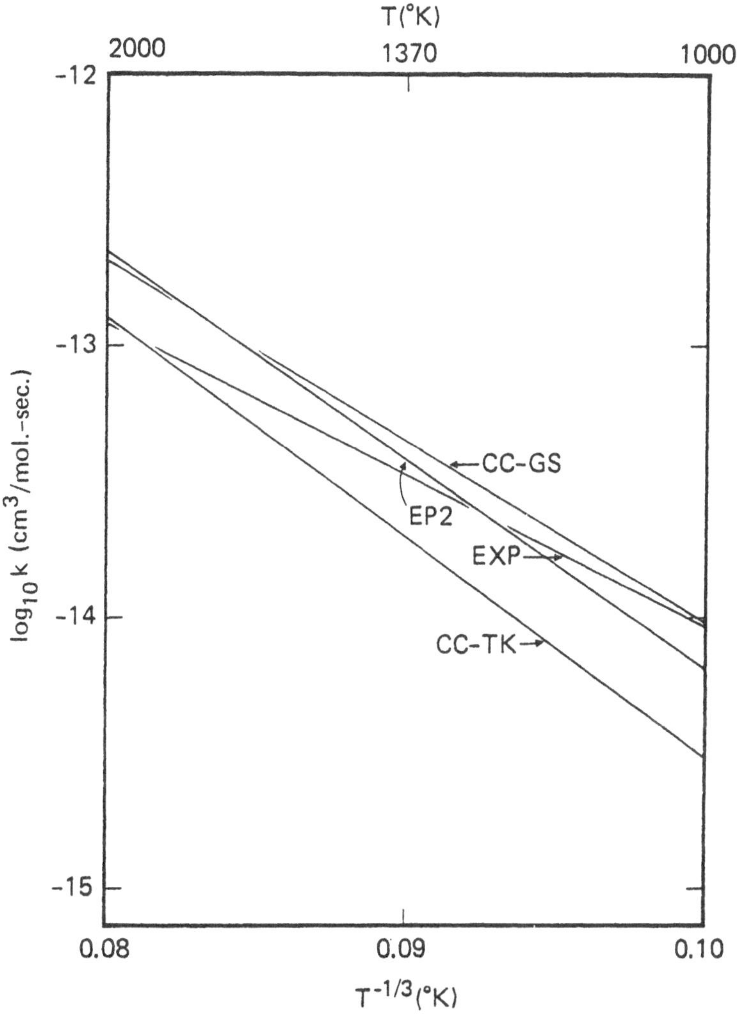

Fig. 6.1 - Temperature behaviour of computed and measured vibrational relaxation rates in the high-T region:

CC-GS: Close coupling calcs on the GS surface [55].
CC-TK: Same as above with the TK surface [55].
EXP : Experimental values [56].
EP2-GS: Effective potential calcs on the GS surface [42].
[from A.W.Raczkowski et al., IBM Research Report RJ2203 (30016), 3/10/78].

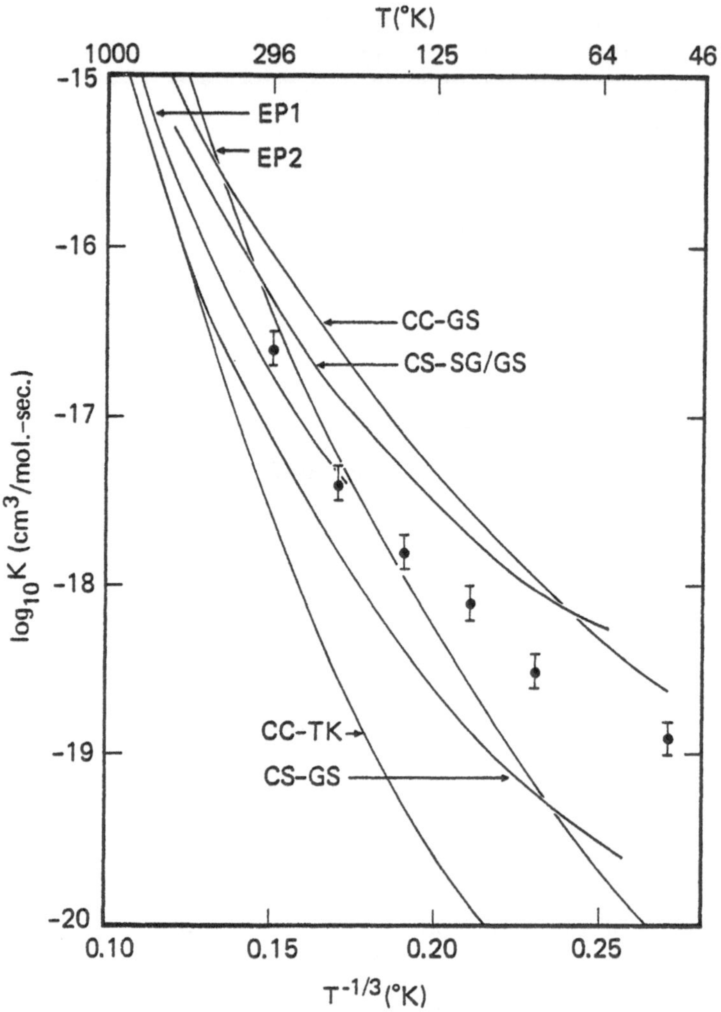

Fig. 6.2 - Temperature behaviour of vibrational relaxation rates in the low-T region:

CC-TK : Close coupling calcs on the TK surface [55].
CC-GS : Same as above on the GS surface [55].
CS-GS : Coupled-state calcs on the GS surface [48].
CS-SG/GS: As above, but with the SG spherical term in the potential surface [48].
EP1-GS: Effective potential calcs on the GS surface [39].
EP2-GS: As above, but using a different fit to GS surface [41].
⧫ ⧫ ⧫ : Experimental values [46].
[From the same Ref. as in Fig. 6.1].

energy dependence. They obviously exclude any simple fitting form valid for large temperature intervals: the molecular HF system and the proton interactions with polar molecules are prime examples of this latter behaviour [33].

The corresponding acceptable level of sophistication that should be used to yield the theoretical inelastic cross sections is obviously related to the above energy behaviours, since the number of channels that need to be coupled rapidly proliferates with increasing collision energy. This rather general behaviour makes mandatory the use of some decoupling procedure when going to high temperatures, unless a simple extrapolation law can be found for the specific case under study. The major factor that appears ultimately to control the reability of computed rates is, however, the quality of the potential energy surface.

Because of its direct bearing on the dynamics, in fact, there are no simple ways for correlating differences in the computed potential surfaces and the corresponding cross sections produced by their use within any given, and same, coupling scheme for the quantal treatment. The careful comparison here reported on the He + H_2 case seems to indicate that the above effects are essentially non-linear in the sense that fairly small and subtle difference in the potential surface calculations employed can shift the low temperature behaviour of rotovibrational relaxation rates produced by the Master equations by orders of magnitude. This is particularly so for those interacting partenrs that exhibit fairly small chemical effects or rather small anisotropic components of their potential multipolar expansions.

Similar results have been reached also in the studies on the CO + He system [31] and on the comparisons on molecular HF [33] or on molecular Hydrogen [34], even when one decides to employ simpler models to solve the coupled equations for the vibrational inelasticity case only [37].

REFERENCES

|1| D.Lambert, *Vibrational and rotational relaxation in gases* (Clarendon Press, Oxford, 1977).
|2| D.I.Rosen, R.N.Sileo and T.A.Cool, J.Quant.Elec., QE-9, 163 (1973).
|3| M.Y.D.Chen and H.L.Chen, J.Chem.Phys., 56, 3315 (1972).
|4| S.Ormonde, Rev. Mod. Phys., 47, 193 (1975).
|5| J.L.Stretton, *Transfer and storage of energy by molecules*, G.M.Burnett and A.M. North Ed.s, Vol.2 (Wiley Interscience, New York, 1969).
|6| C.B.Moore, *Fluorescence*, G.G.Glilbault Ed. (M.Dekker, N.Y. 1967).
|7| C.B.Moore, Accts.Chem.Res., 2, 103 (1969).
|8| J.C.Polanyi et al., Disc.Faraday Soc., 44, 183 (1967).
|9| J.I.Steinfeld and P.L.Houston, *Laser and Coherence Spectroscopy* (Plenum, N.Y., 1978).
|10| J.Reuss, Adv.Chem.Phys., 30, 389 (1975).
|11| H.Rabitz and H.Lam, J.Chem.Phys., 63, 3532 (1975).
|12| H.Rabitz, Ann.Rev.Phys.Chem., 25, 155 (1974).
|13| S.Chapman and T.G.Cowling, *Mathematical Theory of non-uniform gases* (Cambridge U.Press, 1939).
|14| E.A.Mason and L.Monchick, J.Chem.Phys., 36, 1622 (1962).
|15| K.F.Herzfeld and F.O.Rice, Phys.Rev., 31, 691 (1928).
|16| L.Landau and E.Teller, Phys.Z.Sowjet., 10, 34 (1936).
|17| J.S.Dahler and D.K.Hoffmann, in: *Transfer and storage of energy by molecules*, Vol.3, G.M.Burnett and A.M.North Ed.s (Wiley Interscience, New York, 1969).
|18| e.g.: J.H.Ferziger and H.G.Kaper, *Mathematical Theory of Transport processes in gases* Elsevier, N.Y., 1972).
|19| C.S.Wang Chang, G.E.Uhlenbeck and J.de Boer, in *Studies in Statistical Mechanics*, Vol.2, J.de Boer Ed. (Interscience, N.Y., 1964).
|20| L.Waldmann, Z.Naturforsch., 12a, 660 (1957); 13a, 609 (1958); R.F.Snider, J.Chem. Phys., 32, 1051 (1960).
|21| A.B.Callear and I.W.M.Smith, Trans.Faraday Soc., 59, 1735 (1963).
|22| F.I.Tanczos, J.Chem.Phys., 25, 439 (1956).
|23| R.Marriott, Proc.Phys.Soc., 84, 877 (1964).
|24| R.Marriott, Proc.Phys.Soc., 86, 1041 (1965).
|25| M.R.Verter and H.Rabitz, J.Chem.Phys., 64, 2939 (1976).
|26| I.Oppenheim, R.Shuler and G.Weiss, Adv.Mol.Relax.Process., 1, 13 (1967).
|27| R.Gordon, W.Klemperer and J.Steinfeld, Ann.Rev.Phys.Chem., 19, 215 (1968).
|28| T.L.Cottrell and J.C.McCoubrey, *Molecular Energy Transfer in Gases* (Butterworths, London, 1961).
|29| H.Rabitz and S.-H.Lam, J.Chem.Phys., 63, 3532 (1975).
|30| R.Ramaswamy and H.Rabitz, J.Chem.Phys., 66, 152 (1977).
|31| F.A.Gianturco and U.T.Lamanna, Chem.Phys., 38, 97 (1979).
|32| M.H.Alexander and A.E.De Pristo, J.Chem.Phys., 65, 5009 (1976).
|33| F.A.Gianturco, U.T.Lamanna and F.Battaglia, to be published.
|34| J.Schaefer and W.Meyer, J.Chem.Phys., 70, 344 (1979).
|35| E.Dougherty, H.Rabitz, J.Detrich and R.Conn, J.Chem.Phys., 67, 4742 (1977).
|36| M.R.Verter and H.Rabitz, J.Chem.Phys., 64, 2939 (1976).
|37| F.A.Gianturco, U.T.Lamanna and I.Baraldi, J.Chim.Phys., 74, 437 (1977).
|38| W.P.Kraemer, VII Int. Symp.Mol.Beams, June 1979.
|39| H.Rabitz and G.Zarur, J.Chem.Phys., 62, 1425 (1975).

|40| M.D.Gordon and D.Secrest, J.Chem.Phys., 52, 120 (1970).
|41| M.H.Alexander, J.Chem.Phys., 61, 5167 (1974).
|42| M.H.Alexander, Chem.Phys., 8, 86 (1975).
|43| M.Krauss and F.H.Mies, J.Chem.Phys., 42, 2703 (1965).
|44| M.H.Alexander and E.V.Berard, J.Chem.Phys., 60, 3950 (1974).
|45| P.McGuire and J.P.Toennies, J.Chem.Phys., 62, 4623 (1975).
|46| M.M.Audibert, C.Joffrin and J.Ducuing, Chem.Phys.Lett., 19, 26 (1973).
|47| F.A.Gianturco and U.T.Lamanna, Chem.Phys., 17, 255 (1976).
|48| M.H.Alexander and P.McGuire, J.Chem.Phys., 64, 452 (1976).
|49| R.Shafer and R.G.Gordon, J.Chem.Phys., 58, 5422 (1973).
|50| M.H.Alexander, Chem.Phys.Lett., 38, 417 (1976).
|51| M.H.Alexander, Chem.Phys.Lett., 40, 101 (1976).
|52| B.Tsalpine and W.Kutzelnigg, Chem.Phys.Lett., 23, 173 (1973).
|53| A.W.Raczkowski and W.A.Lester Jr, Chem.Phys.Lett., 47, 45 (1977).
|54| M.H.Alexander, J.Chem.Phys., 66, 4608 (1977).
|55| A.W.Raczkowski, W.A.Lester Jr and W.H.Miller, J.Chem.Phys., 69, 2692 (1978).
|56| J.E.Dove and H.Teitelbaum, Chem.Phys., 6, 431 (1974).

Lecture Notes in Chemistry

Vol. 1: G. H. Wagnière, Introduction to Elementary Molecular Orbital Theory and to Semiempirical Methods. V, 109 pages. 1976.

Vol. 2: E. Clementi, Determination of Liquid Water Structure. VI, 107 pages. 1976.

Vol. 3: S. R. Niketic and K. Rasmussen, The Consistent Force Field. IX, 212 pages. 1977.

Vol. 4: A. Graovac, I. Gutman and N. Trinajstić, Topological Approach to the Chemistry of Conjugated Molecules. IX, 123 pages. 1977.

Vol. 5: R. Carbo and J. M. Riera, A General SCF Theory. XII, 210 pages. 1978.

Vol. 6: I. Hargittai, Sulphone Molecular Structures. VIII, 175 pages. 1978.

Vol. 7: Ion Cyclotron Resonance Spectrometry. Edited by H. Hartmann and K.-P. Wanczek. VI, 326 pages. 1978.

Vol. 8: E. E. Nikitin and L. Zülicke, Selected Topics of the Theory of Chemical Elementary Processes. X, 175 pages. 1978.

Vol. 9: A. Julg, Crystals as Giant Molecules. VII, 135 pages. 1978.

Vol. 10: J. Ulstrup, Charge Transfer Processes in Condensed Media. VII, 419 pages. 1979.

Vol. 11: F. A. Gianturco, The Transfer of Molecular Energies by Collision: Recent Quantum Treatments. VIII, 328 pages. 1979.